THE FIRST IDEA

THE FIRST IDEA

How Symbols, Language, and Intelligence Evolved From Our Primate Ancestors to Modern Humans

Stanley I. Greenspan, M.D.
Stuart G. Shanker, D. Phil.

A MERLOYD LAWRENCE BOOK

DA CAPO PRESS
A Member of the Perseus Books Group

Designed by Trish Wilkinson
Set in 12-point Adobe Garamond by the Perseus Books Group

Library of Congress Cataloging-in-Publication Data
Greenspan, Stanley I.
The first idea : how symbols, language, and intelligence evolved from our primate ancestors to modern humans / Stanley I. Greenspan, Stuart G. Shanker.—1st Da Capo Press ed.
p. cm.
"A Merloyd Lawrence book."
Includes bibliographical references and index.
ISBN 0-7382-0680-6 (alk. paper)
1. Evolutionary psychology. I. Shanker, Stuart. II. Title.
BF698.95.G73 2004
153.7—dc22

2004010658

First Da Capo Press edition 2004

Published by Da Capo Press
A Member of the Perseus Books Group
http://www.dacapopress.com

Da Capo Press books are available at special discounts for bulk purchases in the U.S. by corporations, institutions, and other organizations. For more information, please contact the Special Markets Department at the Perseus Books Group, 11 Cambridge Center, Cambridge, MA 02142, or call (800) 255-1514 or (617) 252-5298, or e-mail special.markets@perseusbooks.com.

1 2 3 4 5 6 7 8 9—08 07 06 05 04

Dedicated to our families,
who have and who continue to inspire us.

Acknowledgments

First we want to thank our wives, Nancy and Ginny, for their ongoing support and living the ideas in this book with us, as well as helping us to elaborate on them. We want to express special thanks to Serena Wieder for her many years of wonderful and insightful collaboration in the development of many of the ideas on infant and childhood functioning and the treatment of autism that are embodied in this work. Special thanks are also due to Harry Wachs for his thoughtful reading of the manuscript and his help in describing critical developmental processes, and to Barbara King, whose work on patterned interactions and whose careful reading of Part II was instrumental in our thinking about primatology and evolutionary theory. We are also grateful to Rainer Born, Jerome Bruner, Jeff Coulter, Alan Fogel, Robert Lickliter, Daniel McShea, Raffi, Pedro Reygadas, Sue Savage-Rumbaugh, and Talbot Taylor for their advice and support, or for their helpful comments on earlier drafts of the book. We also want to thank Jan Tunney for her thoughtful work on the manuscript; Sarah Miller for her administrative and clinical support. We especially want to thank Merloyd Lawrence, our editor, who not only raised countless important questions but worked with us to find solutions. Her gifted insights into the material illuminate much of this work. S.I.G. would like to express his thanks to the Interdisciplinary Council of Learning and Developmental Disorders and the Floortime Foundation. S.G.S. would like to express his thanks to his college, Atkinson, and his colleagues in the philosophy and psychology departments, for making it possible to devote so much time to the writing of this book. In addition, S.G.S. would like to express his gratitude for the support he received from the Templeton Foundation and the Unicorn Foundation, which made it possible for him to receive invaluable assistance from the following graduate students in the psychology and philosophy programs at York University: Lisa Bayrami, Maria Botero, Matthew Crippen, Matt Peterse, and Ljiljana Radenovic.

Contents

PART III: THE DEVELOPMENT OF LANGUAGE AND INTELLIGENCE

PART IV: THE DEVELOPMENT OF SOCIAL GROUPS

Introduction

WHAT WAS THE FIRST idea? Was it formed by an early human-like figure, poised with stone in hand, calculating how to fell his prey? Certainly, an intriguing question, but there is an even more important one. How and when did the capacity to create an idea come about in the first place? This question has perplexed ancient and modern philosophers, scientists in the fields of human development and evolution, and most of the rest of us. How do human beings develop their highest mental abilities, the abilities to symbolize and think? And how did these distinctly human abilities arise during the course of evolution? In short, how did we become human beings and how do we maintain our humanness?

We have developed a new hypothesis to address these questions. The key to the evolutionary hypothesis we will present in this book comes from our observations of babies at the beginning of their developmental journey. We have found that the capacity to create symbols and to think stems from what was often thought of by philosophers as the "enemy" of reason and logic: our passions or emotions. While there is mounting evidence that emotions influence the content of our thoughts (see Chapter 11), we have discovered a far more important role for our emotions. We will show how emotions actually give birth to our very ability to create symbols and to think.

However, a very special characteristic of our emotions paves the path to symbols. This characteristic is relatively different in human beings than in other animals. We have observed that it is the capacity to transform basic emotions into a series of successively more complex interactive emotional signals. We will show how emotional signaling enables a child to separate perceptions from fixed predictable actions and, in so doing, free up these perceptions to acquire emotional meaning and become symbols. Something as profound as mastering the word "mommy," or something as basic as the word "apple" or the number "4," comes into being through six initial levels of emotional interaction and signaling in the early years of life. Sensory and subjective experiences, seemingly the enemy of reason and logic, through progressive transformations, therefore, actually become the basis for both creative and logical reflective thought.

These uniquely human abilities are not hardwired into our brains. They must be developed through learning interactions. Humans intensify these types of critical learning processes in the second half of the first year of life and during most of the second year and then continue it throughout their lifetimes. These learning interactions are not instructional, where an adult lectures, shows, or otherwise directly teaches a child. They are natural interactions that result in new learning, such as playful back-and-forth smiles or vocalizations between an infant and caregiver from which the infant learns about relating and interacting. When we use the term "learning," we will be using it in this way, even though at times this type of learning will also involve interactions where a child imitates or copies the adult or another child.

Through our studies of nonhuman primates and a review of the fossil record, we will also demonstrate that what takes a human baby two years to learn took our human ancestors millions of years. Remarkably, however, we can trace the same steps in both.

Based on these studies, we have formulated a hypothesis about the evolution of symbols, reflective thinking, and language skills that challenges the prevailing theory. The prevailing theory asserts that, during the course of evolution, change occurred predominately through changes in genetic structure—through the processes of natural selection, genetic mutation, and random genetic drift. Although modern evolutionary theorists will stress that this process of genetic change was ruled by chance and contingency due to unpredictable environmental events, the basic notion persists that evolutionary change came about predominantly through changes in our genes. The genetic changes that were associated with greater adaptation tended to persist.

We will show, however, that the prevailing theory is incorrect. The origins of symbolic thinking and speaking depend heavily on the social transmission of cultural practices that were not genetically determined but were passed down and thus learned anew by each generation in the evolutionary history of humans: a history that extends far back beyond the appearance of anatomically modern humans, to the early humans, the Australopithecines, and even beyond. These cultural practices are necessary for each generation to master the stages of emotional signaling that lead to symbol formation and reflective thinking. They, therefore, constitute an essential element in the growth of the human mind and human society, and, indeed, in the ongoing development of human minds and human societies.

Ever since the pioneering work of Robert Boyd and Peter Richerson, evolutionary theorists have become increasingly interested in the role of cultural and social factors in human evolution.[1] Indeed, there has been a growing interest in recent years in the so-called Baldwin Effect,[2] according to which learned behaviors operate in much the same way as environmental challenges in classical evolutionary theory, insofar as those individuals who favor that learned behavior are selected (either through mutations or by tapping into pre-existing capacities). But in this theory, the learned behavior is ultimately passed on through the genes, not through new learning in each generation.

In the majority of cases where theorists have begun seriously to explore the role of culture and society in the evolution of the human mind, therefore, the primacy of the genetic perspective persists.[3] For these arguments remain wedded to what Susan Oyama describes as a doctrine that assumes that the effects of the cultural and social environment are passed on to descendants through the genetic structure.[4] The basic principle thus remains that human beings developed—and develop—the capacities to reason, speak, see one another as intentional agents, live in complex rule-governed societies, and so on, because specific genes for each of these abilities were naturally selected.

The strongest challenge to this determinist form of reasoning has come from behavioral scientists who tend to take into account an entire dynamic system rather than to isolate single factors. Because the multiple environments that genes interact with have endless degrees of variation, the developmental outcome of traits or behaviors can be influenced in a near infinite number of ways.[5] According to this way of thinking, evolutionary change involves changes in the developmental system.[6] Therefore, attempts to attribute this or that percent of influence to genetic or environmental factors when looking at intelligence or different types of temperament are not only fruitless, they're inaccurate. Nature and nurture are constantly influencing one another, much like Fred Astaire and Ginger Rogers during one of their memorable dances.

Understanding the dance between nature and nurture (genes and environment) requires more than assertion and vivid metaphor, however compelling. It requires a careful analysis of how each partner actually interacts with or influences the other. For example, Gilbert Gottlieb showed that wood ducklings could only learn the calls of their species if they heard them from their parents or siblings as hatchlings before they were hatched.[7] Nobel prize winning neuroscientist Eric Kandel showed how learning experiences influence regulatory genes, which in turn, influence biological processes involved in the formation of neural pathways that make long-term memory possible.[8]

How does environmental variability interact with genetic variability in the development of a range of human capacities? For some capacities, the genetic structure may set the constraints and environmental experiences may operate more like a switch—turning on or off certain "regulator" genes, which in turn, influence gene expression and behavior. For other capacities, however, environmental variability and learning may play a far more complex role, be far more influential than thought, and be a necessary condition for these capacities to develop. We believe this is especially true for our highest level human capacities, such as symbolic thinking. For example, we have worked with infants who were born with motor and sensory differences, such as low muscle tone and underreactivity to touch and sound. With one type of environment, they have a high probability of becoming self-absorbed and evidencing severe language, social, and cognitive deficits. With another type of environment, one that is geared to their particular physical traits (i.e., we construct a "key" to fit and open their "lock"), we have been able to help many of them master the stages of emotional signaling and become engaged, interactive, and symbolic, with high levels of social, verbal, reflective, and empathetic capacities.[9]

As indicated, in this model, basic biological capacities are a "necessary" but not a "sufficient condition" for an individual learning to construct symbols and to think. That is, our biological *potential* for learning from experience, which includes our rudimentary capacities to perceive, organize, and respond, is the critical substrate for the capacity to learn. The sufficient condition, however, involves a series of learning steps that are the basis for symbolic thinking. In human beings, however, even the tools of learning must be learned and relearned by each new generation.

These include the ability to attend, interact with others, engage in emotional and social signaling, construct complex patterns, organize information symbolically, and use symbols to think. These "tools" enable us to develop knowledge, wisdom, and empathy. They are also the means for effective protection, security, and social and political organizations.

We were able to search for those stable learning processes that have been passed on throughout human evolution by making a critical distinction between two types of cultural and learning processes: There are those that originated millions of years ago and have been passed on through learning from one generation to the next over the evolutionary time period and are therefore quite permanent and near universal in human groups (i.e., the capacities to attend, relate, signal with emotions, etc.). There are also those that are determined by variations in each individual and time period, that is, they express the near infinite variations of human groups (i.e., the specific ways a person attends, relates, and signals with emotions, for example, signaling pleasure, but not anger). The former involve basic learning processes and the latter embrace individual content and behavior stemming from these processes.

Once we could identify these processes in our research on the development of human infants and nonhuman primates and the fossil record, we were able to see how the critical learning steps leading to symbolic thinking were embedded in our cultural learning processes and not in the structure of our genes (as important as these were and are as the necessary foundations for learning). In the chapters that follow we will identify and describe the critical and culturally mediated learning processes that have made us human and have the potential to continue our mental growth.

We will show that this theory can answer the three fundamental questions about the evolution of human beings that challenge current evolutionary theory:

1. What are the factors that promoted the growth of the human mind?
2. What is the relationship between these factors and the processes that enable a child to develop the higher-level reflective capacities that characterize our species?
3. How do these factors relate to the origins of human society?

To be precise, we will show that the same processes we describe in Chapters 1 and 2 that lead to creative and reflective thinking in children were at work over millions of years during human and even prehuman evolution. In fact, we will show that the cultural diffusion of these processes of emotional interaction throughout our evolutionary history constituted the primary engine for the development of our highest mental capacities.

In particular, in what follows we will show how

- certain cultural practices are absolutely necessary for the development of higher-level symbolic and reflective skills, and that without these practices humans cannot develop the species-typical skills that distinguish our species;
- these cultural practices are a necessary factor in the evolution of symbolic and reflective thinking;
- the cultural practices we are discussing date back to prehistoric times and were likely shared at the outset with the nonhuman species that preceded us;
- certain basic practices are also present in nonhuman primates;
- these cultural practices are not self-sustaining nor inherited but passed on only through social learning in each new generation;
- this critical social learning progressed in a specific manner from one generation to the next over millions of years through different species until each of the stages of emotional and intellectual development (to be described in Chapter 2) was mastered; and
- the fossil record, as well as an examination of the behavior and interaction patterns of nonhuman primates, will be shown to support this new theory.

This type of culturally transmitted learning may have been especially important for the evolution of flexible problem solving that uses complex communicative and symbolic processes because these processes needed to be sensitive to changing environments and, therefore, could not be structured into fixed biological pathways as could simpler patterns.

Many evolutionary psychologists believe not only that human intelligence evolved through genetic changes but that the brains of modern humans have remained the same since the time modern humans first appeared. Others believe that human brains have remained the same for at least the last 30,000 to 50,000 years (since the time cave drawings and new tools emerged). Contrary to this assumption, we will consider evidence suggesting that during this time the very structure of the human brain has changed. For to the degree that new generations had access to culturally mediated early interactions that were more "enriched" than those of previous generations, there were opportunities for the brain to develop new pathways. For example, on brain imaging studies musicians have been shown to have more neuronal connections in the area of the brain that regulates the hand movements involved in their musical performance. In Chapter 10, we will present the mounting evidence for a relationship between experience-based learning, including the role of environmentally sensitive regulator genes that turn other genes on and off, and the growth of neuronal pathways in the brain. These enriched, or improved, formative learning interactions likely led to new neuronal pathways. In Chapters 1 and 2, we will discuss how our highest level mental capacities, such as reflective thinking, only develop fully when infants and children are engaged in certain types of nurturing learning interactions. Children who are deprived of critical early nurturing learning interactions tend to have a variety of problems in the development of their social, language, and thinking capacities. But, as indicated, the predominant way these pathways that require learning and development are transferred from one generation to another is through the continuation of culturally mediated learning interactions.

We will also show that a current view of the way in which the brain organizes emotions, advocated by neuroscientists such as Joseph LeDoux, is incorrect. This view, which sees emotions as states of mind that are somewhat separate from and compete with or influence logical thinking, is not consistent with our clinical observations of growing infants and children. In fact, we will show that this view is based on confusion over the difference between pathological and healthy emotional development. In pathological development, the systems that organize emotions and logical thinking may remain separate. In healthy development, these systems become fully integrated and catastrophic emotions, such as rage, become transformed. Emotional signaling will be seen to provide the missing link between the level of the brain that involves basic emotional circuitry (subsymbolic systems, such as the amygdala) and its highest cortical symbolic capacities.

Similarly, we will show, contrary to the views of Chomsky and Pinker on the genetic origins of language, that language and cognition are embedded in the emotional processes that, in our hypothesis, led to symbols.

We will also show that while Piaget and his followers made pioneering contributions in formulating the way a child acts on his world to learn to think, they were not able to figure out the mechanism through which symbol formation and thinking occurs. Piaget described stages involved in thinking and discussed emotionally meaningful behaviors, such as imaginative play. However, he viewed emotions as more of a secondary phenomenon, useful for motivation and, at times, guided by a child's reasoning abilities. He and his followers, however, did not realize that emotions and their transformation into various levels of emotional signaling and mental representation were a critical mechanism in the development of thinking and that at each level of thinking, emotions lead the way to higher levels of thinking. For example, an infant learns causality through his experience of his smiles leading to his caregivers' smiles months before he learns to pull a string to ring a bell (which Piaget believed was the beginning of causal thinking).

Our work with children with autism (discussed in Chapter 11) and other developmental and emotional challenges provides supportive evidence for our developmental, evolutionary model of how symbols are formed. In general, we observe problems when critical emotional interactions are not available, because of either biological or environmental challenges. This is seen in autism (biological) or deprivation of nurturing care in orphanages and multiproblem families, and in experiments with nonhuman primates (environmental), where there are often severe limitations in emotional development, and, in humans, in the capacity for symbolic thinking. In children with autism, biological factors, which vary in pattern and degree from child to child, undermine their capacity to master the stages of emotional signaling that we believe are important for symbol formation. If our theory is correct, helping some of these children find another pathway to master the stages of emotional signaling should enable them to create symbols and think. We have found that a subgroup of these children (where the biological limitations afforded more flexibility) verified this prediction. In fact, their symbolic thinking capacities were proportional to the degree to which they mastered the different stages of emotional signaling. As we will show, children in this subgroup developed the capacity to engage, read, and respond to affect signals, empathize with the feelings of others, form symbols, and think reflectively to a degree formerly thought unattainable in children with autistic spectrum disorders.

Another part of the evolutionary puzzle is how early humans learned to live and work in families and groups and build complex cultures and societies. The concept of natural selection and survival of the fittest has tended to emphasize competition among individuals for the survival of their genes. Unfortunately, since Hobbes, it's been widely accepted that the basis of social and political discourse and organization is language, which is assumed, for the most part, to be genetically mediated.

In contrast, we will show that the growth of complex cultures and societies and human survival itself depends on the capacities for intimacy, empathy, reflective thinking, and a shared sense of humanity and reality. These are derived from the same formative emotional processes that lead to symbol formation. They enable human beings to work together in larger and larger groups. Ironically, even successful competition, beyond the brute force level, depends on cooperative group functioning with a high degree of mutual empathy and trust.

We will trace the growth of these social capacities through nonhuman primates and early human cultures. Then we will show that understanding the types of early emotional processes that lead to symbols can contribute to understanding the developmental levels of different cultural and social groups. This understanding provides a new "lens" to observe both the near-infinite cultural variations and the remarkably stable patterns that have characterized social groups since their inception. The affective processes that orchestrate individual intelligence connect the individual to the social group and characterize the way in which the group functions. The levels of functioning of the social group, therefore, will be seen to be an extension of the intelligence of individuals and a continuation of the evolution of intelligence.

Although most students of human behavior now focus on the interaction between nature and nurture, nonetheless, as indicated earlier, there is an overemphasis on the role of genes and an underemphasis on the role of culture (except as made possible by genes) in the formation of high-level thinking and social capacities. Perhaps there is comfort in assuming that biology, rather than families, communities, and culture, is responsible for and protects our highest mental abilities. Do we somehow feel that biology can preserve our humanity better than learned, culturally mediated processes? Perhaps so! If we believe in "fixed biology" over "experience," we can falsely believe that only these severe mental illnesses, which are characterized by biological aberrations that affect a small percent of the population, can derail these processes and ignore far more common changes in family structure, international relationships, culture, and the environment that threaten the foundations of our humanity and intelligence.

The evolutionary forces that equipped the human baby with the potential to learn made that same baby dependent on a learning environment of emotional, social, and cultural experiences for subsequent development. Remarkably, over millions of years, these basic, learned, culturally transferred processes keep improving and building on one another (at least until now). In Part II, as we trace the "line" of human evolution from prehuman ancestors and many species of nonhuman primates to modern humans, we will show how each new species and group reached higher levels of these core basic processes.

What are these vital learning processes that carry the building blocks of culture and intelligence from one generation to the next? In Chapters 1 and 2, we will identify these stages. They are more basic than the commonly assumed building blocks, such as language, new tools, or symbol use in its own right, and underlie the mechanisms of observing, imitating, and practicing that are commonly assumed to be hardwired. We will show that rather than being hardwired, basic tools, such as social imitation, are themselves learned through earlier emotional interactions.

One reason we assume that our highest human capacities are biological is that, when we look around, many people appear to have at least some degree of them. Most people relate, interact, think, and solve problems. Yet, as we will show, the reason these processes can be found in so many peoples, cultures, and geographical areas across long stretches of human history is that similar culturally mediated learning practices began in our prehistory and, therefore, characterize many cultures, settings, and historical periods.

Similarly, the ability of individuals from across the globe to arrive at an implicit consensus on certain basic elements of reality, in spite of near-infinite cultural variations, emerges in part from these basic shared interactive processes and experiences that originated millions of years ago.

Therefore, although our common DNA creates a potential for learning in helpless newborns and the potential for progression, it is our common ancestor's behavior and cultural patterns that set in motion the development of practices that influence the way newborns of nonhuman primates and human primates are nurtured and master the fundamental emotional learning processes leading to symbolic thinking.

Before we conclude this introduction, we should like to provide a glimpse of the evidence for our hypothesis that a common set of cultural ancestors and interactive emotional patterns led to symbol formation.

Consider two babies engaging with adults. Nathan, a one-year-old, is sitting on the floor, playing idly with his toes, when Sue walks in. The moment Nathan sees her his face lights up, to which Sue responds with a broad smile. Nathan giggles and waves his arms slightly as if beckoning Sue to join him. She sits down on the floor and, laughing, asks, "What are you doing with your toes?" When she tickles his toes, Nathan pulls his feet away sharply, laughing as he does so, his big smile extending from the corners of his mouth all the way up to his eyes. Her eyes sparkling, Sue asks in a gentle voice: "Can I please see your toes? I promise not to tickle them." With a wary look on his face, Nathan slowly extends his right foot, and, just as Sue starts to lean forward, he pulls his foot away and breaks out into a full-bellied laugh. Sue reacts with a crestfallen look on her face and hangs her head slightly, to which Nathan responds by edging closer and slowly extending his foot again. Sue keeps her eyes on the ground and then suddenly darts forward and starts tickling Nathan's toes, laughing loudly. For an instant Nathan looks frightened, but then he starts to smile and begins laughing in perfect unison with Sue while making faint attempts to brush her hands away. After about a minute of this, they both begin to quiet down and their faces and bodies relax; Sue says, "I'm going to go get your dinner ready." As she starts to stand up Nathan suddenly indicates that he wants to be chased and starts to move purposefully across the room, looking back with nodding glances as Sue is in hot pursuit. Nathan then smiles broadly and shrieks in delight as Sue, laughing loudly, shouts, "I'm going to catch you."

Sasha, a one-year-old, is lying on the floor hugging Eeyore, his favorite stuffed animal, when Ginny walks in. When she sees him hugging Eeyore, she smiles warmly and says in a high voice, "Isn't that nice: You're hugging Eeyore." Sasha quickly looks up and struggles to his feet while clutching Eeyore. As soon as he is standing, he holds Eeyore up to Ginny with a look of utter seriousness on his face. Ginny bends over and gently takes Eeyore from his hands, saying as she does so: "Are you giving me Eeyore? Isn't that wonderful. Hello, Eeyore. Would you like a big hug?" As she is saying this, she is giving Eeyore a warm hug, burying her face in his belly. Sasha watches intently and, while she is cuddling Eeyore, he reaches up with his right hand and points. Smiling broadly,

Ginny responds by handing him Eeyore. Sasha takes Eeyore and hugs him in exactly the same way that Ginny has just done, burying his head in the toy's belly. While he does this, he makes a sort of babbling sound that has the same rhythmic contours as the words Ginny has just spoken. The cuddle lasts for a few seconds and then Sasha looks up at Ginny and, his eyes shining, he holds Eeyore up to her again. Ginny, her eyes shining with the same intensity, responds, "Oh boy, can I hug Eeyore again?" As she starts to take Eeyore from him, Sasha gives a huge smile and breaks into a happy laugh and starts to wave his hands, almost as if he were trying to clap. Now laughing herself, Ginny hands Eeyore back and, her mind on something else, turns and begins to walk towards the kitchen. As soon as she has taken her first step, Sasha makes some distinct sounds that have a yearning quality. Hearing this, Ginny immediately turns back to him and, now laughing loudly, asks, "Does Eeyore need another hug?" This time, she bends over and hugs the two of them while Sasha chuckles softly.

These babies are both developing their ability to engage in back-and-forth emotional signaling. As we said earlier, this ability, a critical step in forming symbols, is not genetically mediated, but took millions of years to be learned and was then passed on from nonhuman primate cultures to human cultures. It is not mastered by every baby; these subtle emotional interactions in which Nathan and Sasha are engaging with their caregivers are delicate learned processes that require skillful caregiving. Yet they are not new or even uniquely human.

In fact, one of these two babies is a bonobo chimpanzee interacting with his mom. Which one is it? The answer is in the endnote.[10]

These interactive emotional processes that we share with nonhuman primates and many other mammals reach higher levels of organization in humans. As children transform emotions, through learning interactions with caregivers, from primitive fixed actions, the emotions become part of and orchestrate intelligent problem-solving interactions with the world. For example, we show in Chapter 2 how complex advanced emotions, such as empathy, compassion, respect, and pride, become necessary ingredients in high levels of reflective thinking.

Just as the discoveries of the wheel and fire set in motion enormous technological advances, the learned ability to signal with emotions and progress through various stages of emotional transformation enabled the development of symbols, language, and thinking, including reflective reasoning and self-awareness. In the twenty-first century, however, these culturally mediated learning processes that took millions of years to evolve may be at risk. Because the critical changes in the mind and brain that support reflective thinking depend on the way in which humans interact emotionally with and learn from each other, these patterns are vulnerable. Misunderstanding regarding what's essential about human beings could easily increase this vulnerability.

For example, the human mind is being thought of more and more as a series of neuronal circuits under genetic control. These circuits are viewed as accessible to biological manipulation with everything from "nano robots" to polypharmacy, without sufficient appreciation of the emotional and social contexts within which the mind and brain grow and maintain their subtle equilibrium. There is also insufficient appreciation of the boundary between curing disease on the one hand and tampering with our humanity under the guise of artificially increasing our potential on the other. Events or changes in family or group structure that alter the nature of early interactions between caregivers and their infants and children could alter an evolutionary line that dates back to prehuman and primate cultures.

The first section of *The First Idea: How Symbols, Language, and Intelligence Evolved* will focus on the formation of symbols and reflective thinking. In Part II, we will then show how these insights have helped us know what to look for in human evolution, and present a new model of evolution.

In Part III, we will show how this hypothesis challenges and revises current thinking about the human mind. We will explore its implications for a new view of language, cognition, autism, and the development and functioning of the brain.

In Part IV, we will explore the implications of this model for understanding groups, cultures, and societies. We will show how "groups" evolved through a series of stages and can be characterized according to their developmental needs and functioning. Our developmental model of groups will reveal the new challenges of a world now connected not only by shared communication and economies but by shared technological and environmental dangers as well. It will inform a new psychology of global interdependency that can help us understand emerging social and political forces.

Although the authors speak as one voice in this book, we bring very different training and experience to bear on the questions discussed. One of us (S.I.G.) is a child psychiatrist and psychoanalyst who brings years of clinical work with infants, children, and their families, as well as research in early development and mental health. The other (S.G.S.), philosopher and psychologist, brings intense study of language, both in humans and nonhuman primates, as well as research into Artificial Intelligence and the philosophy of psychology. Although readers may be aware of our individual contributions, the authors have worked together for many years to uncover the developmental pathways leading to language, intelligence, and mental health.

—— PART ONE ——

Origin and Development of Symbols

—1—

Origin of Symbols

HOW DO WE COME to form symbols and learn to think? As we mentioned in the introduction, we have recently found a new answer to this longstanding question. Our ability to form symbols, which enables us to represent our world and reason about it—and all the great intellectual accomplishments that build upon this—has an unexpected origin. In order to develop symbols, we must transform our basic emotions into a series of succeedingly more complex emotional signals. This human capacity to exchange emotional signals with each other begins in early life during an unusually long practice period and leads to symbols, language, abstract thinking, and a variety of complex emotional and social skills that enable social groups to function. The exchange of emotional signals may also play a critical role in the development of the brain, especially that of the higher cortical centers dealing with language and thinking, the prefrontal cortex dealing with planning and problem solving, and the integrating pathways that connect subsymbolic systems that process basic emotions such as fear and anxiety (such as those involving the amygdala) with cortical symbolic capacities.

Symbol formation is a complicated process that has perplexed ancient and modern philosophers and scientists alike. Only human beings, as far as we know, can engage in reflective thinking. They can think about and judge their own thoughts or feelings, saying to themselves, for example, "That's a good idea"; "That's nonsense"; "I want (or don't want) to do that"; "I should or shouldn't feel this way." Many members of the animal kingdom can solve problems, but only human beings, we believe, can use symbols to think reflectively. Thinking means not just having a verbal or visual image, but being able to take that image, manipulate it, combine it with other images, and then organize these at different levels. At the reflective level this involves reflecting on one's own thoughts—"Why am I thinking about that?" We all vary considerably in the degree of reflection we can use. There would seem to be no limit to ever deeper reflections or new insights about many of the eternal questions.

If we are to nurture these vital human talents in childhood and adult society, we need to know how this ability is acquired through development and how it came about during the course of evolution.

How does a baby develop from an action-oriented, global set of responses into a reflective, thoughtful, problem-solving individual? There are various existing theories. In our view, however, they leave this fundamental question unanswered. Although we have traveled to the moon and back, we haven't been able to figure out how we became able to "figure out" how to get there.

A MISSING LINK

The most popular current view held by those in the field of behavioral genetics, neuroscience, and cognitive science is that the ability to symbolize and use language is based on the physical wiring of the brain. This view holds, for example, that there is a language module in the brain. This module is contained in the genetic code and, during the course of evolution, came into being through natural selection. This view was popularized years ago by Noam Chomsky,[1] the MIT linguist who holds that important elements of language (e.g., grammar) are prewired and require only that the environment "turn on" the program. Steven Pinker, also from MIT, in his new book *The Blank Slate,*[2] presents a recent example of this view.

Another theory of how the brain and mind work holds that emotions operate at one level of brain organization and rational, reflective thought at another level. For example, Joseph LeDoux,[3] whose work we will describe in more detail later on, has mapped out the role of the limbic system in the processing of emotions. While this "subsymbolic system" relates to symbolic, cortical systems, it is viewed as a fundamentally separate system within which emotions and reflective thought are often in conflict (passion versus reason).

In both these views the ability to form symbols is mostly genetic (hard wired). There are modules in the brain that are biological givens. If a baby is fed, is not ill, and doesn't have a neurological deficit, the baby develops a certain wiring of his or her brain because it's in the genetic blueprint. While few would argue that a child does not need to hear language to become language-literate, this theory holds that only a very general kind of environmental switch, like hearing the human voice, is required in order to learn language. A sizable group of scholars believe in this viewpoint. Both thinking skills and temperament are viewed as being the product of our genes. The environment isn't discounted entirely; it's just assigned a lesser role.

This view does not discuss the developmental steps in the growth of the mind that would account for the shift from a reaction-driven baby to a reflective adult. The explanation is that somehow this transition is prewired or preprogrammed into our brains and emerges on its own schedule with only more general or global environmental input required.

Since this genetic perspective has not worked out the critical steps and pathways leading to language or intelligence, it can't answer why, for example, we see different levels of thinking among identical twins. Years ago, we did an in-depth clinical study of identical twins and found that many had different levels of reflective thinking.[4] Not only in identical twins but in all children, the emergence of reflective thinking and use of symbols can be derailed by differences in nurture.

Why are some adults impulsive and unable to use the ability to reason and think to control their actions? They just discharge aggressive behavior. Why do children in orphanages who have been given adequate medical care and nutrition and heard language spoken, but who have been deprived of critical social interactions, often fail to develop fully language and social skills? These are questions that would lead us to ask what the mechanisms are that explain (even if you're inclined to believe in minimal environmental influence) these variations in intellectual development.

Another view is that proposed by the field of cognitive development. Jean Piaget and his followers created the most systematic version of this theory. Piaget offers a more developmental and dynamic explanation for how symbols are formed,[5] but misses its most critical elements. He believes that symbols emerge from earlier levels of sensorimotor intelligence. His model sees the baby operating on his environment and learning from experience as a critical factor. He refutes Chomsky's and the genetic and neuroscience view that symbols emerge from a prewired, genetically controlled brain module and emphasizes the role of learning and experience. As we will discuss in more detail later, however, Piaget does not give an adequate explanation for the critical transitions from intelligent action to intelligent thinking with symbols. Piaget's constructivist stance—that is, the child constructs his understanding of the world from active experience with the world—has many helpful elements, including wonderfully detailed descriptions of stages of thinking. His pioneering descriptions of levels or stages of thinking, however, do not explain this critical transition, that is, how symbols and thinking actually come about. What are the mechanisms?

Cultural anthropology, child development, and dynamic systems approaches have also looked at the issue of symbol formation and the genesis of thinking. Researchers in these fields recognize the importance of experience and, like Piaget, believe that children's development depends on human interaction. These fields have also emphasized the enormous variation and the importance of the social and cultural context in the way in which children learn to symbolize and think. Although these fields recognize certain missing pieces in understanding human development, they leave open the question of how symbol formation and thinking are learned.

The question of how a baby progresses from global reactions to reflective thinking not only has perplexed students of human development but has also perplexed those in the field of evolution. As we will discuss in more detail in Part II, evolutionary psychologists have been struggling to figure out how primates progressed from prelanguage to language-based beings—how they progressed from early, action-oriented creatures capable of only a limited degree of tool use to reflective individuals who can build a rocket ship capable of travelling to the moon, create great literature, and construct complex social organizations. The best that current theories of evolution can come up with is that during the last 200,000 years, modern humans, through genetic changes, have developed a new capacity for higher-level thinking and that challenges from the environment have enabled human beings to use this new, genetically mediated mental capacity. The mechanisms by which this happened, and the steps through which it happened, are still a mystery. Therefore, there is no coherent or logical explanation for why, between 30,000 and

50,000 years ago, the fossil record shows a surge of new tools and remarkable symbolic drawings on cave walls.

THE SOCIAL ORIGINS OF COGNITION

In 1929, the Russian philosopher V. N. Volosinov wrote in his book *The Philosophy of Language*[6] that "signs (symbols) emerge only in the process of interaction between one individual consciousness and another. . . . And the individual consciousness is filled with signs." Lev Vygotsky, the well-known Russian psychologist and philosopher, also stressed the social nature of cognition and language.[7] Studies in the middle of the twentieth century established the importance of early emotional engagement between infants and caregivers by showing the massive effects of institutional deprivation on cognitive, language, and social development. The well-known studies of Spitz, Bowlby, and Hunt, as well as the notorious experiments of Harry Harlow on nonhuman primates, showed that even babies who were well-fed and properly cleaned needed emotional contact with a caregiver to grow and develop.[8]

There is considerable current interest in and support for the role of emotions and social interaction in human development. For example, higher-quality child care, including emotionally sensitive caregiving, is associated with stronger cognitive, language, and emotional and social development.[9] On the other hand, recent studies show that insensitive or poor emotional interaction between infants and caregivers, low-quality day care, long hours in full-time daycare (over thirty hours a week), and a mother's returning to full-time work (over thirty hours a week) before her baby is nine months old were all associated with more challenges, for example, increased stress and aggression.[10] In a study of multirisk families, we found that early problems in emotional interaction were associated with a variety of language and cognitive difficulties in children, and that as we improved early emotional interactions through a comprehensive intervention program, competency in emotional and intellectual functioning increased.[11] In another study, we found that the risk factors, such as mental illness and family dysfunction, that undermine caregiver and family emotional relationships are associated with compromises to intellectual functioning. Families with four or more risk factors interfering with relationships were twenty times more likely to have children exhibiting marginal IQ scores and behavior problems at age four.[12] This pattern continued and was again documented when the children were thirteen years old.[13] In studies of emotional rhythmicity between caregivers and babies, it has also been shown that early emotional interactions influence cognitive capacities in the preschool years.[14] A recent study of older children showed that this basic capacity for timing and rhythmicity, which originates during the infant's earliest emotional interactions, is associated with levels in a range of cognitive and academic skills during the school years.[15] A large empirical literature, recently summarized by the National Academy of Sciences, also documents the importance of early emotional interactions for children's later cognitive and social skills.[16] This review includes the well-known studies on the favorable impact of early intervention.[17]

THE ROLE OF EMOTION

Recognition of the social origins of cognition has also been strengthened by neurological research. New knowledge of the pathways that process emotions, the effect of stressful emotional experiences on the brain, and the impact of brain lesions on emotional regulation has illuminated this link (see Chapter 10).[18] In addition, explorations of the types of thinking that are part of skillful social interactions (emotional intelligence) and concepts of multiple intelligences have further supported and increased interest in the role of emotions.[19] Recently, when we worked on emotional interactions with children with developmental disorders, including autistic spectrum disorders, we found that a subgroup of the children could learn to engage with others; think creatively, logically, and reflectively; enjoy peers; and do well academically in regular classes (see Chapter 11).[20]

In spite of this support for and greater interest in the role of emotions in human development, there has not been sufficient understanding of how emotions exert their influence, nor of the ways emotions and emotional interactions affect intelligence and its related cognitive and language abilities as well as many complex social and self-regulation skills. There has not been sufficient understanding of the psychological or neurological mechanisms of action by which emotions influence these different aspects of the mind.

In other words, we have correlational studies showing that emotions are important. We even have neurological data and clinical data showing that interfering with emotional functioning negatively influences cognitive and social capacities. But we haven't had a theory that explains adequately just how during development emotions exert their considerable influence. Because we have not had an adequate understanding of how emotions exert their influence, emotions have taken a back seat to cognition, language, and memory. Therefore, the question remains: What are the developmental steps and pathways through which such distinctly human capacities as symbol formation, language, and reflective thinking emerge in the life of each new infant and child and, furthermore, how did these human capacities emerge during the course of evolution?

A NEW APPROACH TO THE QUESTION

We have had a somewhat unique opportunity to search for answers to this question. Over the years, we have worked with normally developing infants and children and observed them to see symbol formation in progress. We saw thinking skills develop over time. However, because we also worked with children who had difficulties, we had a chance to observe many of the variations in these skills. This research included work with multiproblem families in which there was severe environmental deprivation; indeed, children weren't receiving the basics, and many of them were afflicted with ADHD, ADD, and a variety of learning problems. We worked with children with autistic spectrum disorders and other biologically based, special needs conditions where the biology of the child was expressed through problems with processing information (such as language, motor, visual-spatial processing). The problems interfered with learning to think, relate, and communicate. These children couldn't interact in the way that is needed not because the environment was unwilling to provide essential learning opportunities but because their own biology made it hard for them to learn. Yet when we figured out how to create special learning opportunities that dealt with these processing differences, we saw that many of these children could learn.

Over time we learned which therapeutic and educational experiences worked, even when there were biological limitations, and which didn't. Trying various prevention and intervention strategies gave us a unique window through which to study the developmental processes leading to symbol formation, thinking, and intelligence.[21]

THE MISSING LINK IN SYMBOL FORMATION

These observations have led us to a new understanding of how symbol formation comes about. So that we can understand this process, let's picture two situations. The first situation is that of a baby seeing his mother and crying immediately for food, or a frustrated baby who is biting in anger. In both these instances, there is a link between the baby's perception, a possible emotion (yearning for milk food, anger), and a relatively fixed set of behaviors (such as crying or biting). Consider another situation, that of an adult sitting in a bar. Another patron gives him a dirty look for making a noise and he impulsively hits the patron who gave him the dirty look. Here, too, we see a relatively fixed pattern of action that's closely tied to a perception and a possible emotion (such as anger). The baby and the barroom brawler both have in common this fixed perceptual motor or action pattern in which something that is seen or perceived leads to a possible emotion and an action pattern closely tied to that emotion (anger/biting, anger/hitting, yearning for food/crying).

Contrast this with a young child who can think to himself, "There's Mommy! Maybe she'll get me some milk if I ask her nicely," and then be able to say, "Mommy, I'm hungry. Can I please have some milk?" A more mature and reflective barroom patron might think to himself, "I wonder why that guy glared at me. Was it because I was burping or making too much noise? I wonder if that offended him. Maybe I should say, 'Excuse me.'" The barroom brawler who can think through his own feelings, and the possible feelings of another person, can say, "Excuse me!" and see whether the other person accepts his statement. Or perhaps he could say to himself, "I've seen that look before, and reacting will only get me into more trouble!" In both situations, he is exploring options based on a process of reflective thinking involving the use of symbols rather than taking a relatively fixed action based on an immediate perception and an associated emotion.

To understand how babies or barroom brawlers progress from actions to thinking, we need to look more closely at some critical developmental steps. We have found that there are two conditions that need to be met for human beings to progress from the fixed action level to a level where they can create meaningful symbols and thoughts. The first condition, which we have discussed in earlier works,[22] is that relevant emotional experiences must invest symbols as they form. Images without emotion, or affect—that is, without meaning—create memories without meaning. We see this phenomenon in some children with autism. They just repeat words rather than convey what the words mean. For example, "mama" is one of the first words children use. If we ask, "What does 'mama' mean to you?" and the answer is, "Red hair, green eyes, and tall," we suspect developmental challenges. But if the child says, "Mama is loving and she feeds me" or "Mama is mean and bossy and she doesn't do what I want her to do," then we feel confident about the child's development. Meaningful words involve emotionally rich images. It's the same with "dada" or "no." When children develop words without meaning, they tend to use the words in a scripted way, repeating what they heard on TV or read in a book.

Therefore, to create a meaningful symbol, the image must be invested with emotion. We don't have meaningful symbols (you can have imagery, but not symbols) unless affect invests the image. Even seemingly impersonal objects, such as an apple or concepts such as numbers, we have found, require emotional investment if they are to become meaningful symbols. As we will explain, an apple is not just red and round, it's something you give your teacher with pride and throw at your brother in anger. The understanding of number concepts is based on how it feels to have a "lot" or a "little" of something. To a three-year-old, "a lot" is more than you want and "a little" is less than you expect.

The second condition for creating meaningful symbols is the one that was a dramatic new insight for us. Yet it was so basic in its simplicity that it also surprised us. This second condition is that *a symbol emerges when a perception is separated from its action.* The developmental process that enables a human being to separate perception from action provides the missing link in understanding symbol formation and higher levels of consciousness, thinking, and self-reflection. In Chapter

3, we will also show that this developmental process was a critical step in the course of evolution.

THE SEPARATION OF PERCEPTION FROM ACTION

In most nonhuman animals, very young babies, and impulsive older humans, perception is tied closely to action. A trout, for example, perceives a fly or a larger fish, and what happens next? He moves toward it and attacks, or he flees. The perception triggers an action. There is very little pause or delay in carrying out the action. In a baby, you expect this. In some preschoolers, you expect this (a child sees a toy and he grabs it). Across the animal kingdom, you can see that some are more tied to the perceptual/motor (action) patterns and some a little less because they are more adept at problem solving.

To understand how humans make the huge step from action to thinking, let's look at the parts of this process more closely. First, what is a perception, whether of another person or an object such as an apple? We perceive when the mind registers a sight, sound, smell, or other sensation. Often, a perception involves the registering of several or all these sensations together. Such a multisensory "image" is simply what we see, hear, and/or touch. Without an action, it is a freestanding image or multisensory impression or "picture."

When the baby and the barroom brawler move from perception to action, there are three steps. They perceive something, for example, the baby perceives the mother or the barroom brawler perceives another person giving him a dirty look. They then may experience an emotion, and a tendency toward one or a few actions at almost the same moment. The baby, for example, may experience hunger and then begin crying or sucking to deal with that emotional and physical need. The barroom brawler may experience rage and, in a fixed pattern, immediately strike out. Even verbal adults will often be unaware of this emotional part of the fixed pattern. When asked what happened in a situation, they may reply, "He gave me a dirty look and I hit him." If you inquire, "How did that dirty look make you feel?" the barroom brawler is likely to say, "It made me feel like hitting him." He simply describes his action. Of course, our baby can't verbally describe his actions, but in all likelihood he also experiences his own actions without any self-awareness.

What if this pattern is interrupted? Then a perception may not lead automatically to a fixed action. In other words, our baby would be able to perceive his mother without being driven to some fixed behavior. Our barroom brawler might be able to perceive that somebody is looking at him critically and not be driven instantly to hitting. If the baby or brawler can perceive the other person without taking action, they are left with a *freestanding image.*

The ability to experience a freestanding image (a multisensory picture) is a momentous developmental step. Such a freestanding image is used in a variety of ways; for example, it can be used to accumulate more experience. Rather than the image of mother being associated only with crying and milk, it can become associated with other experiences as well. Over time, this freestanding image of mother can be imbued with many emotional experiences—playing, eating, being comforted, and so forth. The freestanding perception or image acquires emotional "meaning."

As a freestanding image is seasoned with more and more emotional experiences, it is on its way to becoming an internal symbol. Over long stretches of time, a symbol for something as complex and important as "mother" will involve such complex meanings as love, devotion, control, annoyance, and sacrifice. But, to begin this journey toward becoming a symbol, a perception must first be separated from fixed actions.

A freestanding perception that becomes an internal symbol continues to define itself throughout life. This is true for simple images and symbols as well as for complex ones. The internal image of a strawberry enables us to see, and sometimes almost to taste or smell, that berry without actually eating it (taking action). As with the example of the image of "mother," the image of a strawberry can acquire more meaning, perhaps from memories of summer meadows or gardening, or from books and films.

TRANSFORMING CATASTROPHIC EMOTIONS INTO EMOTIONAL SIGNALS

Now to the tough question: How do we separate a perception from the initial tendency or urge to take immediate action? To answer this question, let's first look at instant nonreflective responses, which tend to keep human beings tied to fixed actions. Babies, older nonhuman animals, and barroom brawlers usually experience affects or emotions in a "catastrophic" way. Catastrophic emotions are intense global emotional states, such as massive rage, fear, or emotional hunger or neediness, that press for direct discharge in fixed actions. These intense global feeling states are often tied to the fight or flight reactions, massive avoidance approach, seeking behaviors, or other basic responses. When a caribou sees a wolf, or a baby sees a stranger, fear drives responses. In these situations, there is little subtle nuance or pause to solve a problem.

Human beings under the pressure of intense emotions will often attack or run away. Others may also withdraw, avoid, freeze, or approach. For some, if the emotion of fear is very intense, they are momentarily almost paralyzed. In these situations, we are observing "catastrophic," or global emotional, reactions.

When people say something's emotional, and use the term "emotion," often they are referring to this level of intense (often seemingly irrational) emotionality—fight or flight reactions, panic, avoidance, feeling overwhelmed, being overwhelmed. These catastrophic feeling states are part of a primitive perceptual motor level of central nervous system organization.

In human development, however, infants can learn to "tame" catastrophic emotional patterns. Early to midway in the first year of life, caregivers help babies begin to learn how to transform catastrophic emotions into interactive signals. For example, we frequently observe the following: A mother smiles and the baby smiles back. She coos and the baby coos back. A baby smiles, raises his arms as if to say, "Hey!" and vocalizes emotionally needy sounds that convey "Look at me!" The mother looks his way. Baby then smiles and flirts and the mother flirts back. By seven to eight months, the baby can keep it going, and mothers will often keep responding. Now we have back-and-forth emotional signaling, or "circles of communication." Contrast this pattern with one in which emotional signaling does not occur. A baby smiles and vocalizes and a mother, preoccupied with her own thoughts, ignores the overture. Baby tries again with more "oomph," this time moving his arms, vocalizing more loudly, and smiling even more brightly, but with obvious strain. The mother still ignores his overtures. After three more tries, we observe the baby looking despondent and resigned as he leans back, becomes passive, and stares at his mobile without showing signs of interest or pleasure.

For the type of emotional signaling we described in the first example to occur, the baby needs to have been wooed into a warm pleasurable relationship with one or a few caregivers so that there is another human being toward whom he experiences deep emotions and, therefore, with whom he wants to communicate. He also needs opportunities to be "intentional," to express an emotion or need by making a sound, using a facial expression, or making a gesture with his arm, and to have his efforts become part of a back-and-forth interaction by being responded to. For all this to happen, he needs to have a relatively intact central nervous system so that he can move his facial muscles and arms and begin to recognize patterns.

Human beings are distinguished from most mammals by having a long period of dependence on caregivers. During this time, infants and toddlers can develop, use, and rely on nonverbal, gestural, emotional signals to meet basic needs, interact, and communicate. These years provide continuous opportunities to learn and fine-tune the skills of emotional signaling; these skills will continue to be learned and refined during the course of life even after words and other symbols are mastered. As we will see, these emotional signals are used to communicate, to negotiate, and, ultimately, to govern, or orchestrate, our emotional states, and to build our cognitive abilities.

THE DEVELOPMENT OF EMOTIONAL SIGNALING

Before they learn to use emotions to signal, during these same early months of life, babies are learning about the world outside themselves. As caregivers provide interesting and pleasurable sights, sounds, and calming, regulating interactions, these experiences facilitate interest in the outside world and the use of the senses and motor system. These types of nurturing interactions enable the infant to awaken to the world outside himself. These nurturing, pleasurable interactions also deepen the infant's engagement and relationship to his caregivers. During the early months, this growing intimate relationship enables the infant to become more deeply involved in all aspects of his caregivers' emotional communications.

As his nervous system develops, an infant becomes more able to regulate his facial muscles and give a big smile (or angry glare). Not only can he look where he wants and turn toward pleasurable, interesting caregiver sounds, he can also reach for a caregiver's nose or an enticing object she is holding. Through his relationship to his caregiver(s), he is becoming more intentional.

This budding intentionality can then be harnessed into back-and-forth emotional signaling. For example, the baby turns to look at his mother as she makes an interesting sound, mother gives him a big smile, and he smiles back. She moves her head a little and makes another interesting sound and, with an expectant tone, says, "Can you find me now?" Baby moves his head to find her and greets her eyes with a big smile and sound of his own, as though to say, "What do you think of that?" Mother responds with her own sound and they continue to exchange pleasurable smiles, sounds, and a variety of facial expressions.

As mentioned earlier, a baby wants to communicate because he is so engaged with his caregivers. If his care is too chaotic or neglectful (if he is left alone in his crib all the time), he may either withdraw or become hyper-excitable, vigilant, and/or disorganized. He may remain governed by catastrophic, global emotions rather than enter into a pattern of regulated emotional signaling and interaction.

By nine to eleven months, this process can become very complicated. Babies and emerging toddlers can exhibit a range of emotions—pleasure, disgust, anger, annoyance, surprise. These are not simply expressions but can now be used for emotional signaling. A mother looks annoyed and baby looks shocked. Then she is quiet for just a split second and baby watches, as if to say, "What else are you going to do?" The mother smiles. The baby coos. The subtle facial expressions exchanged can express a vast range of emotions. There is an almost infinite variety of affective variations that can be employed in these interchanges. The exchanges also include sounds, movements, and problem-solving actions (reaching, taking, giving back, crawling, toddling here or there, searching). These exchanges of sound and movement are guided by the toddler's emotional "intent," his purposeful goals.

These baby-caregiver exchanges, or circles of communication, can become quite extended. We refer to long chains of emotional signaling as *reciprocal, co-regulated emotional interactions.* Each of these terms has a slightly different history in developmental psychology and carries with it a slightly different connotation. The term "interaction" is, perhaps, the oldest of the three, and can be traced back to Lev Vygotsky and Jean Piaget, if not beyond.[23] In Vygotsky's usage, which is the sense more commonly associated with child-caregiver interaction, the term primarily applies to social interaction.[24] But by no means does it entail that the two individuals in a dyadic interaction are equal partners, and indeed, more often than not the more experienced partner (the caregiver) monitors the child's abilities and, one hopes, ever so subtly "raises the bar" to promote the child's cognitive, communicative, and social development.

The term "reciprocal," which the developmental psychologists Bateson and Trevarthan began to highlight in the 1970s, draws attention to the back-and-forth nature of preverbal interactions between infant and caregiver.[25] It is not just that the caregiver and infant are interacting, but that there is a certain "conversation-like" pattern to their interactions. Finally, the notion of "co-regulation," which developmental psychologist Alan Fogel has been most instrumental in clarifying, draws attention to how communicating partners continuously influence one another and together create new behaviors and meanings; for example, the toddler and caregiver make interactive sound patterns that become "funny."[26] Combining all three of these constructs—*co-regulation, reciprocity,* and *interaction*—brings us closest to understanding the dynamics and the richness of what is involved when we speak more simply of "emotional interaction" or "emotional signaling." This rich history of emotional interaction, however, never anticipated the truly vital role these interactions would have for forming symbols and higher levels of reflective thinking abilities.

A good, fully engaged conversation between adults has many of the same ingredients. Both partners initiate and respond to the other in a continuing manner. Each partner takes the initiative and influences the other. It's a co-regulated pattern in that the baby is an important initiator and player in the drama. If the baby is not initiating and it's just a mother stimulating baby (tickling the tummy) or a father picking the baby up and getting the baby to give a big smile, that's a one-way street. The baby is not a full, initiating partner. For these experiences to lead to symbol formation, they must be "co-regulated," each party influencing the other, and "reciprocal" in that both participate.

When a baby engages in these co-regulated affective exchanges, he is learning to signal with his emotions to convey *intent* rather than engage in a complete fixed action. For example, the baby who is angry and begins to make an angry expression through a grimace is conveying the intent to bite or hit through the expression. If the caregiver responds before the baby actually bites or hits, that is, responds to the *intent,* the baby is likely to respond in return with another *intent.* For example, the parent responds to the baby's angry looks with a soft soothing look of "what's the matter" and, with hands out, an offer to pick him up and cuddle or feed him. The baby responds with a softening of his grimace and anger and a look of expectation. The parent then responds with another gesture—reaching closer to the baby—and the baby now begins to break into a smile as he reaches his hands towards the parent. A second later, the parent is holding the baby, snuggling, and patting his back, and the baby relaxes. The tension in his body dissipates and he has a look of calm.

Consider different examples. The baby looks angry, signaling his *intent* to bite or hit; the mother, and the father, too, who is right there, ignores the baby's emotional signal. The baby bites the father on the arm. Or the baby looks angry and the mother, instead of responding soothingly, yells angrily: "What do you want?" The baby immediately begins crying and biting his own hand.

In the first example, we saw complex signaling interrupt a fixed action. Mother responded to the baby's signal of intent. He responded back and together they negotiated an outcome characterized by shared soothing pleasure rather than a unilateral aggressive action. In the second two examples, the baby went right from *intent* into actions because his *intent* was either not responded to or responded to in a way that did not interrupt his global response but simply intensified it and triggered fixed action.

As his intentions are responded to, the baby becomes better and better able to signal intent without escalating into direct action. During the middle and latter part of the first year of life, the baby's growing neurological development, in part, makes this possible. It is a necessary but not sufficient condition, however. It must be matched with a caregiving environment that enables the baby to learn back-and-forth emotional signaling. In other words, it won't develop on its own. Without learning and practice, the baby will remain locked into global emotional states (catastrophic emotions) and fixed actions.

There are additional ways in which emotional signaling works to interrupt the fixed action patterns and free perceptions from their associated actions. By having his *intent* responded to, the baby learns to modulate the intensity of his emotions. He is learning to regulate his mood. The baby is learning to show a little annoyance, to negotiate, and to get his needs met. There is, therefore, less of a tendency to explode into desperate action. For example, a toddler is pushing his bottle away with an angry glance at his mom and is beginning to grow agitated. She puts her hand out to take it from him and uses soothing tones to convey empathy and, in a slower rhythm (down regulating) offers her finger to make contact. He squeezes it, feels a bit reassured, and looks expectantly as his mom holds up finger food to see whether he wants it. He waves his hand to convey yes. His anger and agitation is being responded to with soothing actions. If, on the other hand, the mom became agitated and used harsh vocal tones and a fast, abrupt arm movement (further up-regulating), in all likelihood her toddler would become more agitated and upset.

Consider a toddler who is looking a little sad, subdued, and self-absorbed. If his mom energizes up (up regulates) and pulls him into a joyful interaction, he learns to regulate in the other direction. Similarly, when his big smile wins a smile back he knows his joy is being responded to. He has a sense that he can regulate his mood and emotions through the regulating responses he gets from his environment. Infants and toddlers quickly sense that they and their caregivers are regulating one another when there is a back-and-forth, finely-tuned nuance system of affective interaction involving lots of mutual exchanges. It's not unusual to see infants or toddlers sometimes take the lead and flirtatiously pull a preoccupied dad (up-regulate) into a joyful interaction.

In addition, the baby is learning that negotiation involves an interaction between two people. He is not left to take action on the world alone. The perception of mother is not a static one leading automatically to hunger or to the emotion of rage and biting or sucking behavior. Once emotional signaling is learned, the perception of mother is associated with a complex pattern that involves negotiation and the promise of a slightly different outcome each time the negotiation occurs. Rather than a fixed stimulus and response, the child now experiences a variety of problem-solving interactions between two people.

In this way, the baby learns that the person he is perceiving is not a simple "thing." Mother is not simply a feeder or comforter or a juicy person to bite. Now because of the complex signaling, the baby is engaged in discovering that mother has many different characteristics. She feeds, soothes, excites, and limits. This emerging complex picture of mother, father, or others invites equally complex responses (rather than simple fixed ones).

Along the way, the baby learns to exchange emotional signals better for some emotions than others. Our barroom brawler may have nuanced ways of expressing affection, but he remains a brawler when feeling anger.

In the second year of a child's life, caregivers and toddlers signal back and forth to solve more and more problems, which further favors affective signaling over impulsive action. The toddler is hungry, for example. Now he doesn't have to become catastrophic and yell and scream. His mother doesn't have to figure out why he's screaming ("Oh, you must want your milk. I'll go get it for you."). The child can go up to his mother, attract her attention with a little flirtation, and point and gesture. Often, in response, mom smiles, looks expectantly, and gestures as if to say "Where?" The child takes her hand and starts walking over to the cookie jar. When they get there, the child bangs on it and mother looks and asks through her gestures and words, "Are you going to open it or am I?" The child points to her and she opens the jar. The child points to the cookie, and maybe even says "Coo." While the words are helpful, it's the gesturing that's most important. To obtain the cookie, the child has opened and closed many circles of communication with his mother, who has been a very interactive partner.

Little Sally, fifteen months old, wants a book, so she goes and flirts with dad. A dozen circles of communication later, dad's picking her up so that she can take the book off the shelf herself. This is another example of shared social problem solving between a toddler and caregiver. During the latter part of the first year and into the second year of life, this ability is only beginning and there are still many occasions of crying and screaming. Just the same, a huge leap in human development has been taken. Interestingly, it is during this time of rapid improvement in back-and-forth emotional cueing and problem solving that the prefrontal cortex of the brain, the seat of our ability to sequence and plan actions and regulate emotions, is developing.

THE ROLE OF THE CAREGIVER

For the toddler to learn to take the initiative and enter into co-regulated, reciprocal emotional signaling, he needs caregivers who engage him in the steps we have been describing. They need to awaken him to the world outside of himself through regulated, synchronous interactions and pleasurable, interesting sights and sounds, touch, and movement. They need to "woo" him into a pleasurable, joyful, intimate relationship with themselves. They need to read and respond to his emotional signals, support his initiative, and facilitate a continuous flow of back-and-forth expressive signaling.

In this process, caregivers elicit a range of emotions and motor and sensory capacities. For example, they facilitate joy, pleasure, assertiveness, and curiosity, and encourage looking, listening, smelling, touching, and movement all together and all as part of a relationship. (As we will describe later, this enables a child to integrate his different experiences and connect the different parts of his mind.) Because caregivers do all this effortlessly as part of having fun with their babies and toddlers, it's easy to overlook the many steps in this complex learning interaction.

Engaging in co-regulated emotional interactions is just as important for the caregiver as for the child. Through these interactions, the caregiver gradually learns at a very intuitive level how to negotiate with the child through different types of affective interaction patterns, such as when the child is angry, impulsive, sad, withdrawn, aimless, or needy. The more time the caregivers spend with the child, the more time there is to interact and become confident in their caregiving abilities.

We have noted clinically that many caregivers who do not spend sufficient time with their infants and toddlers in direct care may remain tense and anxious when their babies show strong affect. Because they have so little practice in entering into these patterns of co-regulated, emotional interactions, one consequence is that when a child does show strong affects, the caregiver may tend to be dysregulating, rather than regulating. This may lead to a higher likelihood of problems with the regulation of attention, mood, and behavior.

We may be seeing more of these problems in today's busy societies because fewer parents are logging the hours to become confident in their caregiving abilities. When caregivers are confident and early emotional interactions are pleasurable and regulating, communication and symbol use can meet some of the infant's basic needs that were originally met only through direct action. For example, back-and-forth flirtatious babbling and words can feel as good as a hug, and annoyed looks and angry vocalizations or words can satisfy a need to express anger.

Many caregivers do not engage in these learning interactions for a variety of reasons, not simply because of a lack of opportunity or practice. If a father is preoccupied and unavailable (tired or self-absorbed), he might not keep up his part of the dialogue. If a mother is depressed and living in her own world, she's not going to keep up her side. If a baby has processing or attentional problems, these can make it hard for him to initiate signals. He may have a hard time reading the parents' signals as well, and parents may then give up in their efforts.

FROM CO-REGULATED, RECIPROCAL EMOTIONAL INTERACTIONS TO SYMBOL FORMATION

As humans, we're not unique in being able to form perceptions or impressions. Most members of the animal kingdom can form images. Humans, however, can learn to engage in relatively longer, more continuous creative chains of emotional signals. They can then use this capacity to negotiate and solve problems, and, thereby, more fully separate perceptions or images from their fixed actions and construct higher and higher levels of internal symbols. A well-developed facial musculature supports our capacity for emotional signaling because it enables us to express a vast variety of subtle emotional gestures. While we also see many facial expressions of emotions and complex play and mating interactions in animals, it's important to distinguish the degree to which these interactions are co-regulated and creative. In addition, our brains have a relatively more developed prefrontal cortex than those of other animals, which supports the regulating of emotions and planning and sequencing actions. As indicated, however, these "necessary" foundations are not sufficient. Learning interactions of the type we have been describing are essential for the development of complex levels of emotional signaling.

Humans are somewhat unique in their ability to separate perception from action. Most members of the animal kingdom can perceive stimuli (even reptiles perceive a stimulus). For the most part, however, reptiles are tied to very basic perceptual-motor patterns (attack the stimulus). Many mammals engage in some social and emotional signaling. Nonhuman mammals, however, don't tend to engage in the long, affective signaling sequences needed to fully separate an image from its actions. For example, as we will describe in Chapter 3, nonhuman primates can exchange some affective signals, but they tend to do so in short bursts. When observing bonobo chimps, we saw that they looked, vocalized, gestured, and exchanged four to five quick emotional expressions and then went into action. Over time, we were able to help a few become more elaborate signalers. Although nonhuman primates clearly relate to each other and have many forms of covert communication, only some of which we likely have observed, in general, a continuous flow of overt reciprocal affect signaling is not readily observed.

Through learning to transform emotions into long chains of interactive signals, humans can form an image that's less tied to action and, therefore, can acquire meaning and become a symbol. Once the image is separated from its action—once food exists as an image separated from attacking the food source or crying—that image can serve a new purpose. Such an image can be used to plan, solve problems, and think. Human beings can also take a freestanding image and combine it with new and old experiences and other symbols. For example, they can combine the image of parents and siblings and construct a concept called "family."

As we explore the development of symbols, we may wish to return to the question implied in the title of this book—*The First Idea.* What was the first idea? According to our hypothesis, an idea is an image that has been freed from a fixed, immediate action and is invested with affects or emotions (i.e., intent) to give it meaning. Many constructivist theories, including the pioneering work of Piaget, suggest that internal symbols or representations involve getting beyond action pattern problem solving. But these theories haven't been able to explain how this happens or the mechanism involved. They haven't identified the critical role of emotions and affect signaling.

From our developmental studies, we suggest that the first idea is, therefore, the emergence of the capacity to invest a freestanding image with emotional meaning to make it into a meaningful multisensory, affective image (i.e., an idea or symbol). Did the first idea also have specific content? Was the first idea of our ancestors of water, food, mama? According to our hypothesis of its origins, the content of the first idea was as varied as the emotional interests of the many early humans who were constructing it.

In summary, the answer to the question of how human beings form symbols is that symbols come about by separating a perception, which is the ability to form an image, from its action. This is achieved by co-regulated, emotional interactions with other human beings. Ongoing co-regulated emotional interactions provide emerging and later symbols with meaning throughout the course of life.

FROM SYMBOL FORMATION TO LANGUAGE

Symbol formation also makes it possible for language to develop and be used creatively and logically for communication, problem solving, and higher-level reflective thinking. Symbol formation, however, is only one of the foundations for language. The others also, interestingly, stem from co-regulated, affective interactions and, together with symbols, form a confluence of factors that enable language to develop. As we will discuss more fully in the chapters on evolution and the development of the mind, the confluence of these factors occurs in infants and young children during the first two years of life and thereafter. We will show, however, that, during the course of evolution, these same factors took many millions of years to evolve. The critical factors for language that stem from co-regulated, affective interactions include:

- Motor capacities, including increasingly complex vocalizations, facial expressions, affect intonations, postural changes, and problem-solving actions. Co-regulated affective interchanges are the vehicle through which these increasingly complex motor skills are practiced and developed. The tongue and oral motor capacities, more generally, can express an enormous range and subtlety of affective variations and, therefore, were and are an important centerpiece for the motor skills needed to express an almost infinite variety of experience. In comparison, hand or finger variations do not have nearly the motor flexibility of the tongue and oral motor area.
- Increasingly complex social interactions that enable groups to move beyond communicating simply to meet concrete needs, such as food and shelter, and use communication for social goals. Social problem solving enables economies of scale that allow for free time beyond survival needs; it also stirs the creative use of communication.
- The use of communication for communication itself rather than only for meeting concrete needs. Co-regulated affect signaling enables nonhuman and human primates to meet their basic needs for nurturance and closeness with others (a basic need both groups require for survival) through complex, social, affective interaction. Holding, touching, and proximal smelling, and other forms of direct contact, which all rely on proximal and concrete modes of communication, can now be partially satisfied through distal communication across space because of the ability to use gestures to signal over longer distances. Feelings of closeness can now oc-

> cur through presymbolic signal exchanges and, eventually, through symbolic conversations. This means that communication in its own right begins to serve the basic need for intimacy and nurturance. Through serving this need, communication has the motivational system behind it to develop to higher levels (that's one reason why some people just love to talk).

As complex motor skills, including vocalization, combine with symbol formation, complex social signaling, problem solving, and the ability to find pleasure in communication for itself, all the factors needed for language and speech are present. Guided by relationships and the different levels of emotional signaling, these factors work together during development to create the capacity for language and speech. There is no evidence for a specific biological event either during the development of an individual infant and child or during the course of evolution (e.g., a mutation that altered evolution). Rather, there is mounting evidence for this developmental hypothesis, which explains how biology and experience work together. In this hypothesis it is necessary for a child to be engaged in a series of affective interactions that give rise to the development of motor, sensory, and social capacities, which, when combined with symbol formation, lead to language. The symbolic use of language, in turn, creates the foundation for more advanced social and intellectual capacities, including higher and higher levels of reflective thinking.

EXPERIENCE-BASED LEARNING

The theory of symbol formation presented in this chapter allows us to distinguish more clearly the respective contributions of genes and experience to higher levels of symbol formation and reflective thinking. Our genes and the changes in genetic structure over the course of evolution provided *Homo sapiens* with a unique adaptive ability: to learn from experience-based interactions with others and the environment. There is no scientific evidence that genes by themselves, however, can explain complex interactive behavior, problem solving, or thinking. This ability to learn enabled human beings to engage in interactive learning experiences and to acquire the "tools" of learning, including emotional signals, problem-solving capacities, and the use of symbols to organize and reflect on experience. We often take our abilities to think and reflect for granted and ignore all the years of learning that went into their mastery.

We also tend to mix up the learned tools of learning with factual learning, which uses more memorizing than thinking. These learned tools enable *H. sapiens* to master knowledge and wisdom, as well as facts.

Each new learned tool is a basis for learning to construct additional *new tools* for even higher-level learning; for example, symbol formation leads to putting symbols together creatively to form new concepts. In this way, *H. sapiens*' ability to learn leads to various levels of new learning. But this process, that is, the capacity to move from the hardwired ability to learn to acquiring the tools of learning (emotional signaling, problem solving, symbol formation, and thinking) and then on to various levels of knowledge and wisdom depends on each new baby's learning how to signal with his or her emotions rather than taking global actions. This is just as true for helping bonobo chimps reach their highest levels of thinking, as we will see in Chapter 4, as it is for humans. The ability of each new baby to produce these signals depends on nurturing practices that must be provided anew for each new generation. In other words, although our potential for learning is genetically mediated, the development of emotional signaling, problem solving, symbol formation, and reflective thinking, that is, our intelligence, is a culturally dependent process that is as delicate as it is wondrous.

In later chapters, we will draw on the mounting evidence that experience-based learning is an important line of evolution that dates back tens of millions of years. Because they are not fixed by biology, however, not only do these culturally mediated learning interactions have room to improve but they are also vulnerable to regression. In fact, we would postulate that the higher the mental level and the more sensitive it is to its environment (and, therefore, the more adaptive it is), the more vulnerable it is to regression. Perhaps this explains why some lower forms of intelligent behavior, such as that found in the cockroach, may help a cockroach survive a nuclear holocaust better than would a human being. Only a human being, however, can cause such a holocaust and figure out a way to prevent it.

—2—

Intellectual Growth and Transformations of Emotions During the Course of Life

INTRODUCTION

Before we explore the transformations of emotions and intelligence throughout the course of life, let's take a peek back and try to glimpse how emotional interactions and intelligence may have been working together at a critical stage in the history of intellectual development—the Renaissance. On the basis of various biographies and autobiographies, and the paintings of the period, historians of childhood have been able to identify a number of critical changes occurring in the caregiving practices during the Renaissance.[1] Whereas in the Middle Ages a child was swaddled and given relatively little attention when it began to move about on its own, and childhood was not really seen as a distinct stage in human development,[2] childhood was beginning to be seen during the Renaissance as a distinct period in human development: a period in which the child's mind could be developed with toys and special educational practices. But, as we have learnt from the case of the Romanian orphanages in the twentieth century,[3] it takes more than just an interesting toy to stimulate a child's mind. Rather, we can envisage how, through their affect gesturing, Renaissance caregivers were communicating new attitudes and values to their children, even before they could speak.

In fact, the pictures that are available tend to capture the dynamism of human interaction.[4] According to our model of human development, capacities such as critical thinking that characterize the Renaissance could not arise from lecturing children or instilling values in them in a deliberate and conscious way. Rather, it comes from second-to-second and day-to-day interactions which create implicit mastery of this type of core capacity. This type of mastery is embedded in the very processes the child is challenged to use as part of his day-to-day interactions. These processes lead in turn to presymbolic and symbolic capacities and later attitudes and values that define the new trait.

For example, when a 16-month-old toddler explores his mother's ear, looking at it, yanking on it a little bit, making interesting gurgling sounds of fascination, the mother might simply limit the child with a harsh "Stop that!" which throws the child off to do something on their own; or she might be fascinated with her toddler's curiosity and signal back to her toddler a sense of pride with a warm supportive vocal tone suggesting curiosity and admiration for her toddler's curiosity. She might also lower her head to make it easier for him to look in her ear and point to his ear as though to show him the similarity. As he touches his own ear, and then again touches his mother's ear, acknowledging the similarity, she might show him her other ear, and so forth and so on. Through many of these back-and-forth "circles of communication," a mother and her curious toddler might explore ears and noses, knees and elbows, feet and hands, and a symphony of co-constructed affective exploratory interactions. If such a toddler developed an attitude and capacity to try to understand his world with a capacity for joyful but yet systematic exploration, including an uncanny scientific knack for seeing patterns, we wouldn't be surprised.

In a similar way, when her curious toddler shakes his head "no," exercising the critical prerogative of a toddler, the mother could get angry, scaring and squashing a budding critical attitude evidenced in the toddler's experimenting with his first gestural "no"; or she could imitate his "no" with an enthusiastic shaking of her own head and then raise her arms in puzzlement with a matching facial expression and vocal tones that signal "Well, what do you want to do?" As her toddler responds to her welcoming gestures to communicate further his needs and wants, he might point to a different food or toy than that initially offered. His playful mother might then hold up two possible toys, hiding them both in her hand and letting him guess which one is in which hand and eventually select the one he wants. Here we have a pattern of interaction facilitating the toddler's ability to exercise his critical faculties and couple it with discovery as well as intentionality, pattern recognition, and problem solving, all before he can speak.

As the child progresses and learns to utter words and use symbols, the same types of interactions can play out to foster creative, critical, and scientific thinking at the level of ideas. Therefore, while admittedly speculating from limited data on the Renaissance, it is nonetheless a reasonable speculation to consider that changing child rearing practices facilitated, through the processes described above, a much fuller implicit understanding of the world which worked together with the larger social, cultural, and intellectual currents in a synergistic fashion to produce the Renaissance. The thesis of this book is that such a renaissance occurs in the life of each infant and child through the interactions he or she has with responsive caregivers.

We will be describing the different stages through which affective learning interactions operate a little later on. We will also see how these same basic processes of different levels of affective signaling create the necessary capacities to understand and collaborate in complex social organizations during both childhood and adulthood. To understand the relationships between emotions and intelligence throughout the course of life, it's vital to examine how human beings have thought about this relationship over time.

THE NATURE OF EMOTIONS

The emotional transformations that support symbol formation and reflective thinking continue throughout the course of life, resulting in newer levels of intelligence and wisdom. The Western view of human nature and emotions, however, dating back to the ancient Greek philosophers, is that human behavior is governed by two opposing forces: the emotions versus Reason. In the modern version of this doctrine, introduced by Descartes, emotions constitute the animal side of human behavior: the innate feelings, moods, and mental states that are involuntary and automatic. Reason resides in a completely separate realm of the mind: the part that is independent from emotions, and that, one hopes, comes to govern them. To this day, numerous researchers continue to study a child's emotional and intellectual development as if there were a bifurcation between these two elements of the human psyche; that is, as if a child's emotional and her cognitive or linguistic development were independent and autonomous.

Recent attempts to develop a scientific theory of emotions have remained tethered to the Cartesian view about the nature and function of emotions. Influential theorists such as Sylvan Tomkins, Carroll Izard, and Robert Ekman have developed rigorous methodologies for studying the facial expressions of emotions, and cognitive neuroscientists such as Joseph LeDoux and Daniel Schacter have deepened our understanding of the neural processes associated with the primary emotions.[5] But the basic principles underlying these research programs remain those that were enunciated by Descartes. Even Antonio Damasio, a vocal critic of Descartes's bifurcation between reason and emotion, has nonetheless remained committed to a Cartesian model of the biological origins and functioning of the primary emotions. Although he stresses the influence that emotions can have on information processing, he nonetheless sees the primary emotions as predetermined phenomena.[6]

What, then, are the defining features of this Cartesian model of the emotions? Not simply that emotions are "passive," for this thought existed long before Descartes, dating back to the Stoics. Nor that the feelings associated with emotions are some sort of private mental state, for this idea is prominent in the writings of St. Augustine. Nor that there are "basic" emotions; this idea can be found in Aristotle. Nor even that the basic emotions are indexed by facial expressions that serve predetermined communicative functions, for this idea can be found in the Bible. But Descartes introduced a unique slant on all these themes by construing the "passivity" of emotions as signifying that they constitute a distinct class of involuntary mental states.

That is, Descartes viewed emotions as *complex reflexes* that are triggered by internal and/or external stimuli. He believed they consisted of distinctive bodily processes and sensations and are associated with characteristic behaviors and stereotypical facial expressions.[7] In the modern version of this argument, a "basic" emotion is defined as a complex process consisting of neural, neuromuscular/expressive, and experiential aspects.[8] To qualify as "basic," the emotion must be associated with a distinctive facial expression; with certain body movements and postures; with distinctive vocalizations, changes of voice, tone, rhythm, prosody, and stress; and with distinctive sensations and chemical changes in the body.

The two most influential contemporary versions of this argument are elaborated by Ekman and Izard (both informed by the work of Sylvan

Tomkins).[9] Their theories maintain that all the various elements of emotional responses are coordinated and controlled by neural programs. Emotional responses are treated as a composite form of reflex; namely, a stimulus triggers a neural program that controls a neuromuscular/expressive, autonomic, behavioral, and experiential sequence of events. The essential difference between emotions and reflexes proper is thus seen as a matter of complexity, that is, of how many different elements are coordinated. Hence emotional reactions are, as Descartes argued, unconscious and involuntary, because, like other "automatic actions," they operate at a neural/physiological level that is "beneath the threshold" of introspection and conscious planning.

The universality of the basic emotions is explained in psycho-evolutionary terms as resulting from the selection of these affect programs, perhaps sometime during the Pleistocene, and possibly earlier. Cultural and/or cognitive components are "added on" to basic emotional responses to produce the unique emotional behaviors stressed by cultural relativists.[10] These basic emotional programs may also be "blended" with one another to form new emotions, rather in the way that the primary colors can be blended to form new colors.[11] But the key features of the Cartesian theory of emotions are that (1) the basic emotions retain their core features throughout an individual's life, and (2) these basic emotions might either facilitate or impede information processing, but the development and structure of our cognitive and linguistic faculties are seen as autonomous phenomena.

There have, of course, been many important behavioral researchers and clinicians who have challenged this Cartesian tradition and studied the dynamic processes in human development. During the 1940s, '50s, and '60s some of the great theorists of early child development, such as Erik Erikson, Anna Freud, René Spitz, and John Bowlby, looked at the importance of early emotional experiences for subsequent personality functioning.[12] The 1960s, '70s, and '80s saw fruitful research on emotional development by Mary Ainsworth, Jerome Bruner, Berry Brazelton, Myron Hofer, Louis Sander, Allen Sroufe, and Dan Stern, all of whom made important discoveries about the types of interactive experiences that promote strong attachment and healthy emotional functioning and encourage the child's emerging sense of self.[13]

Also highly important was the work done by Robert Emde, Mary Klinnert, and Joseph Campos on social referencing. These theorists showed that a caregiver's facial expressions of affect could have a powerful effect on the infant's emotional responses to strangers and to physical situations.[14] In a particularly telling experiment, Ed Tronick showed that when caregivers assume a still-faced and unresponsive facial expression, the infant typically responds by trying to reengage her caregiver with animated facial expressions, vocalizations, and body movements. When her strategy fails, the infant turns away, frowns, and cries.[15] As important as this research is for our understanding of the socialization of emotional experience, however, it does not explain how the basic emotions arise from early physiological experience, how they then develop and lead to symbols and intelligence, and whether intelligence is separate from the ongoing pathway of emotional development or an intrinsic landmark of it.

Campos and his colleagues David Witherington and Matthew Hertenstein suggest that emotions can best be understood through the adaptive social interactive function they serve for an individual's interactions with her environment.[16] Extending these observations, we have been able to describe the central role and developmental transformations of emotions in the formation of many of the mind's most important capacities.

In the approach we will be describing, an infant (and child) masters a number of *functional emotional (f/e) developmental capacities*. As we explained in Chapter 1, and will explain further in this chapter and subsequent ones, the functional emotional developmental model explains how emotions enable a child to master symbols; the critical role that emotions play in the development of intelligence; how emotions provide purpose to the mind's different processing capacities, for example, motor skills and sensory processing, thus serving as the vehicle through which, like the instruments in an orchestra, all the mind's processing systems work together in integrated patterns; and how, at subsequent levels of transformation, emotions provide the mechanism through which social groups form and societies function. Thus, in our f/e theory, affects play a vastly more important role in the development of the mind and society than has hitherto been imagined. They serve as the source of symbols, the architect of intelligence, the integrator of processing capacities, and the psychological foundation of society.

In recent years, dynamic systems theorists have begun to shed important light on some of these issues. For example, studies by Dan Messinger, Alan Fogel, and Laurie Dickson have found that different smiles become associated with different kinds of pleasurable activities, and indeed, with different kinds of pleasure.[17] There is the pleasure of anticipation; the pleasure of engagement; the excitement of a particular activity; the release from built-up tension (as in tickling) or suspense (as in peek-a-boo); and, of course, the pleasure of enjoyable sensations. Furthermore, the infant experiences the pleasure of observing her caregiver's pleasure and of seeing the effects of her own behavior on her caregiver, including the effects of her facial expressions on her caregiver's facial expressions. And finally, there are the physical sensations that the infant experiences concerning the caregiver's own physical state (relaxed, tense, anxious).

Thus, the significance of an infant's smile cannot be divorced from the context in which it occurs. The primary relevant factors here are the infant's state of physiological arousal; the caregiver's emotional state; the activity in which the infant is engaged with her caregiver; the environment in which this activity is taking place; the direction of the infant's gaze; other muscular contractions in the infant's face; and the type of smile on the caregiver's face.[18] All these factors play an important role in the type of smiles that the infant develops. The emerging stereotypical facial expressions associated with the basic emotions should not, therefore, be viewed as a maturational phenomenon; nor should we assume that discrete emotions are indexed by discrete facial expressions.

What is still missing in these enlightening attempts to move beyond the Cartesian model of emotions, however, is a dynamic account of how the child's basic emotions are formed in the first place and then develop, and what the later forms of emotions are and the mechanisms and steps in their developmental pathway that make symbols and intelligence possible. To answer these questions one of us (S.I.G.) developed, over the past twenty-five years, a developmental model that shows, first, how the basic emotions arise from physical processes and gradually take on signaling and social functions and subjective meaning, and how these emotions are then transformed and become a vital source and architect of the child's cognitive and linguistic development and the development of her higher reflective reasoning abilities. This work is described in the 1979 monograph *Intelligence and Adaptation,* and has continued in subsequent works.[19]

We have observed that infants initially experience a limited number of global states, for example, calmness, excitement, and distress. A caregiver's nurturing, pleasurable, and calming interactions enable the infant to experience soothing pleasure and interest in the caregiver's sounds and sights and in movements such as turning to look at the caregiver. In this way, certain emotional proclivities, such as pleasurable interest in soothing sounds, begin to differentiate from these global states. As their nervous systems develop, in part because of nurturing interactions, and the capacity to discriminate differences and organize patterns develops further, infants begin to further differentiate and elaborate these global states. They do this through continuing interactive experiences with their caregivers if the interactions provide enough subtlety rather than global reactions. For example, as caregivers respond to their infants' interests with their voices and faces using a range of emotional expressions (different types of smiles and joyful sounds), we often observe the infant expressing a range of pleasurable smiles and a deepening sense of joy and security.

Through continuing human interactions of this type, infants associate more and more specific subjective qualities with selective physical sensations. For example, the sound of the voice registers as a sensation, but it's also either pleasurable or aversive. Mother's touch is a tactile sensation that also may be soothing or overstimulating and upsetting.

Every experience that a child undergoes involves this form of dual coding; that is, experiences are coded by their physical properties and the emotional reactions to them.[20] Both these perceptions, the physical and the emotional, appear to be coded together. Similarly, a hug feels tight and *secure* or tight and *frightening;* a surface feels cold and *aversive* or cold and *pleasant;* and a mobile looks colorful and *interesting* or colorful and *frightening.* These emotional reactions can have an almost infinite degree of subtle variation so that each person's sense of pleasure or security is unique and highly textured.

In addition, infants and children differ in their basic perceptions of sensations. Certain types of touch, sound, or smell, for example, may be soothing to one infant and overstimulating to another (i.e., an infant may be hyper- or hyporeactive to a given sensation). The same sound—for example, a violin being played in its upper register—can be stimulating and pleasant for one child and piercing and shrill for another. These physical differences, which we have observed both in typically developing children and in those facing challenges, in turn can also be experienced with a near-infinite range of subjective affective coloring, depending on early caregiver-infant interactions. For example, how a caregiver soothes or overreacts to her infant's hypersensitivity to touch will influence his subjective experience of that sensation.

Thus, as opposed to the Cartesian assumption that every individual responds in the same way to the same stimuli, we see that, for endogenous and exogenous reasons, infants' responses to physical sensations are in part individual, subjective, and emotional.

As an infant constructs a subjective emotional world, "experience" and physiologic expression continuously influence one another. Subtle reading and responding to an infant's emotional cues as part of a reciprocal interaction keeps refining her physiologic and emotional experience and expression. Growing central nervous system organization serves to organize and facilitate the expression of ever more complex and refined interactive emotional experiences. For example, if an infant is hungry and her cries go unheeded, her muscles may tense up, a response associated with sensations of discomfort and tension. Through infant-caregiver interactions, as described earlier, reactions such as discomfort and tension can become more refined or specific. If the caregiver responds to these first signs of anger by scolding the child, or by coldly withdrawing, the physical sensations and the nascent feeling of anger that the child is experiencing may, over time, become bound up with feelings of hopelessness or even shame. These feelings may then serve to organize and give meaning to a variety of interactive experiences accompanied by frustration or discomfort.

The caregivers' continual reading and responding to the infant's experience and expression of an emotion makes that emotional experience more and more "purposeful"—part of a pattern of purposeful interaction. In this way, emotions become social and affective signals. They become the instrument for problem solving. The continuous flow of purposeful interactions leads to complex social patterns. Together with increasing central nervous system organization, these complex social interactions enable the toddler to perceive larger and larger patterns. He uses this ability to perceive his own physiologic and emotional patterns to form organized "subjective" emotional states. At the same time, complex reciprocal emotional interactions are leading to the separation of perception from action and the formation of symbols, as described in Chapter 1. Once the child forms symbols, that is, can create ideas, he is now able to label these perceived emotional patterns. He is, therefore, able to experience the world of feelings at a symbolic level. As we will discuss in the next sections, this process of emotional signaling that leads to symbol formation and the organization of emotional patterns continues to develop throughout life and leads to higher and higher levels of mental organization. We will describe the steps or stages in this process that lead to new levels of symbolic and reflective thinking.[21]

Thus, emotions, which start off as a physiologic system receiving input from the senses, become, through interactive experience, a complex social tool and a vehicle for creating internal mental life. We will see how emotions eventually serve as the architect of intelligence.

In this section, we have described various critical early transformations of emotions:

1. Physical and physiologic reactions become part of the infant's experience of a nurturing, pleasurable, calming relationship with a caregiver (if such a relationship is available).
2. The relationship makes possible a new range of sensations based on human-to-human nurturing interactions. For example, these include a type of soothing, intimate, pleasurable sensation that is not available from impersonal environmental stimuli.
3. These types of sensations, or "emotional" experiences, now make it possible for the infant to *double-code experience* according to its physical and emotional properties; for example, mother's voice registers physically in a certain frequency range and "feels" pleasurable, secure, or frightening.
4. As caregivers respond to the infant's experience and expression of emotion, emotions become more differentiated and purposeful.
5. As simple back-and-forth patterns of emotional exchange become more continuous and complex, they are increasingly used for social signaling and interpersonal communication and regulation.
6. As interactions between caregivers and infants become more complex, the infant experiences a wider range of these unique sensations. A growing central nervous system, in part supported by these interactions and, in part, supportive of them, makes more differentiated interactions possible so that *the infant improves her capacity to recognize patterns.* The ability to recognize patterns enables the infant to "make sense" of these emotional interactions and begin to experience them as patterns and as part of a developing sense of self (as will be described more fully later in this chapter).
7. As these types of co-regulated emotional exchanges continue, they enable the separation of perception from action (explained in Chapter 1); this in turn leads to symbol formation.
8. As symbols are formed (described in Chapter 1), a child not only experiences these emotional patterns but labels them according to their "felt" characteristics.

9. In this way, subjective emotions and feelings differentiate out of early physical and physiologic sensations.
10. The continuation of increasing capacities for emotional signaling and symbol formation creates the basis for a progressive series of emotional transformations and leads to higher and higher levels of intelligence and reflective thinking, as outlined in the stages described below.
11. Through their progressive transformations, emotions, which can be experienced in an almost infinite number of subtle variations, can organize and give meaning to experience. They can, therefore, serve as the architect or orchestra leader for the mind's many functions. At each stage in the pathway to intelligence, emotions orchestrate cognitive, language, motor, sensory, and social experience.
12. The increasing capacities for emotional signaling also provide the means through which social groups form and societies function.

THE STAGES OF EMOTIONAL AND INTELLECTUAL GROWTH

In the pages that follow, we map the mysterious and revealing journey from an infant's earliest emotional interest in sights and sounds to an adult's reflective wisdom. We will observe that, beginning with the dual coding described above, each stage of emotional and intellectual growth involves the simultaneous mastery of what are ordinarily thought of as emotional and cognitive (or intellectual) abilities.

For example, as we discuss in more detail later, a baby first learns "causality" not through pulling a string to ring a bell or other similar behavior, as Piaget thought, but through the exchange of emotional signals (I smile and you smile back). Therefore, this early lesson is emotional and cognitive at the same time. At each stage, new cognitive skills are learned from emotional experiences. Even high-level symbolic and reflective thinking employs emotional awareness as part of its defining characteristics. For example, one of the levels of emotional development we will discuss later involves reflective thinking based upon an internal standard and sense of self. At this level, an individual can reflect and say, for example, "I'm angrier than I should be in this situation." Such an individual can also compare two authors and their treatment of an abstract concept, such as love or justice, in relationship to her own evolving experience of these concepts. She might observe that one author is far more insightful than another in exploring the complexity of "justice" in a situation of competing motives.

To grasp concepts such as justice or love, however, or for that matter to understand the motives of a character in a novel, a person needs to have an internal sense of "self." Without a sense of self, there is no stable internal compass or frame of reference upon which to compare, contrast, or make judgments. It is not by chance that this sense of self becomes more organized and complex as the capacity for reflective thinking and making judgments about one's own behavior and thoughts and comparing them to the perspectives of others is emerging. The *sense of self* is the agent of these judgments. As it further develops during the course of life, it enables an individual to observe, comprehend, and reflect on an ever widening view of the world. In contrast, no matter how strong one's memory, language abilities, or calculating skills, if one has a very rigid sense of self and personality, he will "see" and comprehend only a narrow piece of the world and draw very limited conclusions. As we will see in the sections that follow, our emotions and sense of self influence our "awareness," what we are consciously able to grapple with and attempt to comprehend. It influences the depth of our understanding and ultimately our intelligence and wisdom.

Therefore, memory and selected cognitive skills, which can peak during the early adult years, may be sufficient for mathematical work, but it's difficult to be "fully intelligent" without experiencing deeply and profoundly the full range of many of life's essential, emotional experiences, such as love, disappointment, and competition. These unfold gradually in a variety of contexts throughout life. Therefore, true intelligence, such as wisdom, develops only over time. If we're fortunate, it does so throughout our lives. As we pointed out earlier, our genes are a key element in a biological system that prepares us to respond to and learn from experiences, but specific types of interactive emotional experiences are needed for the development of symbols, concepts, and abstract thinking. There are no known biological mechanisms that can *fully* account for the "meaningful" use of ideas. The description that follows of the stages of emotional and intellectual growth will demonstrate the seamless relationship between emotional and cognitive experience.

We emphasize that as we describe the stages of emotional interactions, we will also be describing and defining emotions in a particular way. We will not be looking simply at when a child smiles, frowns, shows a look of surprise, or expresses feelings such as happiness or sadness. We will instead be looking at the child's overall emotional abilities, such as her ability to engage with others and exchange emotional signals so that she can understand others and communicate her own needs, her ability to elaborate emotions in play and with words and pictures, and her level of empathy.

These overall emotional abilities can incorporate different feelings such as anger, love, and sadness. For example, fully engaging with others involves love as well as sadness or disappointment. Full imaginative play involves happiness and joy as well as anger. These overall emotional abilities are "functional" in that they enable the child to interact with and comprehend her world. Therefore, in earlier writings we have called these abilities Functional Emotional Developmental Capabilities.

These abilities are functional in another sense as well. They orchestrate many of the child's other developmental capabilities. For example, as a child is learning to signal with emotions in the first year of life, she is using her emotions to determine whether to reach—that is, she uses her motor system and muscles—as well as what she vocalizes about: one sound for "I like that" and another for "I don't like that." As indicated, back-and-forth emotional signaling establishes a sense of causality—an early cognitive skill: "I can make my mommy smile with my smile." Emotions also lead the child to search for and find the hidden toy in her mommy's hand. She will search only for a desirable toy. Such searching and finding leads to perceptual motor and visual-spatial problem-solving skills. From early on, therefore, the infant's emotions orchestrate many parts of her "mental team." Her emotions enable the "members" of her mental team to work together, much as the members of a wonderful ballet company or an outstanding basketball team work together. Therefore, the emotional stages we will be describing are the "overall emotional abilities," or Functional Emotional Developmental Capabilities, which are different from specific emotions such as joy or anger. They are fundamental emotional organizations that guide every aspect of day-to-day functioning, unite the different processing abilities, and, as we will show in Chapter 11, orchestrate the different parts of the mind.

As we will see, these emotional abilities build on one another. For example, a baby must be engaged in a relationship with a caregiver for loving feelings to become part of an emotional exchange of signals. Using emotional ideas—"I feel sad"—precedes building logical bridges between emotional ideas: "I feel sad because you didn't play with me."

Emotional abilities and stages may be only partially mastered. When this occurs, emotional development may still proceed, but in a constricted form. Like a house with a weak foundation, constricted emotional development may be more vulnerable to a "strong wind," or less broadly mastered. For example, relationships may be more superficial and less intimate, and empathy for other people's feelings limited only to selected feelings. The functional emotional developmental capacities we will be describing below, which begin early in life and continue through the course of life, are mastered at various times as a human being develops. Through a field study, we have been able to show that the early capacities are mastered for the first time (and then continue to be further developed) during specific, predicted time intervals. Furthermore, we have shown that the mastery of the early capacities is associated with healthy intellectual, social, and emotional functioning. In contrast, compromises in their mastery are associated with developmental and emotional difficulties.[22]

In the next sections, as we describe the stages of emotional and intellectual growth, remember that these are but brief descriptions of more complex processes—the "tip of the iceberg." In addition, in actual development, mental growth is continuous and is "categorized" only for purposes of description and discussion.

Stage 1—Regulation and Interest in the World

Within the first few months of life, babies are learning to transform their emotions from their own inner sensations (e.g., focusing on a gas bubble in the tummy) to the outer world. They are learning to perceive the outer world—a mother's and father's face, voice, smells, and touch. But to perceive the outer world, they must want to look or listen. Although perhaps born with a tendency to perceive some basic patterns, they are enticed by the emotional rhythm of our voices, our big smiles and gleaming eyes—interesting sounds and sights. Rhythmic, almost synchronous patterns between the caregiver's and infant's movements or vocalizations enable the infant to begin relating to and appreciating the outside world. These patterns, which are part of the emotional relationship between an infant and caregiver, begin prenatally. The mother-to-be relates to her baby's movement patterns and responses to sounds and other sensations, not to mention her fantasies about the new baby. The birth process itself and the time immediately after birth can be especially meaningful. Marshall Klaus and John Kennell have described how, when mother and baby are allowed to spend time together with direct physical contact immediately after birth (and an anesthesia has not been used), babies often crawl up their mothers' tummies to find the breast right after birth.[23] Such direct physical and emotional contact appears to have both physiologic and emotional benefit, resulting in less crying and improved mother/infant interaction in the early months of life.[24]

During these early months of parent and baby interactions, babies gradually become more and more interested in sights, touch, and sounds, and begin discriminating between what they see, hear, smell, and touch. To elicit a baby's interest in the outer world, the sensations caregivers provide have to be emotionally pleasurable. If they are aversive, babies tune out or shut down and don't become invested in what is outside themselves. However, each baby has individual ways of responding to sound, sight, touch, smell, and movement. Some babies are very sensitive and require gentle soothing. Some are underreactive and require more energetic wooing. Some babies begin to figure out patterns of sights or sounds quickly; some slowly. Some readily turn toward sound or sights, but others take a while to notice. These responses happen more readily if adults tailor their approaches to each infant based on her individual preferences and abilities. Therefore, even at this first and most basic stage of learning, the baby depends on a caregiver's ability to adapt her gaze, voice, and movements in a pleasurable, emotionally satisfying manner to the baby's unique way of responding to and taking in the world.

Intelligence is forming during the very first stage as a baby is learning to use all her senses to perceive the world and discriminate patterns, such as the difference between mother's voice and father's voice.

The Dual Code. As we shall see, emotions orchestrate this process from the very beginning. A baby can begin the lifelong task of learning about the world only through the materials at hand, which, initially, are the simplest of sensations, such as touch and sound. Years of investigation into initial perceptions and cognition, on the one hand, and emotional development on the other, have left out a vital connection. In our clinical work with many infants—both typically developing and facing challenges—we observed that each sensation, as it is registered by the child, also gives rise, as we said earlier, to an affect or emotion;[25] that is to say, the infant responds to it according to its emotional as well as physical effect on her. Thus, a blanket may feel smooth *and* pleasant or itchy *and* irritating; a toy may be brilliantly red *and* intriguing or boring, a voice loud *and* inviting or jarring. Mom's cheek might feel soft and wonderful or rough and uncomfortable. As a baby's experience grows, sensory impressions become increasingly tied to feelings. This *dual coding* of experience is the key to understanding how emotions organize intellectual abilities and indeed create the sense of self.

Human beings begin this coupling of phenomena and feelings at the very beginning of life. Even infants only days old react to sensations emotionally, preferring the sounds and smells of their mothers, for example, to all other voices and scents. They suck more vigorously when offered sweet liquids that taste good. Older babies will joyfully pursue certain favorite people with their eyes and avoid others. By four months, children can react to the sight and sound of people who have scared them.

However, as we said earlier, a given sensation does not necessarily produce the same emotion in every individual. Inborn differences in peoples' sensory makeup can make the sound of a given frequency and loudness—say, a high-pitched voice—strike one person as rousing and appealing, but another as shrill, like a siren. Though we generally assume that we all experience sensations—such as sound and touch—in more or less the same way, significant variations are now known to exist in the ways individuals process even very simple sensory information. We have explored the emotional consequences of those sensory differences first described by Jean Ayres, a pioneer in occupational therapy.[26] A given sensation can produce quite different emotional effects in different individuals—pleasure, for example, in one person, but anxiety in another. Each of us, therefore, quite unwittingly creates our own personal, and sometimes idiosyncratic, "catalogue" of sensory and emotional experience.

To add further complexity and individuality to the young child's learning, each of her sensory experiences occurs within the context of a relationship that gives it additional emotional meaning. Nearly all her emotional experiences involve the persons on whom she depends totally for her very survival, and who care for her in a manner that can range from expansive nurturing to near-total neglect. Emotions help a child comprehend even what appear to be physical and mathematical relationships. Simple notions such as hot or cold, for example, may appear to represent purely physical sensations, but a child learns them through experiencing baths and bottles. More complex perceptions—big or little, more or less, here and there—have a similar basis in feelings. "A lot" is more than a child expected. "Too little" is less than expected. "More" is another helping of something tasty, "no more" is a dose of nasty medicine. "Near" is snuggling next to mother. "Later" means impatient waiting.

Even abstract, intellectual concepts, those that underlie theoretical scientific speculations, also reach back to a child's felt experience. Mathematicians and physicists may manipulate abstruse symbols representing space, time, and quantity, but they first understood those entities as tiny children wanting a far-away toy, or waiting for juice, or counting cookies. The grown-up genius, like the adventurous child, forms ideas through playful explorations in the imagination, only later translated into the rigor of mathematics. Before a child can count, she must possess this kind of emotional grasp of *quantity* and *extent.* When we worked with children facing developmental problems who could nonetheless count, and even calculate, we found that numbers and computations lacked meaning for them unless we created an emotional experience of quantity by negotiating over pennies or candies.

Each sensory perception therefore forms part of a dual code that has physical properties (bright, big, loud, smooth, and the like) and emotional qualities (soothing, jarring, happy, tense). This double coding allows the child not only to "cross-reference" each experience and subsequent memory in mental "catalogues" of phenomena and feelings but also to reconstruct them when needed.

Anyone who pays attention to the subjective state of her body will almost always perceive within it an emotional tone, though it may often be subtle, elusive, or hard to describe. Our inner emotional tones, tense or relaxed, hopeful or glum, serene or anxious, constantly play out the countless variations that we use to label and organize and store and retrieve and, most important of all, make sense of our experience.

We use our entire bodies to create, express, and bring to life our emotions: the voluntary muscle systems of our faces, arms, and legs for smiles, frowns, slumps, and waves, as well as the involuntary muscles of the gut and internal organs—a thumping heart or a stomach full of "butterflies." Emotions such as excitement and delight reside primarily in the voluntary system; others, including fear, sexual pleasure, longing, and grief, reside mostly in the involuntary system. Some global emotions, for example, "fight or flight," belong to portions of the nervous system formed early in evolution. Those involved in emotional social reciprocity and that make symbol formation and thinking possible belong to more recently evolved parts of the nervous system and rely on many parts of the brain working together, including the highest levels of the cortex.

In the earliest days and weeks, as a baby becomes aware of the world of sensation, a sense of "self" also emerges, but it is not yet a separate entity. The self is part of a global sense of the world of sensation and feelings.

Consciousness is also developing. It is likely that a baby's earliest sense of consciousness is a global state of sensory and affective "aliveness" (i.e., sensations and their registration in the broadest sense define consciousness). The baby's experience of consciousness is linked to the sights, sounds, touch, and, one hopes, pleasurable feelings she is experiencing.

Stage 2—Engaging and Relating

The second stage involves helping a baby use her emotional interest in the world to form a relationship and become engaged in it. With warm nurturing, the baby now becomes progressively more invested and interested in certain people. No longer will just any face or smell do. It has to be the mother's face or smell. From day one, the baby begins distinguishing primary caregivers from others; from two to five months, this ability reaches a crescendo through joyful smiles and coos and a deep sense of pleasurable intimacy.

In addition, higher levels of learning and intelligence depend on sustained relationships that build trust and intimacy. This progress involves more than simply fulfilling concrete needs. The concrete person who just wants "things" never becomes a fully reflective thinker in life's most important areas. For example, if we look at other people just as "things," we will not understand how they think and feel. Understanding others and feeling empathy for them comes from investing other human beings with one's own feelings. This ability, however, begins with first relationships. It depends on nurturing care that creates a sense of intimacy.

When a baby becomes interested in her primary caregiver as a special person who brings her joy and pleasure, as well as a little annoyance and unhappiness, it is not only emotional interactions that begin flowering. A new level of intelligence is also reached. She is now learning to discriminate the joys and pleasures of the human world from her interests of the inanimate world. Her joy and pleasure in her caregivers enables her to decipher patterns in their voices. She begins to discriminate their emotional interests, such as joy, indifference, and annoyance. She begins to figure out facial expressions as well. Thus begins the long journey of learning to recognize patterns and organize perceptions into meaningful categories.

In forming a deeper, more intimate relationship, the baby is also learning her first lessons in becoming a social being, the cornerstone of being part of a family, group, or community, as well as, eventually, a culture and a society. Also, the baby's sense of self and consciousness is moving forward. Now that she is discriminating the human and inanimate worlds, the infant goes from feeling a part of a global world of sensations to a sense of "shared humanity." There is no separate, defined sense of "self" yet, but from shared intimacy with caregivers, a growing sense of special "human" feelings is emerging.

Stage 3—Intentionality

The third stage goes beyond intimacy and engagement. As we saw in Chapter 1, emotions now become transformed into signals for communication. For this to happen, however, caregivers need to read and respond to the baby's signals and challenge the baby to read and respond to theirs. Through these interactions, the baby begins to engage in back-and-forth emotional signaling. We also describe this back-and-forth signaling, which develops throughout infancy, but especially rapidly between four and ten months, as opening and closing circles of communication. The six-month-old smiles eagerly at her mother, gets a smile back, then smiles again. By smiling again, the baby is closing a circle of communication. Different motor gestures—facial expressions, vocalizations—become part of this signaling. By eight months, many of these exchanges occur in a row. As we saw, these emotional interactions help an infant to begin separating perceptions from their fixed actions.

Therefore, intelligence during this stage reaches an important new level. The beginnings of "causal" (logical) interactions, as the baby purposefully smiles to get a smile back, vocalizes happiness to get a happy sound back, and reaches for father's nose to get a funny "toot-toot" sound back, means that from now on, causality and logic can play a role in all new learning. For example, these new lessons in logic are gradually applied to the spatial world as well as to plan actions (motor planning). When the rattle falls to the ground, the baby follows it with his eyes as though he were looking for it. He looks at and touches his father's hand because it just hid his rattle. The beginning sense of causality marks a beginning sense of "reality" because an appreciation of reality is based on understanding the actions of others as purposeful rather than random.

A sense of self also now becomes more defined. There is a "me" doing something to a "not me" or a "you." (The baby smiles [the "me"] to get a smile back from the caregiver [the "not me"].) But the "me" and the "not me" are not yet defined in the baby's mind as full persons. They are defined only in terms of the smiles or sounds being exchanged. In other words, "parts" of "me" (the self) that become involved in causal intentional interactions are forming. Each part of "me," or a self, is experienced as a separate entity. In the next stage, these parts of the self will come together. Consciousness is also growing. The baby experiences her own willfulness and sense of purpose more and more. Her consciousness of herself and of the world is gradually separating the physical world from the emotional world, the "me" from the "not me," and a sense of will, purpose, or agency from a sea of sensations, feelings, and responses.

Stage 4—Problem Solving, Mood Regulation, and a Sense of Self

In the explosive development that takes place between nine and eighteen months, a baby makes momentous strides. As we described in Chapter 1, she learns to engage in a continuous flow of emotional signaling and can use this ability to solve problems. For example, she may take her mother by the hand, gesture with her eyes and hands so that her mother will open the door to the yard, and then point to the swing; or she takes her dad to the car and shows him that she wants a ride. During this stage, true social problem solving emerges. As indicated earlier, however, caregivers need to read and respond to their toddlers' emotional signals and engage in long chains of shared social problem solving for this to occur. She also learns to regulate her moods and behavior and perceive and organize patterns to form a more complete sense of self. All this progress, which is built on increasingly elaborate emotional interactions, leads to higher levels of intelligence and social interaction.

Problem Solving. During this fourth stage, through a wide range of emotional interactions that are part of daily life, a child learns how to predict patterns of adult behavior and act accordingly. She learns, for example, that when her father comes home and looks grumpy, it's best to stay out of his way. Hide behind the couch or he will snap at you. The child learns that before her mother has had her morning coffee, she'd better walk and talk softly. These savvy adaptations are based on, and facilitate, an ability to recognize patterns. Pattern recognition, which ideally is learned first through social interactions, can then be applied to solve problems in the physical world as well.

The child who doesn't interact, however, won't experience or fully learn to recognize a broad range of patterns. The child who is taking a parent by the hand to search for a toy is coming to understand the elements of a pattern. These include her own emotional needs (what she wants), the action patterns involved in finding a toy, the visual-spatial patterns involved in going from ground level to upper-shelf level where the toy resides, the vocal pattern involved in attracting her father's attention (whimpers at not having what she wants, then gleeful exclamations of triumph), and the social patterns involved in working together with parents toward a common goal. In other words, pattern recognition involves seeing how the pieces fit together rather than just being involved in piecemeal behavior. Elaborate negotiations or play with others make it possible to experience the world in larger integrated patterns.

Recognizing patterns helps a toddler predict the behavior of others and adjust her own. She learns when to expect loving responses and when to expect anger, control, bossiness, or limit-setting. The child's moods respond to seeing the gleam in a parent's eye or the nodding approval for something well done. She's learning what respect feels like—as well as what humiliation feels like when she's done something she shouldn't. All these patterns are learned in the second year of life and before language comes in to a significant degree.

Regulating Mood and Behavior. In the daily loving exchanges and struggles with caregivers, the toddler learns to tame such catastrophic emotions as fear and rage (as described in Chapter 1) with the more regulated and interactive use of emotions. Therefore, she learns to modulate and finely regulate her behavior and moods and cope with intense feeling states. Anger is explosive in a very young infant, and sadness seems to last forever and ever. Certain necessary experiences turn these extreme emotional reactions into feelings and behavior that are finely regulated and responsive to the situation at hand.

Once a child is capable of exchanging rapid signals with her caregiver, she is able to negotiate, in a sense, how she feels. If she is annoyed, she can make a look of annoyance or a sound or a hand gesture. A mother may come back with a gesture indicating "I understand" or "OK, I'll get the food more quickly," or "Can't you wait just one more minute?" Whatever her response, the child is receiving immediate feedback that can modulate her own response. The anger may be modulated by the notion that mother is going to do something, even if she can't do it immediately. Just the sound of her voice signals that she is getting that milk bottle ready and it's coming soon. If she can use a soothing voice and gradually calm the baby or toddler, the child will learn not to become so frantic. Adults often do this intuitively when someone they are close to is upset or angry. Some of us get nervous, however, and "up the stakes" by taking the other person's anger personally. If we can slow and soothe and calm via our emotional gestures (as well as our words—the gestures, however, are far more powerful), we can learn to better and better regulate our moods and behavior.

With a fine-tuned reaction rather than one that is global or extreme, the child doesn't have to throw a tantrum to register her annoyance; she can do it with just a glance and an annoyed look. This ability comes gradually. Even if a toddler does escalate to a real tantrum, she will not go from 0 to 60 in one second. Different feelings, from joy and happiness to sadness to anger to assertiveness, can become part of fine-tuned exchanges with patient, caring adults.

A child may not gain this needed experience of nurturing exchanges for a variety of reasons. Perhaps she has a motor problem and can't gesture or signal well, or maybe she has an unresponsive parent who is not signaling back. Perhaps she has a parent who is too intrusive and anxious or too self-absorbed or depressed to respond appropriately. For any one of these reasons, we may see a compromise in this fine-tuned interactive system. For such a child, there are insufficient regulated responses for her emotional expressions. Her expression of feeling is, therefore, not part of a fine-tuned regulated signaling system: It may be simply an isolated expression of feeling.

Without the modulating influence of an emotional interaction, either the child's feeling may grow more intense or she may give up and become self-absorbed or passive. In either situation, the child may be left using the global feelings of anger or rage, fear or avoidance, which are characteristic of very young infants in the early months of life. One of us (S.I.G.) often sees such children in his practice. Not infrequently, when such children continually hit or bite, parents seek help, worried about "aggression," and often ask for "medication." When parents are coached on how to read the child's signals, respond consistently and calmly, and engage in long chains of regulated social problem solving, however, within a few months many of these children can become well-regulated, cooperative, interactive toddlers.

If, on the other hand, caregivers continue to respond inappropriately or not at all, the child can become even more vulnerable. With caregivers who overreact to powerful emotions, a child often tends to become more anxious and fearful. When caregivers tune out, freeze, or withdraw in response to fierce anger or other strong emotions from their infants and toddlers, the child may feel a sense of "loss"; this, in turn, may lead to an increased tendency to depression. When anger or impulsive behavior is dealt with by withdrawal or single intense punitive, rather than regulating constructive, limit-setting responses and opportunities for social problem solving, we tend to see more aggression and impulsivity.

Forming the Earliest (Presymbolic) Sense of "Self." A sense of self begins forming when a baby organizes her emotions and behavior into patterns. As a baby goes from islands of intentional behavior, such as a few vocalizations or one or two hand wavings or a few smiles, to a whole pattern of dozens of exchanges, which she uses to solve problems, she is learning that she and others can operate in larger chunks or patterns. This enables her to be even more intentional and to negotiate, rather than take piecemeal, episodic action.

This process happens step-by-step. When an adult responds reciprocally, the baby makes a discovery: "I can make something happen." This teaches a baby to take initiative (do something and something happens in return; smiling gets a smile from mom or dad). As we indicated, from this process the infant is beginning to gain a sense of purpose and will and, very importantly, a sense of "self" (it's "me" making something happen, "me" getting that smile or getting that little red rattle by reaching out "my" hand). As a toddler's repertoire of emotional signaling grows richer and she begins to discern patterns in her own and others' behavior, she adds these observations to the map delineating herself as a person. Her mother usually responds when she makes friendly requests, but not when she's cranky. Her father loves to roughhouse, but not to sing lullabies. Grandmother is a good deal less strict than either parent. Which actions get affection and approval? Which yield only rejection or anger? Is she worthy of care, attention, and respect? Are those around her also worthy?

With the growing capacity to perceive and organize patterns, these types of experiences continue to define a developing sense of self even before words or ideas are used to a significant degree. In a similar way, the child is discovering how the physical world works—turning this little plastic thing causes a funny animal to pop up, or pushing this big, smooth, see-through object makes a loud noise and maybe even produces a splatter of little pieces and yelling from adults. Seeing the world in patterns increases understanding of how it works and leads to expectations and mastery, a scientific attitude.

The child uses this ability to discriminate, to distinguish among many patterns of emotions; she knows the difference between those meaning safety and comfort and those meaning danger. She can tell approval from disapproval, acceptance from rejection. Life's most essential emotional themes are identified and patterns of dealing with them formed. The child also begins to use this new ability in increasingly complicated situations. Is her mother's tense face a signal that she is angry with her daughter? The child starts to use this awareness to respond to people according to their emotional tones, for example, and to pull away from a situation that seems undermining.

The intuitive ability to decipher human exchanges and pick up emotional cues before any words have been exchanged becomes a "supersense" that often operates faster than our conscious awareness. In fact, it is the foundation of our social life.

Therefore, long before an infant can speak, personality and expectations are already being molded by the countless interactions between caregiver and child. However, no child's family or daily life affords equal interactions in all areas of experience. We may have many different responses to love and pleasure and only two to anger, or vice versa. Some families avoid certain emotions entirely. No child's environment is perfect, and parents who try too hard to provide one often quash the emotional spontaneity that is so critical to the entire process of development.

It is also important to remember that the developments of this formative period, though influential, are not definitive. Many elements of personality form early in life, but daily interaction continuously redefines it.

The basic skills that enable a child to read caregivers' and family members' signals also enable her to learn about her culture. The responses of her caregivers provide an unspoken but expressive running commentary along a scale of approval/disapproval, anger/happiness, curiosity/fear. Is defiance permissible? Is aggression or passivity sanctioned? How do people greet each other? Picking up cues from this subtext, the child learns before she has gained significant amounts of verbal language and knows what is good and bad, what is done and not done, what is acceptable and unacceptable.

During this period, co-regulated emotional signaling also has a role in gender differences. Our work observing very early interaction suggests that these differences may not be based solely on hormones or different brain structures. Learning experiences vary early in life and lead to important questions. For example, we have observed that many caregivers tend to engage female infants and toddlers in longer preverbal, affective "conversations" than boys. Does this contribute to girls developing earlier language skills, and perhaps empathy, than boys? The longer we gesture and signal back and forth to babies, the more we enable them to signal and negotiate with a large range of our emotions. Of course, individual girls and boys vary considerably in their early experiences. Each one will have his or her own unique early interactions and eventual personalities. As a large group, however, boys tend to be more active as babies and, therefore, invite shorter bursts of affective signaling. In our culture, we often play differently with boys and girls, more roughhousing with boys and more back-and-forth dialogues with little girls.

Is it any surprise that a child with more extensive early experience in navigating her emotional terrain will grow up better able to express how she feels? Or that a child who tends to experience shorter emotional interchanges might develop some of the characteristics we think of as typically "male," such as an inability to acknowledge his feelings, the strong desire to separate his emotional world from his rational one, or even the disturbing "all-or-nothing" discharge mode for these feelings, such as rage or withdrawal? At the same time, the action orientation and roughhousing may provide an opportunity to feel secure with his body and learn to be assertive and able to overcome obstacles.

It's important to emphasize, however, that as we speculate about the role of early interactions in the behavior we commonly think of as "male" or "female," each child and his or her caregivers will negotiate their own unique patterns. Therefore, each boy and girl can be understood only in terms of his or her unique history. During Stage 4, the toddler progresses to significantly higher levels of intelligence. Because of her emerging ability for long exchanges of signaling, she is becoming a better and better problem solver. Also, because growing problem-solving abilities involve signaling between herself and others, she is involved in an increasingly creative endeavor encompassing the input of two or more parties. For example, she may want to solve a problem one way and her father may gesture with a point, a puzzled look, or a big smile, suggesting either confirmation or an alternative. Through these types of interactions involving longer and longer chains of communication, new approaches are constantly being learned.

The toddler is breaking new ground in all her different intellectual domains. More complex vocalizations are emerging. A private language may be forming as a prelude to learning the family's language. With her caregivers as interactive partners/explorers, a more elaborate sense of physical space and an ability to engage in visual-spatial problem solving (such as finding hidden objects or figuring out a new way to get to a toy on the shelf or mother's favorite jewelry) is rapidly emerging. This occurs because physical space is now invested with emotional meaning through the pursuit of emotional goals.

Similarly, the ability to plan and sequence actions—conducting a five-step maneuver with a new truck (loading and unloading it, moving it to one side of the room and then the other)—is also rapidly learned because of interactive play where emotional goals are used to guide actions. As discussed earlier, the toddler is also learning to regulate her mood and behaviors better because of interactive emotional signaling and, in this way, is also learning to modulate sensations. She is no longer as likely to become sensory overloaded or underaroused because she is now able to participate actively in the sensations modulating her. For example, she can seek out just a bit more sound or touch. She can slow down an interaction through her expressions, hand gestures, or body posture if it's becoming overloading.

As the toddler becomes an interactive problem solver in all these domains, she is literally becoming a multilevel, scientific thinker, figuring out and implementing new solutions all the time. This progress, in turn, makes her a better and better "pattern recognizer" and organizer, an ability that will underlie all her future academic skills.

Many infant observers and researchers focus on the first year of life and then jump to a focus on language development in the latter part of the second and third years of life. In fact, it's in the second year of life that emotional signaling becomes more and more complex and sets the foundations for language and higher levels of intelligence through the accomplishments just described. As we described in Chapter 1, it is during this time that a toddler is learning to use affect signaling to separate perception from action. This leads to freestanding images that the toddler can invest with emotional experiences. Such emotionally meaningful, freestanding images become meaningful ideas or internal symbols.

New social skills are also developing at this stage. Social signaling enables the toddler to handle multiple relationships at the same time, signaling a mischievous grin to her father and an annoyed look at her mother. Reading these emotional signals is also part of this process and it helps the toddler inhibit aggression, cooperate, and copy altruistic behaviors.

The ability to imitate also advances significantly. Now the toddler can copy large patterns, such as putting on her father's hat, lifting his briefcase, and imitating him as he walks about the house with a confident stride. As can be readily imagined, these abilities for social negotiation, multiple relationships, and rapid learning of whole patterns through imitation are the foundations for participating in groups. We have observed toddlers forming friendships by copying each other, following each other, enjoying some rough-and-tumble play together, and eventually, by eighteen months of age or so, hamming it up together and laughing at and with each other. Shared humor communicated via facial expressions and movements can become quite organized.[27] Initially, there is the family group and then a community, a society, and a culture. Everything from simple tool use to attitudes towards aggression or closeness and intimacy are learned through complex interactive signaling and imitative learning.

Also, as we mentioned, long chains of signaling enable the presymbolic sense of self to become more integrated. Consciousness is, therefore, not simply made up of a "part me" (a smile getting a smile or a sound getting a sound). It's now made up of a more integrated sense of "me" as a whole person interacting with a more integrated sense of another (mother or father) as whole people. In other words, the happy "me" and the angry "me" are now part of one person, as are the "nice mommy" and the "frustrating or mean mommy." This greater integration of the parts of "me" and others occurs very gradually during the second year of life. During this time, islands of "me" gradually come together through longer and longer interactions with caregivers that embrace such polarities of feelings as anger and love. Consciousness also expands to include larger and larger physical spaces, as well as language and motor domains.

This stage of development figures importantly in the observations we have made regarding the developmental pathways leading to autism. As we will discuss in Chapter 11, children with autism (even those who develop verbal abilities, score above average on IQ tests, and do well on school-based academic work) have difficulties with making inferences; using higher-level, abstract, reflective thinking; empathizing with others; and dealing with their own and other people's emotions. In studying such children over time and exploring their histories and videotapes of their interactions during their formative years, we have found that the vast majority, even those who seemed to be doing well and "regressed" only at age two or later, did not fully master these emotional interactions of the second year of life and the skills that are based on them. Although some of the children could engage with caregivers and signal a little with emotions, they did not reach the point where they could take a caregiver to find a toy or engage in long exchanges or wordless dialogues of affective interaction to regulate their behavior and mood. They were, therefore, unable to develop the full range of higher-level abilities. For example, using symbols meaningfully and negotiating emotional and social challenges require investing symbols with regulated emotions ("Mom" is understood as the total of one's emotional experiences with one's mother). This development can occur only through many emotional interchanges with mother. Similarly, empathy requires a full sense of another as an emotional "other" and can only be learned through the same process. In contrast, math or history facts can be learned largely by memorizing them.

We have formulated a theory, the Affect Diathesis Hypothesis,[28] which suggests that in autism co-regulated affect signaling is difficult because of a unique biological challenge. Children with autism, we believe, have a biologically based difficulty in connecting emotion to their emerging ability to plan and sequence their actions. Therefore, complex interactions that require many steps are not guided by needs or interests (emotions). As such, they stay simple or become repetitive. Fortunately, however, we have also found that this important developmental pathway involving emotional signaling, although a challenge for children with autism, is often not completely blocked. As we will show in Chapter 11, extra practice with meaningful emotional interactions can often help children with this type of challenge develop more fully. Perhaps there are "side" pathways that can be mastered when the main one is blocked. Through a comprehensive program that worked with this experience and ability, we found that we could help most children become engaged and interactive and a subgroup of children with autistic spectrum disorders become meaningfully verbal, empathetic, creative, and reflective, and engaged in solid peer and family relationships (see also Chapter 12).[29]

Stage 4 is an important stage that develops over several levels and according to how complex and broad the interactive emotional signaling and problem-solving patterns become. These include:

- Action Level—Affective interactions organized into action or behavioral patterns to express a need, but not involving exchange of signals to any significant degree.
- Fragmented Level—Islands of intentional, emotional signaling and problem solving.
- Polarized Level—Organized patterns of emotional signaling expressing only one or another feeling state, for example, organized aggression and impulsivity; organized clinging; needy, dependent behavior; organized fearful patterns.
- Integrated Level—Long chains of interaction involving a variety of feelings: dependency, assertiveness, pleasure. These are integrated into problem-solving patterns such as flirting, seeking closeness, and then getting help to find a needed object. These interactive patterns lead to a presymbolic sense of self, the regulation of mood and behavior, the capacity

to separate perception from action, and investing freestanding perceptions or images with emotions to form symbols.

Stage 5—Creating Symbols and Using Words and Ideas

The emergence of formal symbols, of words and ideas, involves a momentous transformation. By this time, if there have been many opportunities for emotional exchanges, the child can now more easily separate action from perception and hold onto freestanding images and invest them with emotions. As children learn to regulate their tongues, other mouth muscles, and vocal chords, they can begin forming words to talk about these meaningful images or internal representations. If they have had lots of emotionally relevant experiences, they can create a broad range of meaningful symbols.

When children haven't learned to create emotionally meaningful images but are neurologically capable of speaking, the effect is very different. A child may see a picture of the table and say the word "table." She can label and perform rote memory tasks. But she won't be able to say "Mommy, play with me!" or "I don't like that!" Such a child won't have meaningful language later on. She won't fully comprehend written language, either. The child might learn to read and parrot back "red ball, green ball, blue ball," but won't be able to tell you the meaning of a story or the motives of the characters.

It is through emotional interactions that images acquire meaning. The child is learning what an apple is, what love is. She can use words or pictures to convey the feeling of giving mom a big hug and by saying, "Love you." She can symbolize hitting and screaming by saying, "Me mad!" She can also use pretend play to symbolize real or imagined events, such as tea parties, monster attacks, and the like. In addition, a child can now use symbols to manipulate ideas in her mind without actually having to carry out actions. This allows her tremendous flexibility in reasoning and thinking because she can now solve problems in her own mind.

To the degree that they refer to lived emotional experiences, the new words a child acquires become meaningful. The twelve or eighteen months of exchanges the child has already experienced with caregivers and the available world provide a foundation for the emergence of meaningful language. Continuous emotional interaction with others and the world maintains progress throughout life.

This stage of developing ideas and language, which grows rapidly between eighteen and thirty months and continues thereafter, also moves through several levels based on the complexity of the ideas used and how the ideas are used to express wishes or actions or feelings. These include:

- Ideas or words and actions are used together (ideas are acted out, but words are also used to signify the action). Ideas or words are not yet used instead of actions.
- Action words are used instead of actions, and these action words convey intent ("Hit you!").
- Feelings are conveyed through words, but are treated as real rather than as signals ("I'm mad," "I'm hungry," "I need a hug," as compared with "I feel mad" or "I feel hungry" or "I feel I need a hug"). In the first instance, the feeling state demands action and is very close to action; in the second one, the words are more a signal for something going on inside that makes possible a consideration of many possible thoughts and/or actions.
- Words are used to convey bodily feeling states ("My muscles are exploding," "My head is aching").
- Words convey feelings, but they are mostly global feeling states ("I feel awful," "I feel okay."). The feeling states are generally polarized (all good or all bad). These polarized uses of words can also characterize the next stage, when logical bridges are created to link ideas together; if they persist, however, they often indicate a constriction or limitation in the full mastery of using words and connecting ideas together logically.
- Words begin to convey more differentiated feelings ("I feel sad" or "I feel angry") and, therefore, are beginning to represent more fully a specific feeling that is not tied to action. This more differentiated use of words characterizes the relative mastery of this stage and the next one.

Intelligence has now reached the symbolic level. This is when we ordinarily think of intelligence as truly beginning. As we have shown, however, intelligence has already been on a long developmental journey and is now simply reaching a new level. Although we emphasized the acquisition of verbal symbols, which is a cornerstone of many intellectual endeavors, the ability to construct symbols actually occurs in many domains and gives rise to higher levels of intelligence in all of them. This development includes the formation of visual-spatial symbols (the preschooler can build a house and elaborate about what goes on in each part of it) as well as planned actions, which serve symbolic goals (taking the toy bus from the house to the school to pick up some children).

Now the child is able to solve problems in her mind. She can explore creative and novel possibilities through the manipulation of symbols (i.e., ideas). This new foundation for intelligence, like its antecedents, will be further developed throughout life.

The ability to construct symbols also enables individuals to share meanings. This includes the common use of words and emerging concepts—not only what's "nice" and "not nice," "fair" and "unfair," but a sense of justice and other concepts that can unite groups socially. Symbols also enable new levels of social negotiation. Basic needs, such as dependency, curiosity, assertiveness, and aggression, can be dealt with by larger and larger groups. Preschoolers are still usually better at sharing meanings and symbols with one other person than with large groups of people. At big birthday parties, preschoolers tend either to organize into smaller groups or play on their own in a more parallel way. Over time, however, the ability to use symbols in larger and larger groups emerges.

The sense of self is also reaching a higher level. A sense of "me" and "not me" is forming, now at the level of internal images rather than simply integrated patterns of behavior, as was true at the prior stage. In other words, there is now a "symbolic sense of self" beginning to form. Consciousness is, therefore, reaching an important new level as a symbolic awareness of the world is beginning to complement the presymbolic one, which had been materializing for some time. This symbolic awareness of the world builds on and incorporates the awareness of the world that had already existed. In other words, language does not create conscious awareness. It provides a new way of labeling and expanding consciousness. It builds on a sense of the self and the outer world that is already well established. It now becomes possible to create new realities and new levels of consciousness through manipulating and creating symbols. Fantasies and imagination blossom.

Stage 6—Emotional Thinking, Logic, and a Sense of "Reality"

Another momentous transformation occurs when children learn to connect symbols together logically, making possible logical thinking and reflection. We just described how, in addition to constructing new meanings, the child has infused her formal symbols with the meanings already established, to some degree, in earlier and ongoing emotional experiences. In the sixth stage, from approximately thirty to forty months, the child learns to connect these symbols together. She says, "I want to go out and play!" and you say, "Why?" and the child says, "Because it's fun," or "Because I want to go down that slide." Now the child offers reasons for her behavior. "Why are you so mad?" "Because Sally took my toy!" She can combine symbols together to think causally.

The child learns to connect symbols in a variety of contexts, including an understanding of how one event leads to another ("The wind blew and knocked over my card house"), how ideas operate across time ("If I'm good now, I'll get a reward later" or "He was mean to me yesterday; I bet he'll be mean again"), and how ideas operate across space ("Mom is not here, but she is close by"). Ideas can also help explain emotions—"I got a toy so I'm happy"—as well as organize knowledge of the world.

Connecting ideas logically is also the basis for reality testing, because the child now connects experiences inside herself with those outside and categorizes which are which (fantasy versus reality). Her ongoing emotional interactions support this ability to form a category of reality because they continuously put a "me" in contact (through the interactions) with a "not me or you." This ongoing contact with someone who is "not me" provides constant contact with an external reality outside oneself. The emotional investment in relationships enables the child to recognize the difference between her fantasies and the actual behavior of others. While "reality testing" might appear to be a purely cognitive capacity, as we are describing, it requires an ability to organize an emotional sense of self that is distinct from one's sense of others. Such "reality testing" is a critical foundation for logical thinking. Without it, facts are often used to support irrational beliefs.

Logical thinking leads to an enormous flowing of *new* skills, including those involved in reading, math, writing, debating, scientific reasoning, and the like. The child can now create new inventions of her own, such as a new "game," and play games with rules.

General "reasoning" emerges from understanding emotional interactions and is applied to the more impersonal world. For example, cause-and-effect thinking with symbols comes from dealing logically with someone else's intentions or feelings: "When I'm mean, my mom gets annoyed with me." Once a sense of causality has been established at the symbolic level, a child can understand how the light from the sun causes "day time." A "reality sense" enables us to "think" realistically about many different things. Of course, purely cognitive skills enhance this core ability, but it is founded upon emotional interactions.

The ability to build bridges between ideas leads to a new level of intelligence. The child can apply her new reasoning ability not only to arguing about "why I should watch more TV" (because it's fun) but also to why certain letters make up a word (reading) or why adding numbers together can help you figure out how many apples you have. In other words, the concepts behind most academic abilities depend on this type of logical thinking. We often think about school readiness, intelligence, and academic abilities without regard to a framework for understanding the stages in thinking. However, this fundamental level of thinking is a vital component of intelligence because it is a precondition for all higher levels of intellectual functioning.

New social skills emerge from the ability to connect ideas. The child can now understand why it's important to follow the rules of the group. The child can also participate in forming group rules. These new skills eventually provide the basis for participating in larger groups, communities, and societies, where individuals need both to follow and help define the rules that will enable a large number of people to live safely and securely and solve challenges together. For example, the child can now understand why it's important not to hit (it will hurt someone else or because "I will be punished in the following way"). The child can suggest a "new rule": She gets to watch her show on TV and then her brother gets to watch his show. This type of social problem solving, as indicated, not only enables siblings to get along but enables entire groups and societies to negotiate ways to live and solve problems together.

The sense of self is now defined at a higher level. Understanding the logical bridges between different feeling states enables the child to connect the different parts of what she considers "me" or "you" together at a symbolic level. The mischievous or angry "me" and the nice and happy "me" are understood logically: "When you don't let me do what I want, I'm mad. When you're nice to me, I'm happy and nice back."

Consciousness is also moving to a higher level because there is now a symbolic awareness of one's own feelings: "I'm sad because you were mean to me." This is the beginning of what will be a monumental journey towards higher and higher levels of reflective self-awareness.

Stage 7—Multiple-Cause and Triangular Thinking

From simple causal thinking children progress to recognizing multiple causes, often experiencing a rapid growth in this capacity between four and seven years of age. If someone won't play, instead of just concluding, "She hates me," the child can say, "Maybe she has someone else she wants to play with today. Maybe her mother is making her come home after school." She can set up multiple hypotheses. Or, "Maybe she doesn't want to play with me because I have always played Nintendo. Maybe if I offer to do something else she'll want to come over." The child is now becoming a multiple-cause thinker in many contexts. In school, she can now look at multiple reasons for the Civil War or why a storybook character is upset. With peers, she can compare two friends—"I like Sally better than Stephanie because she has great toys." Multiple-cause thinking enables her to engage in "triangular" thinking. At home, if mother is annoyed, the child can try to make her mother jealous by going to her father and being coy with him. She can become friends with Sally so as to get to know Sally's friend Judy. She can figure out how a character in a book pretended to like her vegetables so that she could get dessert.

The child becomes a more flexible thinker as a result of multiple-cause thinking. Eventually she comes to understand more intricate plots in literature, the multiple causes for historical events, and a physical phenomenon that requires a scientific explanation. Therefore, multiple-cause thinking constitutes a higher level of reflective thinking in all spheres.

To learn multiple-cause thinking a child not only needs to have learned the earlier levels discussed above, she needs to be able to invest emotionally in more than one possibility. For example, she may not be able to consider a second friend as a possible play partner if she is too dependent on the first friend. She may believe that she will "lose" that friend unless she plays only with her. She may not be able to woo her father into playing with her, or even consider this possibility if she is too anxious about losing her mother.

The child can now also understand her own family dynamics through relationships among different people rather than just by whether her own needs are met. For example, it's not simply a question of whether her mother is paying enough attention to her but how she competes with a sibling when they both want mother's attention. Similarly, she can negotiate social triangles and figure out that another child may prefer to play with someone else on a given day without necessarily disliking her.

As the child's intelligence and social skills are advancing, her sense of self is expanding to include new horizons as well. She can now begin to look at herself as competitive and needy, mischievous and funny (multiple dimensions) all at the same time. Her consciousness is gradually developing an awareness of how she can employ these different facets of herself in a variety of contexts—whether it's to compete with her sibling or to recruit a new friend at school.

Stage 8—Gray-Area, Emotionally Differentiated Thinking

The ability to engage in emotionally differentiated thinking enables a child to understand the different degrees, or "relative" influence, of different feelings, events, and phenomena. For example, often between ages six and ten she is rapidly learning that she can like other kids or be angry with them (and vice versa) to different degrees. In school, she not only looks at multiple reasons for events but can weigh the degree of their influence as well. "I think opinions about slavery were a lot more important than where people lived (the North versus the South) in causing the Civil War." With peers she can compare feelings in a graduated way: "I like Sally a lot more than Stephanie because she is much nicer to me when I'm upset."

Gray-area thinking enables children to comprehend their roles in a group and deal with increasingly complex social systems ("I'm third best at spelling and fifth best at telling stories"). All future complex thinking requires mastery of this stage, whether it involves looking at the relative influence of variables in science and math or understanding one's social group and society.

Intelligence now expands to include a more gray-area understanding of the world. She now not only can look at multiple reasons for an event but can also weigh how much each factor contributed. She can do the same with her own feelings (a little, a medium amount, or a lot of anger). Her reflectiveness, therefore, includes a new appreciation of both the world and herself.

Socially, the child is now truly able to negotiate the politics of the playground. She can figure out and participate in multiple social hierarchies involving everything from power or dominance and submission to athletic skills, academic abilities, and likeability. New ways to solve problems, especially group problems that involve multiple opinions, are now possible because the child can compromise in the "gray area." The ability to operate in social hierarchies and employ gray-area negotiation strategies creates the foundations for participating in the larger social reality of one's community, society, and culture.

The sense of self is now expanding to include a sense of being a member of a social group. If there is sufficient security in the family, so that the child can take her "chicken soup" for granted, she can move out into the social group with full vigor and begin defining herself more and more through her peer relationships and these newly understandable social hierarchies. Her sense of self, therefore, is achieving a new level of organization and a truly social self (in comparison to the earlier family-defined self) is emerging. This emerging sense of self in the social group, however, is often initially quite rigid and polarized: "I'm the worst soccer player (or dancer)!" Interestingly, many social groups, as we will discuss in Chapter 13, are organized around rigid rules and hierarchies rather than reflective processes and institutions.

Similarly, consciousness is also broadening. The comprehension of social hierarchies and relativistic intellectual concepts leads to new levels of awareness and self-awareness. The child is conscious, not only of herself in the group but also in a relativistic sense in the world. It's not surprising, therefore, that we see greater concern with life and death at this time as the child becomes aware of the cycles of life. As indicated, children may attempt to handle this broadening of their consciousness by becoming temporarily more rigid and compulsive. However, this is a temporary phase, adaptive in a sense, a way to slow down the progress a bit and so digest the new awareness.

Stage 9—A Growing Sense of Self and an Internal Standard

By puberty and early adolescence, more complex emotional interactions and thinking build a ninth level—an internal standard based on a growing internal sense of self. This process, which builds during the adolescent and adult years, can begin as early as ten or twelve years as children gradually create a more defined sense of self. That sense of self, in turn, constitutes an inner standard by which to judge experience. For example, they can say for the first time, "I shouldn't be so angry because the insult wasn't that great." Similarly, they can look at peers who are doing something naughty and say, "I shouldn't do that because it isn't the right thing for me to do. It may be okay for them." A child aged from ten to twelve and beyond can have a hard time on the playground and still feel like a good person. In contrast, an eight-year-old might have felt like a bad person instead of being able to apply her inner standard. Using that internal standard, a child can now look at history and say, "I agree with the North (or the South) for the following reasons." Or, "In the First World War, I think that when people behaved that way, they were exercising bad judgment."

At this stage, children become able to make more inferences. Inferences mean thinking in more than one frame of reference at the same time, or creating a new idea from existing ones. To use more than one frame of reference, one needs to have an "agency" that can do the looking and relating. One of the two frames of reference that are being compared must be based on an organized sense of self that is the product of meaningful experience. Intelligent inferences involve all the developmental levels we have been describing as well as emotionally meaningful experience in the sphere of knowledge where the inference is being made. Both are required. Experience and knowledge in the area where the inference is made are necessary for sophisticated inferences rather than naïve ones.

Cognitive, educational, and learning theory has not been able to develop an adequate model to figure out or explain how to promote the highest levels of reflective thinking because it has not focused on the role of emotion. In other words, the highest levels of thinking require combined emotional and cognitive development because they involve comparing frames of reference, which have to be based on ongoing emotional experiences and a complex internal sense of self.

The new ability to think in two perspectives—objective reality and personal opinion—at the same time separates individuals who remain somewhat more concrete (they haven't yet mastered this level) from those who have. For those who have, the door is now open to the higher levels of intelligence and reflectiveness that will be characteristic of adolescent and adult thinking.

During the early adolescent years, the ability to consider the future as well as the past and present broadens the sense of self, an internal standard based on it, and the hypothesis-making abilities that lead to inference and other types of creative thinking. Probabilistic thinking about the future, however, can be more readily applied to areas such as math and science than others, such as one's personal plans. Emotionally investing in the future comes in the next stages.

From a social point of view, values and ideals can now be constructed, debated, and argued. To be sure, there will be all kinds of trial balloons as teenagers experiment and argue for cultural norms that are different from those of their parents. Nonetheless, such debates signal a whole new level of social and cultural understanding. It creates the foundations not only for understanding values and ideas and pursuing them but also for investing institutions with stable personal beliefs that can sustain whole societies; for example, concepts and institutions that support justice.

The ability to reflect an internal sense of self and standard may be viewed as one of the basic capacities required to support values and institutions commonly associated with representative democracies. Such forms of government require a certain percent of the population to invest in abstract principles such as justice and the institutions that support it ("the consent of the governed," according to Jefferson). Only individuals who can think from an internal standard and sense of self can invest in abstract principles and institutions. Individuals unable to do this are more likely to adhere to a specific leader or a concrete belief.

The sense of self now embraces a new agency as values and ideals gradually become an integral part of the self. The self has also expanded to include the two dimensions we've been discussing—the one that operates in the emotions of the day-to-day events, and the one that can evaluate these events against a longer term based on a more integrated sense of the self over time. This creates an entirely new level of consciousness, one in which self-observation is possible. Individuals can now evaluate how they feel, what they've done, and plan to change and improve. They can also use this new level of reflective awareness for self-directed criticism and anger as well as to praise or invigorate themselves.

If children remain mired in all-or-nothing thinking, they are likely to take extreme positions about themselves. On the other hand, if they've progressed adaptively through the different stages, through gray-area thinking, they can use this new self-observing ability to take an honest look at their behavior and then plan accordingly. Because this process is just beginning (as are values and cultural norms), there will be lots of experimentation. We will see a mixture of harsh or extreme attitudes alongside what appear to be very reasonable ones. Nonetheless, an entirely new level of self-reflection and, therefore, conscious awareness of the world is now possible.

THE STAGES OF ADOLESCENCE AND ADULTHOOD

The nine stages we have been describing constitute basic functional emotional development and create the very "structure of thinking." From childhood on, these skills are applied to an ever broadening range of experiences. In adolescence and adulthood, we take the abilities for multiple-cause and gray-area thinking, and an internal sense of self and standards, and apply these to an even more ever-widening world. In adolescence, this means biological changes, new emotions and social networks, new interests, and future-oriented probabilistic thinking about education and work. In adulthood, marriage or intimate relationships, having children, and/or career challenges create new demands for reflective thinking. Some of us can reflect maturely in some areas, such as our work problems, but not in family issues. Some of us can apply complex thinking skills around family, but not around work. Some of us can do it in scholarly work, but not in politics. We vary in how well we apply our thinking abilities because they are dependent on the range of our emotional experiences, the breadth and complexity of our growing sense of self, and the scope and depth of our reflective abilities.

As each one of us is simultaneously increasing the complexity and level of integration of a sense of self—for example, a self that can embrace intimacy with one's own children or spouse—we also broaden and further integrate our internal standards. This includes looking at actions from the perspective of not just oneself and family but community, culture, and society, and ultimately the world community. Each of these broader applications leads to a higher level of reflective thinking.

Thinking, as well as its content, changes. For example, the ability to empathize simultaneously with one's children and their feelings, one's spouse and his or her feelings, and, at the same time, to be aware of one's own feelings without overidentifying with those of one's children or spouse is not simply a very high level psychological achievement but a high level of abstract, reflective thinking. Just as thinking of a sense of self and internal standard is at once an emotional and cognitive ability, similarly at each higher level of reflective thinking the building blocks are at once emotional and cognitive. At these higher levels both are required for true reflective thinking. The reason we see so much pseudo-reflective thinking at seemingly high levels, and why truly reflective thinkers are so rare, is because of the difficulty in mastering these combined cognitive and emotional skills.

The next seven stages will be described very briefly. Although a great deal of interest has been shown in the stages of adolescence and adulthood—for example, the work of Erik Erickson, as well as many others[30]—there is a great deal more to learn about these phases of human development.

Stage 10—An Expanded Sense of Self

During this stage, which begins in early and middle adolescence, the ability to reflect by using an expanding internal standard and a growing internal sense of self encompasses new learning experiences, including physical changes, sexuality, romance, and closer, more intimate peer relationships, as well as new hobbies and tastes in art, music, and dress. Some adolescents become fragmented or rigid and constricted in an attempt to cope with all these challenges. They may attempt to return to an earlier, narrower sense of self. In other words, as the complexity of new challenges and experiences expands, an individual may lose the ability to use gray-area, multiple-cause thinking, or an internal standard. When an adolescent is able to incorporate new challenges into an expanding sense of self, however, he or she gains broader reflective abilities. No matter how good cognitive skills are (such as mathematical reasoning), a person's world can be narrowed by naïve or rigid thinking that ignores the full complexities of life.

As the self is expanding to include these new experiences, intelligence is also broadening to include a greater appreciation of the world along with the new levels of reflection. For example, literature can be understood as an exploration of complex relationships between the characters in a novel or drama. Motives for historical events can be more fully understood. A broadening social reality facilitates an appreciation of culture and society.

As a consequence of these changes, one's consciousness and reflectiveness is reaching higher levels. For example, one can *think about thinking and observe one's own patterns of thought and interaction* ("I tend to be quick-witted with the guys but slow-thinking with girls."). This is no small feat. Used adaptively, thinking about thinking creates new insights about oneself and others. It can also, however, lead to overload and fragmented thinking. Nurturing support and the successful negotiations of earlier stages support the adaptive pathway.

Stage 11—Reflecting on a Personal Future

A significant development in late adolescence and early adulthood is an ability not only to think about the future but also to become emotionally invested in one's personal future. Although the ability for probabilistic thinking often begins in early adolescence, the ability of a teenager to apply this skill to her own life becomes more firmly established as she becomes emotionally invested in her future: leaving home, going to college, getting a job, and other such concerns. Without such an investment, probabilistic thinking may not fully develop. Too much anxiety about the future will discourage reflection and restrict cognitive and emotional development. The need to invest emotionally in the future to develop probabilistic thinking in full is another illustration of how emotional and cognitive development work together.

Investing in future-oriented probabilistic thinking is not only needed for mathematical and scientific reasoning, it also enables an appreciation of social patterns. One can look at the implication of social, political, economic, and cultural patterns for the future in relationship to the past and present and future. This not only helps one plan but also leads to a more sophisticated and intelligent analysis of history, culture, and society. It enables one to be a more intelligent voter and economic participant. One's sense of self now embraces a sense of the future and a more integrated sense of the past as well as the immediate present. At the same time, consciousness expands to include this new perspective on time as well as a continuously growing perspective on one's personal history and future.

This perspective will continue growing as the psychological investment in the future (one's children's future, one's retirement, the future of the world) becomes more important.

Stage 12—Stabilizing a Separate Sense of the Self

This stage of early adulthood involves separating from the immediacy of one's parents and nuclear family and being able to carry the warmth, security, and guidance of those relationships inside oneself. It involves stabilizing the sense of self and one's internal standards so that one can always set internal limits, use good judgment, and care for oneself, even when operating outside the immediacy of one's own nuclear family and/or related institutions.

What was earlier a variable or relatively unstable sense of self and internal standard now must be more stable to hold firm in the face of growing independence at college or work. It also must be applied to exploring intimate relationships more deeply.

Without such a stable, broad inner framework, intellectual judgments and reflective thinking tend to be naïve, limited, or fragmented. For example, young adults who retreat back into earlier adolescent modes of being may not be able to appreciate the nature of long-term, intimate relationships with a partner or a future family, let alone reflect upon them. A whole realm of life, literature, and culture that deals with intimate adult relationships, careers, and independence from parental figures may remain outside of, rather than under, the perspective of one's reflective thinking.

Young adults can now often make judgments that "thoughtfully" incorporate and accept or reject the standards of their caregivers. The standards of one's caregivers, however, are not simply their values and judgments, but, through their good offices, the history of their culture as well as one's own—that is, one's heritage. There is, therefore, greater independence from daily reliance on one's nuclear family, greater investment in the future—mobilized in the prior stage—and greater ability to carry one's past inside oneself as part of a growing sense of self and internal standard. This stage ushers in the beginning of a long process that involves reflective thinking and that can use the past, present, and future in a relatively more independent manner.

Stage 13—Intimacy and Commitment

The ability for commitment and intimacy now builds on all the earlier stages of emotional development. It includes taking the initial steps involved in life's major decisions. It calls on all the prior stages as well as new depth in the ability to reflect upon relationships, passionate emotions, and educational or career choices. This challenge can deepen and further stabilize an expanding sense of self and broaden one's thinking (for example, with new levels of empathy). For example, the challenge of loving another person over a long period of time involves engaging in a relationship with deepening intimacy and growing respect for unique differences. This is not an easy feat, and it can lead to a narrowing of emotional investments, rigidity, and fragmentation or new levels of reflectiveness.

Reflective thinking achieves a yet higher level as a new set of time and space dimensions are incorporated into educational, career, and personal relationships. For example, involvement with a potential mate and having a family of one's own inspires a shift from relative states of emotional immediacy to increasingly longer-term commitments. Decision-making involves greater lengths of time and more stable long-term commitments to different types of interpersonal space (work and school environments, setting up homes as opposed to living in dormitories or apartments). With this new level of reflection we may also begin seeing longer-term political and religious values consolidate, although these will often form and consolidate for some time.

Stage 14—Creating a Family

For those who choose to create a family of their own that includes raising children, the ability to reflect broadly and wisely is challenged by the experience of raising children, without losing closeness with one's spouse or partner. An even harder challenge, however, is empathizing with one's children without overidentifying or withdrawing. At each stage of a child's development there is an opportunity for caregivers to overidentify, pull away, or empathize with a balance of caring, understanding, and guidance. Meeting this challenge can significantly expand, deepen, and ripen one's reflective skills and sense of self. At each stage in the child's life it enables one to rework issues in one's own development, as well as construct new empathetic capacities at a level of intimacy and depth, perhaps not attained in any other relationship. On the other hand, it can make a person pull back, wall off parts of the self, and become fragmented. As with all new demands and challenges, there is the risk that thinking will become concrete, narrow, or rigid when challenges are too great.

An adult with all the early stages in place can now develop a new level of consciousness and reflective thinking because of the growing ability to view events and feelings from another individual's perspective, even when the feelings are intimate, intense, and highly personal. In other words, the empathy learned through taking care of children opens up new dimensions of feelings that were not possible at earlier stages of empathy. As this ability develops, one is able to generalize it and look at and empathize with the goals, needs, and perspectives of other communities and cultures while maintaining a strong sense of one's own cultural heritage, social values, and commitments.

Stage 15—Changing Perspectives on Time, Space, the Cycle of Life, and the Larger World: The Challenges of Middle Age

Middle age brings with it new perspectives and the need for an expanded, reflective range. Often, the experience of accompanying one's child through various stages of development, including possibly grandchildren, has brought new insights into one's own developmental journey. Ideally, one has deepened one's relationship with a spouse or partner. During this stage, one is propelled into having to think about the next steps in work and family life. Unrealistic or wishful expectations and earlier fantasies about attainments are tempered with an appreciation of accumulated reality-based experience and wisdom.

One's perspective of time is also changing. As Jaques[31] has pointed out, during the middle age years, the sense of time changes. The future is no longer infinite. Relative to one's own life, time appears to pass more quickly. In addition, one's allegiance often extends more and more into the world community and global concerns. When emotional investment moves beyond family, local community, or even nation, both the sense of self and consciousness further expand. Most important, however, this stage creates an ability to appreciate a new social reality, the global or world group. As we will discuss in Chapters 13 and 14, technological changes are challenging us all to embrace the global group as the new unit of survival.

Therefore, at this stage, the elements of time and space take on an important new dimension that leads to yet a higher level of reflective thinking. If these adaptive processes do not occur, however, varying degrees of depression, pessimism, anger, rigidity, and/or "escapes" back to earlier themes are possible.

As part of this stage, individuals frequently (either at a conscious or intuitive level) have a sense of where they are in their life's journey, including their goals. Implicit in this appraisal is a sense of one's own patterns in relationships to others, such as family and career. Most individuals operate within identifiable patterns related to their own prior experiences. For example, one gifted scholar shared that she had always tried to please a demanding, competitive, and "even more gifted" father and had operated that way throughout her seemingly successful adult life. In trying to please authority figures, however, she had inhibited her own creativity for fear of revealing her true competitive feelings towards those authority figures, as well as revealing her fear of failing. She secretly fantasized that after absorbing all the wonderful guidance from these "knowing" authorities, she would have her day in the sun. Now, in midlife, however, she was sensing that the very pattern she had elected to reach her goals was not leading her there and that the prize she was seeking was not available from the strategies she had elected.

To be sure, there were many other underlying dynamics related to this pattern. Illustrative of the point being made here, however, is the fact that this talented individual was now becoming depressed over the realization of the limitations of her own strategies and the related sense that the prize she had sought would not be forthcoming. This individual experienced at least certain elements of her pattern consciously. Others experience it at more intuitive or less conscious levels. Regardless of the way it's experienced, however, it often contributes to a sense of loss, which can also lead to sadness and depression.

Alternatively, however, this "awareness," which can be particularly poignant in midlife because the future is now finite rather than infinite (in a relative sense), can lead to a reappraisal and a decision to find an adaptive pathway outside one's "pattern." Interestingly, this type of adaptive solution often involves a reappraisal of one's goals as well, since the original goals, like the pattern associated with them, may have been partially colored and limited by a variety of previous experiences, including conflicts, and childlike solutions to family dramas. The relatively new pattern and goals, based on a reappraisal, can lead to a surprisingly fresh and robust direction for life's next stages. The reflective skills involved in such a reappraisal—that is, the ability to understand one's own patterns and make a "midcourse" adjustment—is an important component of an adaptive resolution of this particular stage.

During this stage, the cycle of life, including death, can also now be thought about in a new emotional manner. This further definition of the sense of time and space and one's emotional investment in community, society, and the world community can lead to an entirely new perspective of one's place in the world.

Stage 16—Wisdom of the Ages

The later stages of aging can be a time committed to true reflective thinking of an unparalleled scope or a time of retreat and/or narrowing. As Erik Erikson has pointed out, aging is a time of potential generativity.[32] It brings with it the possibility of broader wisdom, free from many of the self-centered and practical worries of earlier stages.

However, preoccupation with one's changing physical status, or a narrowing of interests and perspectives, accompanied by fear, anxiety, and depression, can lead to limited thinking. The decline of physical abilities, including memory and the ability to sequence actions and information, and the fear of terminal illness can either overwhelm or lead to further growth.

If memory loss and sequencing problems are not severe, the aging process opens up new vistas. Life is much more finite. Goals have been either met or not met. Grandchildren or great grandchildren may be a part of one's life or on the horizon. A spouse or partner may be an even deeper ally in life's travels. One may be able to comprehend the cycle of life in a richer, fuller manner.

When the aging process and changes in one's own body become dominant, the appreciation and acceptance of the life cycle is juxtaposed with the possibility of depression and withdrawal. New, almost impossible to anticipate feelings and experiences are generated. Time, space, person, and self have new dimensions and meanings. In other words, aging can bring not just a new insight but what some have called wisdom, an entirely new level of reflective awareness of one's self and the world.

This quick journey has described the transformations of emotions throughout life. During each of the sixteen stages, we observe the relationship between emotions, intelligence, and reflective thinking (see Table 2.1 for a brief overview).

But even more intriguing than attempting to figure out how each individual develops emotions and intelligence is the question of how this process first evolved millions of years ago. In the next section, we will present support for our hypothesis that cultural patterns dating back to prehuman societies and that support the developmental steps just described have been passed on by each generation teaching the next generation. We will trace how different species of primates, prehumans, and, eventually, humans, over millions of years, achieved mastery of progressively higher levels of emotional transformation and intelligence.

TABLE 2.1 Overview of the Transformation in Emotional and Intellectual Growth

Functional Emotional Developmental Level	*Emotional, Social, and Intellectual Capacities*
Shared attention and regulation *(from birth on)*	Pleasurable interest in sights, sound, touch, movement, and other sensory experiences. Leads to looking, listening, calming, and awareness of the outer world and simple patterns.
Engagement and relating *(from 2 to 4 months on)*	Pleasurable feelings characterize relationships. Growing feelings of intimacy.
Two-way intentional, emotional signaling and communication *(from 4 to 8 months on)*	A range of feelings become used in back-and-forth emotional signaling to convey intentions (e.g., reading and responding to emotional signals); the beginning of "cause and effect" thinking.

(continued on next page)

TABLE 2.1 *(continued from previous page)*

Functional Emotional Developmental Level	*Emotional, Social, and Intellectual Capacities*
Long chains of co-regulated emotional signaling, social problem solving, and the formation of a presymbolic self *(from 9 to 18 months on)*	A continuous flow of emotional interactions to express wishes and needs and solve problems (e.g., to bring a caregiver by the hand to help find a toy): a. Fragmented level (little islands of intentional problem-solving behavior) b. Polarized level (organized patterns of behavior express only one or another feeling state, e.g., organized aggression and impulsivity or organized clinging, needy, dependent behavior, or organized fearful patterns) c. Integrated level (different emotional patterns—dependency, assertiveness, pleasure, etc.—organized into integrated, problem-solving emotional interactions such as flirting, seeking closeness, and then getting help to find a needed object)
Creating representations, symbols, or ideas *(from 18 months on)*	Experiences, including feelings, intentions, wishes, action patterns, etc., are put into words, pretend play, drawings, or other symbolic forms at different levels: a. Words and actions used together (ideas are acted out in action, but words are also used to signify the action) b. Somatic or physical words are used to convey feeling state ("My muscles are exploding," "Head is aching") c. Action words are used instead of actions to convey intent ("Hit you!") d. Feelings are conveyed as real rather than as signals ("I'm mad," "Hungry," "Need a hug" as compared with "I feel mad" or "I feel hungry" or "I feel like I need a hug"). In the first instance, the feeling state demands action and is very close to action and in the second one, it's more a signal for something going on inside that leads to a consideration of many possible thoughts and/or actions e. Global feeling states are expressed ("I feel awful," "I feel OK," etc.) f. Polarized feeling states are expressed (feelings tend to be characterized as all good or all bad)

(continued on next page)

TABLE 2.1 *(continued from previous page)*

Functional Emotional Developmental Level	*Emotional, Social, and Intellectual Capacities*
Building bridges between ideas: logical thinking *(from 2½ years on)*	Symbolized or represented experiences are connected together logically to enable thinking. This includes the ability for: a. Differentiated feelings (gradually there are more and more subtle descriptions of feeling states—loneliness, sadness, annoyance, anger, delight, happiness, etc.) b. Creating connections between differentiated feeling states ("I feel angry when you are mad at me") and logical thinking ("The letters 'C,' 'A,' and 'T' spell CAT")
Multicause, Comparative, and Triangular Thinking	Exploring multiple reasons for a feeling, comparing feelings, and understanding triadic interactions among feeling states ("I feel left out when Susie likes Janet better than me").
Emotionally differentiated gray-area thinking	Shades and gradations among differentiated feeling states (ability to describe degrees of feelings around anger, love, excitement, love, disappointment—"I feel a little annoyed").
Intermittent reflective thinking in relation to a sense of self, and an internal standard	Reflecting on feelings in relationship to an internalized sense of self ("It's not like me to feel so angry" or "I shouldn't feel this jealous").
Reflective thinking with an expanded self; the adolescent themes	Expanding reflective feeling descriptors into new realms, including sexuality, romance, closer and more intimate peer relationships, school, community, and culture, and emerging sense of identity ("I have such an intense crush on that new boy that I know it's silly. I don't even know him").
Reflective thinking with an expanded self; considering the future	Using feelings to anticipate and judge (including probabilizing) future possibilities in light of current and past experience ("I don't think I would be able to really fall in love with him because he likes to flirt with everyone and that has always made me feel neglected and sad").

(continued on next page)

TABLE 2.1 *(continued from previous page)*

Functional Emotional Developmental Level	*Emotional, Social, and Intellectual Capacities*
Reflective thinking with an expanded self; the adult years	Expanding feeling states to include reflections and anticipatory judgment with regard to new levels and types of feelings associated with the stages of adulthood.
Reflective thinking and the separation, internalization, and stabilization of the self	The ability to separate from, function independently of, and yet remain close to and internalize many of the positive features of one's nuclear family and stabilize a sense of self and internal standard.
Reflective thinking and commitment, intimacy, and choice	Intimacy (serious long-term relationships).
Extending the self to incorporate family and children	The ability to nurture and empathize with one's children without over-identifying with them.
Middle age	The ability to broaden one's nurturing and empathetic abilities beyond one's family and into the larger community.
	The ability to experience and reflect on changing perspectives of time and space and the new feelings of intimacy, mastery, pride, competition, disappointment, and loss associated with the family, career, and intrapersonal changes of mid-life.
The aging process	The ability for true reflective thinking of an unparalleled scope or a retreat and narrowing of similar proportions. There is the possibility of true wisdom free from the self-centered and practical worries of earlier stages. It also, however, can lead to retreat into one's changing physical states, a narrowing of interests, and concrete thinking.

A TIMELINE FOR HUMAN EVOLUTIONARY DEVELOPMENT

In recent years, paleoanthropologists have made striking discoveries that suggest the highest capacities we associate with human functioning, such as symbolic and logical thinking, emerged gradually and earlier than was previously thought. The saga is still unfolding, and evidence is mounting that supports this picture. For example, it has just been reported that an animal bone between 1.4–1.2 million years old was found in the Kozarnika cave in northwest Bulgaria. The bone has a series of parallel lines engraved in it, which, according to Dr. Jean-Luc Gaudelli of the University of Bordeaux, "were not from butchering; in this place there is nothing to cut. It can't be anything else than symbolism" (BBC News Online, March 16, 2004).

Discoveries such as this support the hypothesis regarding symbol formation during the course of evolution presented in this work. We have suggested that symbol formation results from a series of stages of affective transformations, which we have described as functional/emotional developmental levels. We have further suggested that although biological foundations were necessary for this progression, the *sufficient* condition for the steps leading to symbol formation was culturally transmitted caregiving practices from one generation to the next, across species, over millions of years. These culturally transmitted learning processes enabled new generations to master more complex stages of affective interaction leading to symbol formation. If this hypothesis is correct, the fossil record should reflect a gradual process of symbol formation, the signs of its early stages discovered over time. The alternative hypothesis, that genetic mutations were the *sufficient* condition for symbol formation, would lead one to expect the sharp emergence rather than the gradual unfolding of symbolic artifacts.

Therefore, the current findings support our hypothesis of a continuously unfolding biological/cultural process, in which formative caregiving practices were passed down from one generation to the next. In the future, we can expect to hear of further discoveries that show the development of our higher capacities was even more gradual and emerged even earlier than the argument presented in this book, which is based on current paleoanthropological data. On the basis of this existing data, however, we can now outline, in general and schematic terms below, and discuss more fully in the next section, the timeline of the broad trajectory of f/e developmental capacities and related abilities for symbolic thinking, with a few examples that characterizes human evolution:

(I) and Elements of (II) and (III) Attention and self-regulation and with elements of engaging and signaling	**tamarins, marmosets, and other mammals** —e.g., marmoset and tamarin infants spend a lot of time looking at their caregivers' faces and staring into their eyes, and they engage in much back-and-forth vocalizing. As the infants grow older, they spend more and more time looking at their caregivers' faces and staring into their eyes. Much vocalizing goes on during these episodes of shared gaze, and when it is time to be weaned, the infant can be seen calling for food; the caregiver responds by slowly extending something to eat.
(II) and Early (III) Engaging and relating and early signaling	**rhesus monkeys and other mammals** —e.g., a rhesus baby spends a lot of time snuggling into its mother's body and looking keenly at her face; it visibly relaxes and vocalizes happily while being rocked, and rhythmically moves its arms and legs. As adults, rhesus monkeys use back-and-forth emotional signaling to deal with danger, aggression, assertiveness, etc.
(III) and Early (IV) Two-way purposeful affective interaction and communication	**baboons and other mammals** —e.g., baboon infants engage in back-and-forth emotional signaling with their caregivers. An infant might happily vocalize, receive the same vocalization from its caregiver, and then make the vocalization again. Recent research has shown that, as adults, baboons engage in extremely subtle back-and-forth signaling in both male-male and male-female greetings whereby they work out such matter as friendship, mating, and coalitions.
(IV) Co-regulated affective signaling and shared social problem solving	**H. erectus (2–.4 mya)** **H. habilis (2–1.5 mya)** **Australopithecines (5.3–1.4 mya)** **Ardepithecines (5.8–4.4 mya)** **chimpanzees, bonobos and other primates** —e.g., chimpanzees in the wild engage in organized hunting parties, demonstrating a sophisticated understanding of how the other hunters will behave in any given situation as well as how the different species of prey will behave. Particularly striking is the amount of complex emotional communication that is involved. Through gestures, body movements, head nods, and facial expressions, the hunters coordinate their actions and signal to one another who is to do what and when.

(V)
Creating ideas or internal representations; symbolic and linguistic abilities

Archaic H. sapiens and Early Moderns (600,000–60,000 years ago)

—e.g., over 300 fragments of pigment, believed to be between 350,000 and 400,000 years old, were found in a cave at Twin Rivers, in Zambia. These pigments had been ground up into a powder and ranged in color from brown to red, yellow, purple, blue, and pink. Paleoanthropologists believe that they were used for ritual body painting, and perhaps, for cave painting. That is, current evidence points to the fact that *Archaic H. sapiens* were mastering, through long and complex affective interchanges, various elements of presymbolic communication and were even beginning to use symbols to convey abstract thoughts and feeling states.

(VI)
Connecting ideas together; Logical Thinking

H. sapiens sapiens (Approximately 130,000 years ago)

—e.g., evidence that early modern humans were at least at the stage of beginning to build bridges between their ideas and to think logically, if not higher, is provided by the first colonization of Australia that took place about 60,000 years ago, which required them to build some sort of crude transport to cross the sea. Operating in this kind of complex group endeavor would have required the ability to perceive connections between natural events and plan their actions accordingly.

(VII-VIII)
Multi-causal and gray-area differentiated thinking

Magdelenian Period (12,000–8,000 B.C.)

—e.g., the Magdelenians were figuring out how to adapt to their hostile environment with more complex kinds of shelters and more protective types of clothing. Not only did they likely experiment with different kinds of structures, but they also experimented with different materials and techniques for fashioning cooking pots and for making protective clothing out of animal skins. Such technological advances reveal a flexible form of abstract reasoning. The emergence of settlements at the end of the period also indicates that they were mastering the sort of gray-area thinking in which one comprehends one's role in a group and can deal with increasingly complex social systems

(IX-X)
Thinking according to an internal standard and growing sense of self

Ancient Civilizations (3,500–72000 B.C.)
—e.g., individuals in the Sumerian state were starting to think of their family's rights, and what sorts of public actions were available to them to avenge transgressions. Laws invested with our internal standards were instituted to ensure that women and vulnerable members of the society were protected by the State. Accomplishing this level of social development requires that the individuals involved master thinking from an internal standard and acquire a more integrated sense of self. There are different degrees here. At a fairly concrete level, the shared sense of reality that binds together the members of a group may simply be based around common words and concepts. At a more abstract level, a shared sense of reality may be grounded in institutions and practices that support a shared system of values.

(XI-XII)
Reflective thinking on the future and an expanded concept of the self

Ancient Greece (6th–3rd century B.C.)
—e.g., in sixth century Miletus, thinkers such as Thales and Anaximander began to develop the first concept of *science*: i.e., the first attempt to explain natural phenomena empirically. These early scientific advances were later followed by remarkable advances in abstract thought, e.g., in the development of metaphysics, logic, mathematics, and geometry, as well, of course, as the elaborate forms of ethical, aesthetic, and political theories found in Plato, the Phytagoreans, and Aristotle. These f/e advances are also reflected in the creation of the *Polis*: the city-state that was based on the idea that people should live in a community of individuals who were all conscious of their shared interests and common goals, and who would manage collective concerns by debating possible courses of action in a public setting. In order to reach consensus through public decisionmaking forums, individuals have to be able to recognize that there are different degrees of satisfying their interests and must also reconcile their personal wishes with the larger needs of a community that now includes members that one is not tied to by blood or possibly even customs.

—— PART TWO ——

A New Direction for Evolutionary Theory

Introduction

> Asking what humans would be like without culture is like asking what a duck would be like if it had lips instead of a bill. It wouldn't be a duck; it would be something else.
>
> —JONATHAN MARKS, *What It Means to Be 98% Chimpanzee*

THE EVOLUTIONARY HYPOTHESIS we are presenting in this book challenges a long determinist tradition in Western thought that dates back at least as far as the Enlightenment, if not beyond. According to this tradition, all human beings, apart from those who have suffered some serious biological disorder, have the same mental capacities; only human beings have these capacities; and the nature and maturation of these capacities is somehow fixed in advance. It is important to recognize just how deeply ingrained this determinist paradigm is in Western thinking about the mind. Long before Charles Darwin boarded the *Beagle,* Western scientists were thinking about evolution; and long before James Watson and Francis Crick discovered the structure of DNA, Western thinkers were thinking about evolution in determinist terms.

According to the great seventeenth-century philosopher Nicholas Malebranche:

> We may say that all plants are in a smaller form in their germs. By examining the germ of a tulip bulb with a simple magnifying glass or even

> with the naked eye, we discover very easily the different parts of the tulip. It does not seem unreasonable to say that there are infinite trees inside one single germ, since the germ contains not only the tree but also its seed, that is to say, another germ, and Nature only makes these little trees develop. We can also think of animals in this way. We can see in the germ of a fresh egg that has not yet been incubated a small chick that may be entirely formed. We can see frogs inside the frog's eggs, and still other animals will be seen in their seed when we have sufficient skill and experience to discover them. . . . Perhaps all the bodies of men and animals born until the end of times were created at the creation of the world, which is to say that the females of the first animals may have been created containing all the animals of the same species that they have begotten and that are to be begotten in the future.[1]

Malebranche's views were enormously influential and played a key role in the emergence of the doctrine known as Deism: the view that God meticulously designed all Nature at the beginning of time and then allowed His creation to operate independently. On this mechanist outlook, the universe is like an extraordinarily complex clock that God fashioned and then left to run on its own.

Twentieth-century evolutionary theorists mounted a vigorous attack on Deism, *not* because they objected to the mechanism but because they objected to the teleology that earlier mechanists had implicitly embraced. Perhaps the most influential of the neo-Darwinian critics of this teleological outlook was Stephen J. Gould. Gould documented considerable evidence throughout his career to show why evolutionary change cannot be viewed as progressive. Thus Gould placed great emphasis on the nonadaptive consequences of genetic variation. In their famous article, "The Spandrels of San Marco and the Panglossian Paradigm: A Critique of the Adaptationist Programme," Gould and Richard Lewontin argued that biological adaptation is analogous to the spandrels in the cathedral of San Marco, which Venetian artisans decorated with detailed mosaics portraying biblical rivers. Just as the architectural space for these mosaics arose as the side effect of an engineering requirement, so too, according to Gould and Lewontin, certain biological structures are nonadaptive consequences of the demands of a local environment.[2]

The anti-Deistic implications of Gould's neo-Darwinism have been most vigorously pursued by Richard Dawkins. According to Dawkins:

> All appearances to the contrary, the only watchmaker in nature is the blind force of physics, albeit deplored in a special way. A true watchmaker has foresight: he designs his cogs and springs, and plans their interconnections, with a future purpose in his mind's eye. Natural selection, the blind unconscious, automatic process which Darwin discovered, and which we now know is the explanation for the existence and apparently purposeful form of all life, has no purpose in mind. It has no mind and no mind's eye. It does not plan for the future. It has no vision, no foresight, no sight at all. If it can be said to play the role of watchmaker in nature, it is the blind watchmaker.[3]

But despite the considerable difference between the teleological view of Nature embraced by eighteenth-century Deists and Dawkins's "blind watchmaker" scenario, what is even more striking is how Dawkins remains committed to the same determinist view of the human mind that the Enlightenment philosophers embraced. Indeed, modern genetic determinists in general have remained committed to the view that human beings' higher reflective capacities are *universal, unique,* and *innate.* Sociobiologists and evolutionary psychologists insist that there is "a universal human nature [that] exists primarily at the level of evolved psychological mechanisms, not of expressed cultural behaviours," and that "the evolved structure of the human mind is adapted to the way of life of Pleistocene hunter-gatherers, and not necessarily to our modern circumstances."[4] According to this outlook, if we want to understand how the human mind works, we have to investigate the "set of environments and conditions [that] defined the adaptive problems the mind was shaped to cope with: Pleistocene conditions, rather than modern conditions."[5]

Such an argument rests on the same assumption as that made by eighteenth-century Deists: that the structure of the human mind evolved at the dawn of our existence—according to sociobiologists and evolutionary psychologists, sometime between 200,000 and 30,000 years ago, as opposed to Bishop Usher's calculation that the world was created on October 26, 4004 B.C., and has stayed the same ever since. Civilizations may rise and fall, but this basic cognitive structure remains unchanged. It is as if Nature rather than God had somehow designed a mechanism—in its latest mechanist incarnation, a computer (the human brain)—together with a means for reproducing exactly the same software over and over again in each individual (the human genome). Hence, the sociobiologist or evolutionary psychologist sees himself as resembling an archaeological computer scientist whose task is to strip away the layers of culture—the applications of those programs—and expose the bare bones of "How the Mind Works": a mechanist metaphor first coined by the nineteenth-century logician George Boole and repeated by mechanist thinkers ever since.

There are, no doubt, many reasons why such a mechanist outlook has been so prominent in Western culture. One of the most important factors was the impetus that seventeenth-century mechanism received from the Newtonian revolution. Suddenly not just the movement of the planets and the stars but even the operations of the human body, and matter in general, could all be given a mechanical explanation. Perhaps the chief reason why this mechanist view of "how the mind works" has been so influential dates back to the Cartesian bifurcation between Reason and the emotions. Convinced that the mind is a "rational machine" that is guided by Aristotle's Fundamental Laws of Logic, generation after generation of philosophers and psychologists have sought to explain the operations of the mind in mechanical terms.[6] What could be more natural than the assumption that there must be some innate mechanism, namely, "genes," that contains blueprints for these mechanical processes that were selected in the Pleistocene and then passed down from one generation to the next in the same form?

Sociobiology and evolutionary psychology provide us with some penetrating insights into how the evolution of our biology, for example, the structure of our sensory organs, is intimately bound up with our evolution as social beings. But although natural selection may in part explain the evolution of our basic biological capacities, it cannot explain how we develop our higher-level reflective and social skills. The evolution of our higher-level capacities, which involve flexible problem-solving skills that use complex communicative and symbolic processes, depends on a very different process; for flexible problem solving needs to be sensitive to changing environments and cannot be structured into fixed biological pathways.

As we remarked at the beginning of the book, the assumption that natural selection alone can explain how such flexible problem-solving capacities evolved is based on a narrow paradigm that was originally applied to simple organisms. The reasoning here is quite similar to what was seen in the field of Artificial Intelligence (AI) forty years ago. Early

AI scientists were successful at modeling simple motor-visual and logical problems. They then assumed that the same computational procedures used to simulate the simple processes involved in these "toy domain" problems could be applied to higher-order cognitive processes; that is, the latter are somehow "built up" out of these "atomic units."[7] One problem with this assumption, however, is that, for deep historical reasons, AI ignored the role of emotions in the growth and functioning of the mind. Indeed, it was for precisely this reason that AI scientists assumed that higher cognitive functions could be simulated on a computer. But as we saw in Chapters 1 and 2, emotions serve as the critical architect for the mind's higher-order cognitive processes. Hence it is not surprising that AI failed to live up to its early expectations and was unable to advance beyond the level of modeling toy domains.

The same problem has surfaced again in modern evolutionary psychology. Given the assumption that the fixed behaviors of simple organisms are genetically determined, evolutionary psychologists then assume that the same mechanistic explanation must apply to the much more complex behaviors displayed by higher social organisms. These modern determinists assume that there must be special genes governing our social capacities, for example, genes for "mindreading," for seeing one another as intentional beings. Thus, just as AI scientists sought to account for the marked differences between following an algorithm and creative thinking in terms of the kind of heuristics embodied in a computational system, evolutionary psychologists seek to account for the difference between the "social behavior" of lower organisms—say, ant colonies—and the social behavior of humans in terms of the "programs" that are encoded in each species' genes and that dictate the manner in which they "process" information.

But there is no algorithm, computer program, or scientifically validated biological mechanism to explain complex human symbolic and reflective thinking. The problem here, however, is not that technology has still not advanced far enough to answer this age-old question of how we develop our higher-level reflective skills. The problem is that for over four hundred years, Western thought has been drawn to a hypothesis that, at present, not only has no scientific data behind it, but is not consistent with any known biological or physical mechanism.

For a generation raised on determinist principles, the idea that the evolution of the human mind and of human societies was the result of formative cultural practices that guide caregiver-infant interactions during the formative periods of development, and that these critical cultural practices were not genetically determined but, rather, were passed down and thus learned anew by each generation in the evolutionary history of humans, may come as something of a shock; indeed, it may seem not just a heretical but, in some ways, even a sacrilegious idea. For such a hypothesis radically alters our view about the forces that brought about the evolution of human beings: It changes our understanding of the ongoing importance of those cultural practices that enable a child to develop his "species-typical" capacities, and it changes our understanding of the basic developmental processes and needs that enabled prehumans and early humans to form groups, societies, and cultures.

Species-typical capacities—the differentiation of cortical and subcortical processes; our basic cognitive capacities, such as attention, the ability to inhibit nonsalient information, and short- and long-term memory; the emotions we feel and are capable of feeling; the thoughts we have and the beliefs, desires, and intentions we form; our ability to think logically and reflectively or imaginatively and creatively; our ability to communicate with others, both nonverbally and verbally; and our ability to understand what others are thinking and feeling—emerged, and can only emerge, in the context of the close nurturing relationships that a child experiences with his caregivers. No matter how much potential his brain may have, unless a child undergoes very specific types of interactive affective experiences that involve the successive transformations of emotional experience and that are the product of cultural practices forming the very core of our evolutionary history, that potential will not be realized in a traditional sense. For that potential does not reside in the physical structure of the brain, but is defined only in the types of complex interactions between biology and experience that we will be describing.

—3—

The Early Stages of Emotional Regulation, Engagement, and Signaling: Nonhuman Primates and the Earliest Hominids

IN THIS CHAPTER WE look at the operation of nonhuman primates and early nonhuman hominids at different stages of functional/emotional development. We will show how tamarins and marmosets are just starting to master the first f/e stage, that of self-regulation; rhesus monkeys are at a higher level, where they are beginning to engage and relate; and chimpanzees and bonobos, which may serve as a possible model for the behavior of the first hominids, are at a still higher level and able to engage in collaborative problem solving.

Before we embark on this fascinating theoretical journey, it is important to note that the term "hominid" has recently gone through an important transition. Traditionally the term referred to any member of the Hominidae family of erect bipedal primate mammals, and thus put recent humans together with extinct ancestral forms. But in the last twenty or so years, we have learned that apes did not all descend from one ancestor, that chimpanzees and gorillas share a more recent ancestor with humans than do orangutans. This means that, at the strict taxonomic level, chimps and gorillas are actually hominids. To avoid this issue, scientists are beginning to use the term "hominin," which refers to those on the human lineage and once again excludes apes.[1] We have chosen in this work to retain the more familiar term "hominid" in the belief that the reader might find the newer term more confusing. But we do so with an eye on how fluid the science of paleoanthropology is, given the major new discoveries that are constantly being reported.

We saw in the opening chapter how a baby learns to take in the world around him, and that, for this to happen, the baby has to *want* to engage with his caregivers and explore his surroundings. To promote the development of their child's capacity to attend to the outer world and to regulate his sensations, caregivers entice their babies with animated facial expressions, lively gestures, and soothing vocalizations. Right from the beginning, caregivers and infants are engaged in rhythmic, co-regulated patterns that enable the infant to begin attending to the outside world. As the baby becomes more interested in his world, he not only begins to discriminate what he sees, hears, smells, and touches, but, in the next stage, becomes more and more interested in interaction. In particular, he starts to develop a strong attachment to his primary caregiver, which provides the foundation for the development of the child's ability to engage in long chains of co-regulated emotional communication.

The primates that we will look at in this chapter, especially chimpanzees and bonobos, have mastered these first two stages of f/e development. But it is the third stage that appears to have been particularly important in early hominid evolution, for it marked the onset of a new range of social behavior that went significantly beyond preprogrammed biological templates. At this stage, we see the emergence of co-regulated emotional interactions, which involves flexible reading and responding to a great variety of multisensory cues (such as facial expressions, sounds, hand gestures, touching, body posture, and the like). The first two stages, having to do with regulation and engagement, while in part social, can be highly influenced by known biological mechanisms. But whereas engaging or bonding are affected, in part, by known biological factors (such as the effects of oxytocin on nurturing behavior),[2] emotional communication is a sufficiently complex social process that must go beyond its biological substrates. It can be developed only in the context of early nurturing experiences. In whatever species uses this process, it must be carried from one generation to the next, through the social transmission of learned caregiving practices, if the species is to survive.

The caregiving practices that nurture all three of these early stages of f/e development are not uniform but must be tailored to the individual strengths and weakness of the infant. Newborns vary significantly in how they respond to sounds, sights, touches, smells, and movements. Some are very sensitive and require soothing, others are underreactive and require energetic wooing. Some quickly turn towards the source of a sound, others take more time to develop this skill. Some begin to figure out visual or auditory patterns fairly quickly, others more slowly. Thus caregivers have to tailor their interactions to their baby's individual preferences and abilities.[3]

Caregiving practices that promote f/e levels are not unique to humans. Complex primates demonstrate comparable caregiving behaviors to a lesser or greater extent.[4] According to our model of intelligence, the greater and more complex the emotional communication in which a species engages, the higher its intelligence. At the lowest end of the nonhuman primate spectrum, in New World tamarins and marmosets, we see some emotional communication, but it is highly constrained, both in the number of modalities involved and the length of reciprocal interactions. At the highest levels, chimpanzee and bonobo infants become able to engage in long bouts of co-regulated, reciprocal interactions with early back-and-forth facial expressions, vocalizations, and other motor gestures.[5] There is also some evidence that chimpanzees and bonobos are just starting to establish conventionalized patterns of behavior that serve as the foundation for problem solving and, perhaps, even some symbolic communication.[6] One can situate nonhuman primates on a similar type of *f/e trajectory*, therefore—although with different quantitative and qualitative features—that we have described for human development. For example, the "higher" the species the more complex its affect signaling. This "complexity" can be seen both in the subtlety and the control that an organism has over its affect signaling apparatus (e.g. facial musculature), and in the number of modalities and the way these modalities are orchestrated in two-way communication (e.g., vision, auditory processing and vocalization, and motor movements, including facial expressions).

Our earliest hominid ancestors, the Australopithecines, may have been similar in varying respects to what we see in chimpanzees or bonobos today.[7] Thus, studying these nonhuman primates may afford us an insight into the caregiving practices that set us off on our evolutionary trajectory.[8]

INSIGHT FROM APE LANGUAGE RESEARCH

The belief that only humans have language, and indeed, that it is the possession of language that demarcates humans from animals, is one of the most entrenched of all modern philosophical ideas. While undergoing formal training in the philosophy and psychology of language at Oxford in the late 1970s and early 1980s, one of us (S.G.S.) saw no problem in this assumption. The field at the time was deeply divided between two warring camps: the "Chomskeans" and the "Wittgensteinians." According to the former, children do not so much *learn* language as have it *grow* in them. According to the latter, children learn language in much the same way that they learn other forms of complex social behavior. Despite the glaring opposition between these two positions, however, the one point on which both sides agreed was that only human beings are capable of acquiring language.

So deeply engrained was this idea that, regardless of the side on which one stood in the language-development debate, few of us at Oxford were prepared to treat seriously the claims coming out at the time that chimpanzees had acquired primitive linguistic skills. Certainly that was true for me. Asked by a colleague to write a review of some of the papers on ape language research by Sue Savage-Rumbaugh, a primatologist who began working with bonobos in the 1980s, I was skeptical of the claim that these apes could have acquired language skills. Rather, I approached Savage-Rumbaugh's papers to see whether her research could shed light on the development of those communicative skills that enable a child to learn language.

Sue responded to my article with an invitation to come down to the Language Research Center (LRC) in Atlanta and see for myself the work being done with the bonobos. It was an invitation that was to turn out to be one of the most momentous events in my academic career. It marked the start of my long and fruitful collaboration with Sue.[9] It was the catalyst for the exciting work I am now engaged in with Barbara King on the nature of nonhuman primate communication as viewed from the perspective of dynamic systems theory.[10] And, most important for this book, it was the source of my deep interest today, both as a philosopher and as a psychologist, in the study of emotional gesturing in nonhuman primates and in the insight this research brings to our understanding of the growth of the human mind.[11]

My arrival that first day at the Language Research Center was eventful. The LRC is located on the outskirts of Atlanta in a heavily wooded area. The entrance to the center is hidden off a small side road with no signs announcing its location. After driving around in circles for some time, my taxi driver stumbled upon it quite by accident. Facing us was a high chain-link fence, a speaker off to one side. These were the days when NASA was funding some of the research being done at the LRC and insisted on fairly strict security. When I announced my presence, we had to wait for several minutes while my credentials were checked. Then the gate mysteriously swung open and we proceeded slowly down a forest road to a second gate, where we had to go through the same procedure. I will never forget the moment when my driver nervously asked me what sort of place this was and, on cue, as I began to answer that it was a research facility where they studied apes, we were passed by two researchers in a golf cart with a bonobo sitting in the front seat. My driver was in such a rush to drop me off that I had to argue with him to wait while I paid for the trip.

After signing in at the office, I was directed to the back compound where Sue was working with the star of her research, Kanzi. The compound was a huge enclosure equipped with an oversized jungle set, various climbing ropes, a child's plastic house, and a large wading pool. I could see Sue off in the distance; she was working with what looked like a large black chimpanzee wearing a bright orange sweatshirt. As soon as she saw me, she came running over and told Kanzi to follow. The first thing that shocked me was that Kanzi lumbered towards me bipedally, and, when he was just a few feet away, stood facing me gently swaying his arms.

I was unsettled by his size, by his stare, and, most of all, by the fact that he was so comfortable standing upright. But even more unsettling was what Sue said: "Kanzi," she announced, "this is Stuart Shanker. Remember the airplane that went over our heads an hour ago, and I told you there was a scientist on board that was coming to visit you? Well, this is the person I was telling you about." This brief introduction left my mind spinning. Did this woman really believe that this ape knew what a scientist or an airplane was? I suspected that, to ease me into Kanzi's presence, Sue was doing the same sort of thing that Tracy Hogg describes in *Secrets of the Baby Whisperer:* In that work, Hogg counsels parents to speak naturally to their newborns and describe what's going on as if they were speaking to an adult. Today, I'm not so sure; Kanzi's comprehension of spoken English is so extraordinary that I would say exactly the same sort of thing to him and expect him to understand.

After these introductions had been made, Sue turned to Kanzi and said, "I'm going to take Stuart around the lab. Could you please water the tomato plants for me while we're doing this?" And sure enough, I watched as he trundled over to an outdoor water faucet, picked up a bucket that was lying beside it, turned on the spigot and filled the bucket, turned off the faucet himself, and then walked down to a vegetable patch at the far end of the compound, carrying the bucket in one hand. When he reached the vegetables, I watched as he poured the water on a small patch of tomato plants growing in the corner of the vegetable garden. Nothing at Oxford had prepared me for this sort of scene.

Shortly after this, Sue led me into the group room where Kanzi had been joined by his sister, Panbanisha, and Tamuli, one of the controls at the LRC who had not been exposed to language instruction. My first experience of being in the midst of a group of bonobos involved a steep learning curve. Sue was orchestrating my every move: "Crouch down," she ordered; then, "Lower your eyes," "Stop waving your hands," "Look up slowly," "Smile." I was oblivious of the problems involved in my introduction into the group. To begin with, I am male: an immediate threat to Kanzi, the alpha male of the group. To make matters worse, I am quite tall; I towered over Kanzi, and, for that matter, over Sue. And worst of all, because I had no experience with bonobos, I was blind to the subtle emotional gesturing that was going on around me as the group sized me up.

Afterwards, Sue asked me about my perception of how this initial meeting had gone. I replied, honestly, that the animals seemed indifferent to my presence. Fortunately, we had taped this initial encounter and I was able to see just how mistaken my first impression was. When I viewed the tape in slow motion, it was clear to me that a virtual flurry of emotional communication had been going on. All the apes were studying me carefully out of the corner of their eyes. Even more striking was the way they were expressing their anxiety through slight fear-grins and raised hair, and then how they showed their acceptance of me with the slightest of head nods and subtle hand gestures. Their body posture and the hair on their backs gradually relaxed as they returned to the grooming activities that I had interrupted.

Over the years I have learned that bonobos are constantly communicating with each other in a variety of ways: gestures, facial expressions, gaze, vocalizations, arm and leg movements, posture, and proximity. Some of the time they may communicate through one modality only, but more often than not this communication is multimodal. Moreover, their emotional gesturing in each of these modes can be even more subtle than similar forms of gesturing amongst humans: so subtle, in fact, that it has led many casual observers to assume, mistakenly, that they are relatively passive and incommunicative creatures. Nothing could be further from the truth. To be sure, it takes time to tune in to their subtle gestures; therefore, for someone who has not spent time studying these apes, it can be difficult to pick up on the emotional currents that are constantly swirling around a group. But then, this problem is not confined to human observers.

A couple of years ago I witnessed a fascinating encounter when an adult male who had been raised in Japan was introduced into the colony. P-Ske was having a great deal of trouble integrating into the group. At first it was thought that this was simply because he had been the alpha male in his colony back in Japan and was having trouble adjusting to Kanzi's role as the alpha in the LRC group. But after a while, it became clear that the trouble went far deeper than this. Once, I was watching as Panbanisha and Tamuli were playing with a gorilla mask. Panbanisha had put the mask on and was chasing Tamuli around the cages, squeals of delight emanating from both. This is fairly unusual behavior for bonobos, and certainly P-Ske had never encountered anything like it. As he entered the cages, his hair immediately began to bristle, his nostrils flared, and he had a pronounced fear-grin on his face. His behavior became aggressive as he responded to what he apparently regarded as threatening behavior. Moreover, he failed to pick up on the significance of the gestures and vocalizations that Panbanisha and Tamuli were making to try to involve him in the game. Like a stranger who has been raised in a totally different culture with completely different social conventions, he had trouble picking up on the emotional cues of the group and for a long time had to be somewhat isolated until he began to grasp the significance of the gestures, vocalizations, and facial expressions of the bonobos at the LRC.

Seeing this, I realized just how critical it is for babies to be raised within a group if they are to absorb the group's distinctive manner of emotional communication. I have seen this over and over again with Nyota. When Nyota wants me to chase him he gestures in exactly the same way as Panbanisha. When he is happy, his smiles and soft vocalizations remind me of Kanzi. Over and over again, I am reminded of that passage in *Philosophical Investigations* in which Wittgenstein describes how "the various resemblances between members of a family: build, features, colour of eyes, gait, and temperament, etc. overlap and criss-cross in the same way."[12] But the family resemblances that one senses so strongly at the LRC are not so much physical as behavioral.

The very first time I met Nyota, I could see how strong these family influences are. Sue started sending me regular reports about Nyota's development shortly after his birth. I couldn't wait to fly down to Atlanta to observe him first-hand. I knew from Sue that Nyota was a sheer delight; what I wasn't prepared for, however, was the family scene that greeted me the evening I arrived. Panbanisha was quietly grooming Nyota, who was lying sprawled across her lap while Uncle Kanzi was sitting close by munching on some sugar cane. Every once in a while, Nyota would languidly reach over to touch Kanzi, perhaps in the hope that he'd be offered some of the sugar cane. All this time, Panbanisha was gently stroking Nyota's arms and abdomen and making a low chirping sort of sound, to which he would respond by nuzzling deeper into her arms and occasionally looking up and gently touching her face. Both of them had the most serene look on their faces. As I sat in the corner watching this tranquil domestic scene, I quickly began to share in the emotional sense of well-being that filled the room.

Sue has written extensively about her own experiences with Kanzi and his mother, Matata. One vignette that I particularly like is Sue's description of "nest-building." Just like any human child, Kanzi was rarely ready to go to bed when Matata wanted. Still eager to play, he would flop on Matata with a full play face, smiling and laughing while waiting for Matata to play-bite and tickle him. Matata would indulge him in this way for fifteen to twenty minutes, at which point Kanzi would nurse and go to sleep while Matata groomed him.[13] The reason I like this vignette so much is because I have watched my wife perform exactly the same ritual with our own son every night at bedtime. She would go through exactly the same routine, playing with him in the same way that Matata played with Kanzi and making the same sorts of vocalizations and facial expressions. And just as with Kanzi, these play bouts would always end with Sasha nursing and then suddenly falling asleep.

The primatologist who has, perhaps, added most to our understanding of the dynamic process whereby the affective behavior of nonhuman primate infants is shaped in co-regulated interactions with their caregivers is Barbara King. The following are wonderful examples from her research of the co-regulated patterns that develop within primate families:

> Elikya, two months old, sits with her mother Matata. Her mother hands her over to Neema sitting nearby. From Elikya's facial pout, we can tell she is distressed by this transfer. Three times in succession, she extends

her arm and hand, palm up, back toward her mother. After the third gesture, her mother takes her back. Neema pats Elikya as Elikya relaxes against her mother.

When eight months old, Elikya moves toward Neema; she may lightly touch Neema's outstretched leg, it is hard to tell. Neema lowers her leg, then begins to stomp her feet on a platform as Elikya stands bipedally facing her; Elikya has a playface and raises her arms. Immediately, Neema moves to Elikya and hugs her, covering her with her whole body, then quickly moves back and resumes her previous position.

Eleven months old, Elikya climbs up a chain-link fence outdoors and approaches Kanzi, her brother, who rests on his back in a hanging tire. Elikya stops, then extends one leg and foot to Kanzi. Kanzi opens his foot, spreading his big toe apart from the other toes, in Elikya's direction. Elikya climbs over Kanzi's body up to his face. Kanzi wraps his arms around Elikya and pats her.[14]

King uses material such as this to illustrate how infant bonobos (and in other examples, gorillas) begin to produce and comprehend communicative gestures as a means of coordinating their actions with their parents and older siblings. Her work constitutes an invaluable analysis of the co-regulated process whereby infant great apes come to participate fully in their family groups.[15]

CHIMPANZEES AND BONOBOS

The first primatologist to investigate systematically the caregiving practices of nonhuman primates was Jane Goodall. Goodall has spent a lifetime studying an extended family of chimpanzees in Tanzania, called the "F" family. Through a series of books, films, and photographic records, she has enabled us to witness for ourselves the warm emotional bonds and rich emotional gesturing that characterize chimpanzee family dynamics. Presiding over the F family was Flo, an extraordinary matriarch who inspired great love in her infants Faben, Figan, Fifi, Flint, and Flame. The simple act of baptizing a chimpanzee with a proper name aroused a surprising amount of wrath amongst primatologists at the time. One of the first rules of studying nonhuman primates was that they should be referred to only by numbers and letters, never by names, for fear of committing the cardinal sin of "anthropomorphism." But

Goodall insisted on this practice. What Goodall most wanted was for us to recognize the "humanity" in these chimps: to see them, not as "savage brutes," but as creatures similar to ourselves who are capable of feeling deep love and affection for one another, display loyalty and trust, perform acts of kindness and self-sacrifice, experience sorrow and grief, and even feel awe at the mystery of life.

Flo was between thirty-five and forty years old when Goodall started studying her in the 1960s. You can easily tell from the films that Goodall made in the 1970s that this aging mother is finding child rearing exhausting.[16] And yet she constantly pushes herself to care for her newest infant, Flint. It is clear that his needs always come first: He feeds before Flo even though she is hungry; she soothes him when he is frightened; is patient when he is fussy; tolerates his playful antics; and is always responsive to his overtures. In a way, one might almost say that she spoils Flint. It appears that she is simply too tired to fight with him when it is time for him to be weaned, and instead she allows him to continue nursing long past the point when it was no longer permitted for her other offspring. Sadly, this decision may have had dire consequences, for Flint was stricken by grief when Flo died. The bond between them was, Goodall speculates, just too strong for him to survive on his own. The scenes that show Flint pining over his mother's body, trying to tease a caress out of her lifeless arm, and then wasting away after her death, are profoundly moving and change forever one's perception of these creatures.

What we have learned from Goodall's research is that, in chimps, infant development in all its aspects depends on a long-term caregiver. In the first eight months after birth, the mother satisfies the infant's physical and emotional needs. She remains the infant's primary source of security for at least the first five years. The infant suckles for around three minutes every hour and does not switch to solid foods until around his third year. At two months the infant is as helpless as a human infant, but his muscles for clinging and grasping develop quickly. In the early years, Flint would cling to Flo's ventrum when he was being carried. As he gained weight, he began to travel on Flo's back; later on, he crawled along Flo's body, which appears to have provided him with a secure playground for acquiring early motor coordination. Flo not only put up with all this but even encouraged it with gestures or by physically guiding his movements.

At twelve weeks, Flint was strong enough to dangle from Flo by one hand, and soon after this he began to break contact with her. At four months, he could pull himself into a standing position beside Flo. He would attempt to walk on all fours but stumble, at which point Flo would quickly gather him up. At six months, he would venture a few steps away from Flo; at seven months, he started to move about on his own. Throughout these initial forays, Flint would constantly monitor his mother's face to provide himself with a sense of security. She encouraged him in his efforts and warned him when he began to stray too far through gestures, head nods, vocalizations, and facial expressions.

Over the past thirty years, developmental psychologists have become fascinated with the phenomenon of attachment between caregivers and their infants and the factors that inhibit its healthy development. Watching these scenes of Flo caring for Flint, one realizes how important it is that John Bowlby's original paradigm be revised. Bowlby's studies on attachment were informed by his readings in ethology: in particular, Lorenz's work on imprinting and Hinde's work on attachment behaviors in rhesus monkeys.[17] This led Bowlby to formulate his famous hypothesis that attachment behaviors (in both infants and caregivers) were naturally selected in order to protect the infant from predation.[18] On this outlook, both "secure" and "avoidant" infant attachment behaviors can be seen as serving some predetermined evolutionary function. To be sure, Mary Ainsworth contributed a critical developmental dimension to this paradigm with her extensive research on the connection between a child's attachment behavior in the "strange situation" and prior measures of maternal sensitivity.[19] But to this day, Bowlby's relatively narrow picture of the evolutionary origins of attachment behavior has continued to inform research in the area, as can be seen, for example, in the numerous attempts to reduce attachment to a psychobiological phenomenon.[20]

Given the climate in which he was writing, it is certainly understandable that Bowlby should have come up with his ethological view. After all, one need only think of the impact of Harlow's experiments on separation and attachment to appreciate the reason why security should have been so uppermost in everyone's mind. But Bowlby extrapolated, from the behavior of care-deprived primates in fairly hostile conditions, to what conditions must have been like in our distant evolutionary past when attachment was "naturally selected." But Goodall's extensive research on the "F" family presents a very different picture of the role and development of attachment.

As with humans, chimpanzee mothers show a strong affection for their offspring and this bond persists throughout the life span. There is just as much physical contact between Flo and Flint as between a human caregiver and her infant, and a considerable amount of emotional exchange involving gestures, facial expressions, body movements and posture, and vocalizations. Time and again we see Flo look at Flint with a gentle smile, and Flint responding with a wide grin. Flo smiles back, and then the two of them are off on a sustained bout of smiling, gesturing, and vocalizing. Time and again we see the two of them engaged in gestural exchanges which Flint initiates as much as does Flo.

Particularly interesting, as far as the present chapter is concerned, is how Flo dealt with her older daughter Fifi, who was fascinated by the new baby and was constantly trying to hold and handle him, to touch his face and gently shake him. In the beginning Flo would discourage this with a stern look on her face. But over time she allowed Fifi to have more contact with Flint, to hold him for longer periods, and to carry him while traveling with Flo. However, if Flint cried Flo would take him instantly. By playing with her baby brother, Fifi was learning and practicing maternal behavior, imitating what Flo did with Flint. When Fifi had her first baby her maternal style proved to be somewhat more relaxed than Flo's, as she allowed uncle Flint to touch baby Freud at only two weeks; but overall Fifi's maternal behavior, and even her facial expressions, were strikingly similar to Flo's.

Goodall's research, together with that of Savage-Rumbaugh, King, and De Waal,[21] provides us with a much richer understanding of the role of attachment behavior in nonhuman primates, and thus assuming that chimpanzees and bonobos may serve as a possible model for the behavior of the first hominids—a deeper understanding of the role of attachment in human evolution.[22] Even more profound than its role in negotiating security is the role that attachment plays in the co-regulated affect signaling that takes place between infants and caregivers. As we explained in Chapter 1, by the time we come to test infants in the "Strange Situation," they have already undergone a considerable amount of f/e development. It is through these attachment behaviors that nonhuman primate as well as human caregivers help their babies begin to learn to tame their catastrophic reactions and transform them into interactive signals. In both species we see the same sorts of vignette: For example, a mother smiles and the baby smiles back. She coos and the baby coos back. The baby smiles, raises his arms as if to say, "Hey!" and vocalizes emotionally needy sounds that convey "Look at me!" The mother looks his way. Baby then smiles and flirts and the mother flirts back.

For this type of emotional signaling to occur, a baby needs to have been wooed into a warm pleasurable relationship with one or a few caregivers so that there is another being toward whom he experiences deep emotions, and, therefore, with whom he wants to communicate. Herein lies the deeper function of attachment. The baby needs opportunities to be "intentional," to express an emotion or need by making a sound, using a facial expression, or making a gesture with his arm, and to have his efforts confirmed by being responded to. Of course, one key respect in which human beings are distinguished from nonhuman primates is the much longer period where the infant is fully dependent on his or her caregivers; for during this time, human infants continue to develop, use, and rely on nonverbal, gestural, emotional signals to meet their basic needs, interact, and communicate, which enables a child to develop language, reason, and higher states of consciousness (see Chapter 2). But the important point here, as far as concerns the phenomenon of attachment, is that in nonhuman primates as much as in humans, attachment provides continuous opportunities to learn and fine-tune the skills of emotional signaling that are used to communicate and negotiate emotional states.

By no means, then, is attachment confined to the sort of intense, even catastrophic emotions that we observe in nonhuman primate and human infants when they are exposed to stressful situations. But even in terms of survival, the role of attachment is more profound than is suggested by Bowlby's narrow ethological point of view. For in order for both great apes and humans to survive, they not only had to protect their infants from predators; more fundamentally, they had to develop the capacity to live in groups. And, as we will see in Chapter 13, the very growth of this ability, and thus human survival itself, is derived from the same formative emotional processes that lead to symbol formation and logical problem solving in the infant.

Our work with nonhuman primates thus provides us with a striking insight into how critical the function of attachment must have been for the survival of early human groups, for it is through attachment with its caregivers that nonhuman primate babies, as much as human, come to absorb their group's distinctive manner of emotional communication. The upshot of this argument, as far as concerns our understanding of the nature of attachment, is that the full depth of this phenomenon cannot be gleaned from conditions that simply stress an infant's sense of security. Bowlby's attachment paradigm was based on the assumption that such conditions must have prevailed in our evolutionary past and that attachment developed as an adaptation to this challenge. To the degree that these situations prevailed, Bowlby's ideas represented a truly pioneering contribution to our understanding one of the foundations of human relationships. However, modern studies of primates suggest that the original paradigm, as important as it was, was too narrow; that is, nonhuman primate dyads operate in a broad range of environmental contexts and ecologies, and under less threatening conditions we observe a broad array of attachment-type or relationship-type behaviors, as described above.

In other words, even nonhuman primates are able to transform emotions from the catastrophic level, which would be operative under conditions of threat, to more differentiated affective expressions in a variety of patterns of co-regulation that typically permit the complex social negotiation of subtle affects dealing with themes of dependency, independence, and in between the two. What we glean from our f/e approach to the study of attachment, therefore, is that, in the evolutionary sense, attachment had to serve other purposes besides simply that of survival, of being able to deal with catastrophic or threatening conditions. Attachment also had to serve as the foundation for co-regulated affect signaling, leading both to higher cognitive and communicative functions and to the formation and coherence of groups.

It is not surprising, given the sorts of caregiving that we have looked at in this section, that chimpanzees are solidly established at the third level of f/e development (see Chapter 2). They engage in sustained exchanges of emotional communication, and caregivers are sensitive to their infant's overtures, thereby developing the infant's capacity to initiate and respond. In Chapter 4 we will see how, through co-regulated interactions, juveniles move into higher levels of f/e development, including problem-solving and recognition, and have a wide range of emotions.

THE DIFFERENT STAGES OF FUNCTIONAL/EMOTIONAL DEVELOPMENT

Tamarins and Marmosets

As one would expect, we see far less co-regulating communication between caregivers and their infants in the so-called "lower orders" of nonhuman primates. A perfect example of this can be observed in the Callitrichids, the New World tamarins and marmosets found in the tropical forests of Central and South America. Because of their small size and beautiful coloration, these creatures are perennial favorites at zoos all over the world. They can be as small as 5 inches (130 mm), their tails as long as 6 inches (150 mm). They look like other primates who cling vertically to trees, their forelimbs shorter than their hind limbs, but their hands and feet resemble those of squirrels and they have very sharp claws that they use to dig into the bark of trees.

What is most interesting about the Callitrichids for our purposes is their style of caregiving, which is often a family enterprise. In species that routinely twin, fathers aid with the birth and carry the infants around fairly soon after birth while the mother forages, only transferring them back when the infants need to nurse. Pygmy marmoset infants weigh only around 0.5 ounces (15 g) at birth and are completely helpless. They require constant care for the first two weeks and are nursed for up to three months. They generally become sexually mature between twelve and eighteen months and reach adult size by the age of two. The same thing is seen in the Emperor tamarin. A newborn Emperor is completely helpless. It weighs around 1 ounce (35 g) and has a coat of short hair. The mother needs to feed her baby every two or three hours for about a half hour each time, after which she returns the baby to the father. The babies ride around on the backs of their parents for six or seven weeks, and at around two or three months they go through a weaning period. Most tamarins become sexually mature from around sixteen to twenty months and the life span of the species is between ten and twenty years.

Such a short infancy allows only a brief period in which the infant can develop its ability to engage in emotional communication. Immediately after birth, the baby clings tightly to its mother or father's back, its head buried in the crook of the parent's neck, or it is held tightly to its caregiver's body while being licked all over. As it grows older, the infant spends more and more time looking at its caregiver's face and staring into its eyes. A lot of vocalizing goes on during these episodes of shared gaze, and when it is time to be weaned, the infant can be seen calling for food, to which the caregiver responds by slowly extending something to eat. Tamarin and marmoset infants demonstrate a keen interest in sensory experiences (sights, sounds, touch, and movement), poking their head up and looking around with very quick, darting movements. And they can be seen to calm down when held or nursed. We would thus situate Callitrichids as mastering the first stage of f/e development; that is, infants learn to be relatively calm and focused. But there is little evidence that they are capable of mastering the second stage of f/e development; that is, of experiencing deeper feelings of engagement.

Rhesus Monkeys

Rhesus monkeys, which are found throughout India, Nepal, and western Afghanistan, are up to 25 inches (62 cm) long with a 12-inch (30 cm) tail. They live in troops of as many as a couple of hundred individuals. Life in these troops is highly interactive: full of play behavior, grooming, and even, according to some, the first vestiges of gossip. The troops have a matrilineal hierarchical structure, even though the males are dominant. In recent years, primatologists have come to view these monkey societies as hives of political intrigue: To survive in this environment, an infant has to acquire street smarts at an early age so that it can form strong political alliances.

Much of our knowledge about the caregiving behavior of these monkeys comes from Carol Berman's studies of a free-ranging population on Cayo Santiago, a small island off the coast of Puerto Rico.[23] Scientists collected data on the animals from 1974 to 1990 and found a direct correlation between caregiving styles and the size of the group. The smaller the group, the less time mothers spent with their infants and the more those infants were allowed to interact with other adults; the larger the group, the more the infant's social network was restricted to kin members and the more time mothers spent with their infants in close proximity. This finding suggests to us that the ability of rhesus monkeys to live in large troops is grounded in strong caregiver-infant ties.

Although a baby rhesus doesn't express its positive affects with the same sorts of wide joyful smiles that we see in human infants between the ages of two and five months, in other respects it behaves in a manner similar to that of a human infant. The rhesus baby spends lots of time snuggling into its mother's body or looking keenly at her face. It visibly relaxes while being rocked, and vocalizes happily when the mother plays with it. We can even see the baby rhythmically moving its arms and legs and vocalizing in time to its caregiver's movements and vocalizations.

Perhaps the greatest source of our understanding of the similarity between a rhesus and a human infant mastering the second stage of engaging and relating with a caregiver comes from Harry Harlow's infamous experiments on rhesus monkeys in the 1950s and '60s. Harlow's experiments have been widely vilified as representing the cruelty to animals that scientists are capable of if their research is not carefully monitored. Pictures of baby rhesus monkeys clinging to a cloth "mother," or, worse still, one of the wire contraptions that Harlow built, are a standard feature in textbooks on child development. Deborah Blum's *Love at Goon Park* clarifies this picture, however, by situating Harlow's work within the context of mid-twentieth-century psychology.[24] It would appear that Harlow set out to challenge the then-prevailing behaviorist hostility towards any mention of the role of affect in a child's development.

Harlow was trying to build up a population of rhesus monkeys for experimental purposes—something that had never been attempted before—and was raising infants in very much the sort of sterile conditions that behaviorists were then prescribing for human infants. Despite the meticulous care they received, the animals were growing up afflicted with severe emotional, social, and physical disorders. Harlow introduced his "cloth mothers" into the infants' cages to see whether the infants might thrive if they had something comfortable to cling to. Harlow introduced the infamous wire "mothers" as a control to satisfy his critics that it was not the feeding that was drawing the animals but the comfort of the cloth "mother." Harlow showed that even when the food was placed only on the wire surrogates, the monkeys still spent the majority of their time on the cloth ones.

Although it was clear that the infants were deriving great comfort from the cloth-covered surrogates, they still suffered from striking social and emotional disorders. They would rock back and forth to soothe themselves, stare into space, and compulsively suck their thumbs. Worse still, the appearance of other monkeys would startle them and cause them to stare at the floor of the cage. The latter behavior was disastrous, given Harlow's goal of building up a captive population for research purposes, so he set the surrogates in motion to see whether movement would further encourage the development of a healthy nervous system. This modification did result in some improvement, although still not to the extent of creating animals capable of breeding.

What is so interesting about the latter modification is that, in addition to stimulating the rhythmicity of the infant's movements, these swinging surrogates may also have started to stimulate the infant's capacity for emotional communication. One of the most telling of all comments about these experiments came from a colleague of Harlow's: "I've come to realize that the mobile surrogate was more like a real monkey than we had expected. . . . [The moving surrogate] provided a limited simulation of social interaction, which the stationary surrogate did not."[25] Harlow's team may have inadvertently discovered just how important such interaction is—even the caricature of interaction that occurred with the moving surrogate—for enabling a rhesus infant to develop the capacity to manage its way in a troop.

For the model of f/e development that we are presenting in this book, we would argue that the outcome of Berman's studies and Harlow's experiments reveals that, to survive in the complex world of a rhesus troop, infants develop the capacity to experience deeper feelings of engagement (the second of our stages) only through warm emotional relationships with their caregivers. We would thus hypothesize that the longer period of infancy that we see in rhesus monkeys than in tamarins and marmosets—rhesus infancy lasts up to eighteen months—allows for more emotional communication. The infant thereby comes to develop the particular emotions and behaviors that are favored by the group, and, indeed, that define it. That is, the enculturation of the infant is inextricably involved in the growth of its mind. The types of gesturing that cultivate shared assumptions in a group, and that underpin the group's implicit processes for dealing with danger, aggression, and assertiveness, are literally bred into the infant in these formative early stages of development. Without these experiences, an infant rhesus will display the sort of stunted, autistic behavior that was observed in Harlow's lab.

THE CONNECTION BETWEEN LENGTH OF INFANCY AND FUNCTIONAL/EMOTIONAL DEVELOPMENT

Building on Kathleen Gibson and Dean Falk's work showing that longer periods of dependency facilitated the extended learning that is associated with the evolution of bigger-brained and more intelligent primates,[26] we would hypothesize that a direct connection exists between the length of infancy and the f/e development of a species: The longer the infancy, the more opportunity there is for an infant to engage in and thereby develop its capacity for emotional communication.[27] With an extended infancy, infants and their caregivers spend more time in close emotional as well as physical contact with one another; more time communicating with each other through co-regulated gestures, body movements, vocalizations, and, in the higher primates, facial expressions; more time engaging in joint activities such as food sharing, or acquiring problem-solving and tool-solving skills; more time playing with each other; and more time mastering the social behavior of their group. It is because of the impact these exchanges have on the growth of the mind, and, indeed, on the development of the brain, that comparative psychologists have found such a striking correlation between a species' length of infancy and its scores on the sorts of intelligence tests that are administered to young infants. This correlation is demonstrated in the following chart and tables:

CHART 3.1 Primate Infancy Periods Including Estimates from B. H. Smith

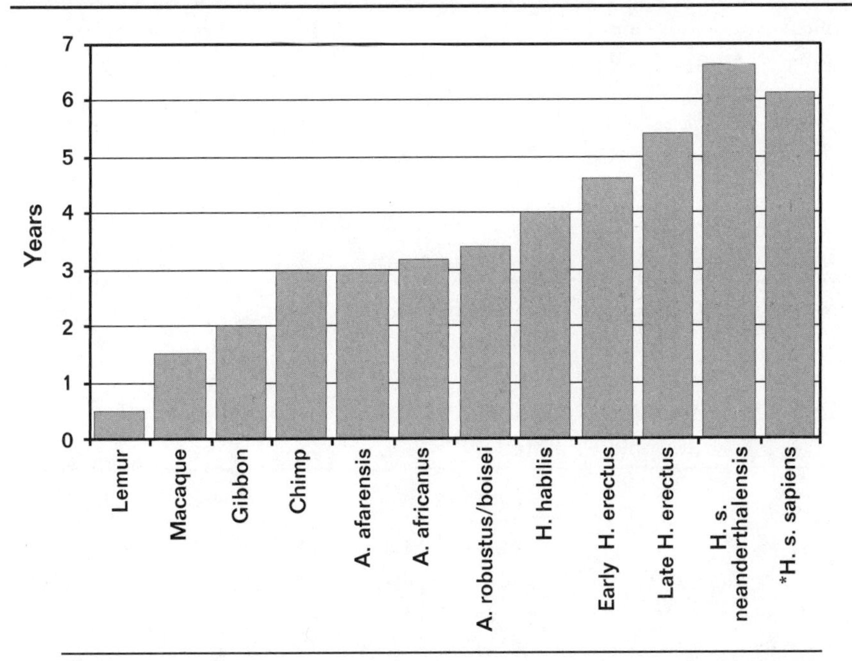

Adapted from Smith (1991). Smith's early hominid infancy periods (defined as the period from birth up to the first molar eruption) are based upon equations relating life history to brain and body weight in living anthropoid primates.

*H. s. sapiens score is averaged from Smith's two estimates based on different data sets.

TABLE 3.1 Molar Eruption Sequence for Selected Anthropoid Species in Years of Age

Species	Molar 1	Molar 2	Molar 3
Macaques	1.4	3.2	5.5
Baboons	1.7	4.0	7.0
Orangutans	3.5	5.0	10.0
Gorillas	3.5	6.6	10.4
Chimpanzees	3.2	6.5	10.7
Homo erectines	4.5	9.5	14.5

From Parker and McKinney (1999) *Origins of Intelligence*, p.227.
Original source: B. H. Smith (1993), B. H. Smith, et al. (1994).

TABLE 3.2 Cognitive Development Relative to Dental Development

Species	*Highest Level Achieved*	*Occurrence of Highest Sensorimotor Achievement*	*Occurrence of Highest Preoperational Achievement*
Monkeys	Sensorimotor up to fourth or fifth stage	After last deciduous tooth	Before first molar
Great apes	Preoperational up to symbolic	Long after last deciduous tooth	Coincident with third molar
Humans	Formal operational	Coincident with last deciduous tooth	Coincident with second molar

From Parker and McKinney (1999) *Origins of Intelligence*, p.227.

TABLE 3.3. Ontogenetic Characters for Cognitive Development in Anthropoid Primates

Taxon	*Ontogeny*
Humans	ESM – LSM – EPO – LPO – ECO – LCO – FO
Homo erectines	ESM – LSM – EPO – LPO – ECO
Great Apes	ESM – LSM – EPO
Macaques	ESM – LSM

Note: ESM = Early sensorimotor, LSM = Late sensorimotor, EPO = Early preoperations, LPO = Late preoperations, ECO = Early concrete operations, LCO = Late concrete operations, FO = formal operations.
From Parker and McKinney (1999) *Origins of Intelligence*, p.198.

As interesting as this Piagetian framework is, however, it provides us with little insight into the developmental factors that promote these kinds of cognitive advances. There has been something of a tendency—which runs contrary to Piaget's own constructivist outlook—to view these charts in predetermined, linear terms; that is, as registering how the different primate species have a biologically fixed cognitive capacity that increases, for strictly genetic reasons, as the ratio of a species' brain-to-body-size increases.[28] For our purposes, the significance of these charts is the relation between length of infancy and levels of cognitive development. Thus, rhesus monkeys demonstrate causal understanding, are curious and exploratory, and are purposive in their actions because of the time they spend as babies snuggling into mother's body, looking keenly at her face, and playing with and moving rhythmically in time to her movements and vocalizations. Apes have been able to count and sort and combine things, and even acquire symbolic skills, because, as we will see in the next chapter, their infants engage in long emotional exchanges that involve flexible reading and responding to multisensory cues (such as facial expressions, hand gestures, body posture, and the like). Furthermore, our model allows for a more complex profile of a species' cognitive skills than this Piagetian framework suggests. For example, monkeys may score on the fourth or fifth substage of sensorimotor development for problem-solving tasks but considerably higher on tasks involving spatial memory. Intelligence as such cannot be reduced to a single metric, nor can its development be explained as purely maturational.

A MODEL FOR EARLY HOMINIDS

Sue Savage-Rumbaugh and I (S.G.S.) have often discussed how the more time we spend with the bonobos, the more we feel we are glimpsing the dawn of our own species.[29] It is so easy when you are with them to imagine that here before you is a family of Australopithecines bound together by all sorts of rich attachment and communication.

The truth is that we know very little about the earliest hominids, the Ardepithecines and Australopithecines. Indeed, after the remarkable discovery of the Chad skull and the so-called Millennium Man in 2000, paleoanthropologists are now seriously debating whether the hominid lineage can be traced back even further, to *Sahelanthropus tchadensis,* a creature that existed between 7 and 6 million years ago, and *Orrorin tugenensis,* which existed approximately 6 million years ago. The former displays ape-like features (e.g., a small brain size) and some hominid characteristics (e.g., a brow ridge and small canines). There is no evidence that *Sahelanthropus* was bipedal, but there is evidence that *O. tugensis,* which was about the same size as the chimpanzee, was bipedal. The debate over whether either of these creatures constitutes the first hominid is likely to be quite lively over the next few years, but what is already becoming apparent is that, because *O. tugenensis* was still living in a forested area, we need to reconsider our views about why hominids became bipedal *(infra).*

The current thinking amongst paleoanthropologists is that the Ardepithecines existed between 5.8 and 4.4 million years ago and the Australopithecines between 5.3 and 1.6 million years ago. The fossil evidence indicates that both *Australopithecus anamensis* and *A. afarensis* were partly arboreal creatures, conclusions drawn from the shape of their feet, hands, and shoulders. Their skeletons reveal the same sort of strong sexual dimorphism as is seen in chimpanzees and bonobos, and their first permanent molar erupted at the same time as that of chimpanzees and bonobos, which suggests that they reached sexual maturity at the same time and had approximately the same infancy and life span. Paleoanthropologists have concluded that the Australopithecines are best described as bipedal human-like apes. For this reason, we speculate that they had cognitive skills, communicative behaviors, and social organizations similar to those we see today in chimpanzees or bonobos.

A few paleoanthropologists have suggested that some of the crude stone tools that have been attributed to *Homo habilis* may, in fact, have been created by Australopithecines. If this were so, Australopithecines must have possessed a fair amount of foresight: Because only specific types of stone could be used to manufacture a hand-axe, they would have had to learn where these deposits were located. The archaeologist Nick Toth has shown that a considerable amount of practice and skill is required to produce the kind of stone tools that the earliest hominids left behind.[30] One particularly interesting experiment that Toth performed on Kanzi certainly seems to bear out the possibility that Australopithecus possessed this ability. Curious to see whether Kanzi could learn how to make stone tools, Toth baited a clear plexiglass trap with some food. The trap was bound by stout cord that was too thick for Kanzi to bite through, but he could easily sever it with a stone knife. Toth then demonstrated the art of stone knapping for Kanzi. After a few clumsy attempts to imitate Toth's actions—difficult for Kanzi, because of the shape of his hands—Kanzi hit on his own technique of throwing one stone against another to produce a flake with a sharp edge. Kanzi immediately proceeded to use this to cut the rope. Kanzi has since become a proficient stone knapper, unerringly choosing the sharpest of the flakes that he produces to use as his knife.

Whether or not Australopithecines were capable of producing stone tools, it is almost certain that, like chimpanzees, they were constructing tools out of more perishable materials, such as wood and leaves. Chimpanzees in the wild have been found making tools out of twigs to fish out termites, or using stones and large branches to crack open hard nuts, or using special types of leaves to collect or soak up water. For a long time it was thought that chimpanzees respond only to environmental contingencies, but chimpanzees have been observed choosing appropriate tools well in advance of using them and transporting them to a tool-working site. It is highly likely that Australopithecines would have shared these abilities, which for our purposes suggests that they would likely have had the same sorts of nonhuman primate caregiving practices that support the development of such skills.

Both chimpanzees and bonobos continually assess relationships and maintain them at some cost, for they use relationships for various goals and negotiate over different aspects of the expression of their relationships. Chimpanzees cultivate tense male-male relationships based on shifting dominance ranks, but females play a subordinate role. Amongst bonobos, however, the females form strong alliances, which, according to the primatologist Richard Wrangham, accounts for the more harmonious and stable social existence that bonobos enjoy.[31] We are only just starting to understand the subtlety and importance that gestures, body movements, facial expressions, and posture play in creating and maintaining individual social relationships and group cohesion amongst chimpanzees and bonobos.[32] It is clear, however, that the social organization of chimpanzees and bonobos is far more complex than was thought as little as a generation ago.

Savage-Rumbaugh and others have reported that bonobos in the wild may follow rules as a means of maintaining their stable group behavior.[33] Such a phenomenon would be of the utmost importance insofar as it means that bonobos are starting to reach our fourth level of f/e development, for rule following provides an essential foundation for the sorts of problem-solving abilities which, as we will see in the following section, are the hallmark of the fourth stage. When we talk about *rule following* we are, of course, talking about more than merely patterned or conditioned behavior. To claim that bonobos are capable of following rules, we need evidence of instruction directed to young or new members of the group about the nature of the rule and when and where it applies, as well as evidence of sanctions against those who break the rule, and perhaps even awareness on the part of the rule breaker that a rule has been broken. The bonobos studied by Savage-Rumbaugh and her colleagues at Wamba, in the Democratic Republic of the Congo, present us with some intriguing signs of such behavior.

The scientists working at Wamba regularly provide sugar cane at designated feeding sites so they can observe the bonobos close-up. It seems that when it arrives at one of these sites, the group has to wait for some sort of signal from the alpha male before his subordinates can commence feeding. In one sequence captured on a film made by the Japanese production company NHK, a juvenile male can be seen taking a piece of sugar cane before the signal has been given by the alpha. As the juvenile surreptitiously reaches, he is caught in the act: An infant standing upright beside the alpha sees what the juvenile is doing and begins to vocalize excitedly. The second the infant starts to vocalize, the alpha looks directly at the offending party. He then stands fully erect and, with a stern look on his face, slowly raises his right arm. The juvenile responds immediately by approaching the alpha submissively. As he approaches the alpha, he turns around and presents his back in a crouching position. In what looks like a symbolic act of punishment, the alpha steps on the juvenile's exposed back and then turns to the others. The juvenile retreats to the outer limit of the group, where he sits dejectedly waiting to be given permission to join the others in feeding.

Does this small vignette offer us a glimpse into the kinds of rudimentary social rules that bonobos in the wild might follow? Evidence to confirm such an interpretation has been documented by Sue in her work with the bonobos at the Language Research Center. For example, in one particularly striking scene in the video *Kanzi: An Ape Genius,* we see Kanzi's sister Panbanisha playing a little too boisterously with the family dog. At one point, she jumps with all her weight on its back, which sends the poor dog yelping into the bushes. Sue is not going to stand for this sort of behavior and immediately begins to berate Panbanisha. Panbanisha adopts a facial expression of utter contrition and then hits the lexigram key for "milk" on the board that Sue is carrying a sign that she intends to be good. Then Panbanisha quite deliberately goes over to the dog and gently pats his back. This isn't just an animal who has learned to appease Sue's anger; this is a social creature who, whether or not she is genuinely sorry, has clearly learned how to apologize for her actions.

There are also accounts in the primatological literature of deception by a subordinate chimpanzee, demonstrated when a group member takes a piece of food or furtively mates while monitoring the movements or attention of a more powerful conspecific. Again, in the film of the bonobos at Wamba made by NHK, there is an intriguing scene in which a juvenile male, eager to initiate sex with one of the alpha male's females, looks at her and gestures towards a tree at the edge of the clearing. She responds to this overture with a facial expression that can best be described as "demure." Shortly after, the male moves to the tree while she follows as unobtrusively as possible and, sure enough, the two mate, both of them clearly monitoring the movements of the alpha all the while.

Does such behavior indicate that apes follow simple rules? Of course, one can always try to explain it as the result of some type of conditioning. But by categorically dismissing the possibility of nonhuman primate normative behavior, we rule out the possibility that the great apes participate in a community that follows social conventions. Moreover, we have all sorts of evidence from ape language research that apes can grasp and follow rules. Once again, a personal experience vividly illustrates this point.

This question of whether bonobos are capable of following rules was one of the very first that I asked Sue when we began working together, for it make little sense to investigate whether bonobos can acquire linguistic skills, however primitive these might be, if they cannot grasp simple rules. Sue's response was that Kanzi absolutely loved playing games that involve complex rules. Incredulous, I asked for an example, so she offered to have Kanzi play against me in a "Pac-Man" face-off. For readers who are not familiar with Pac-Man, it is a computer game in which the player operates a joystick to avoid being captured by aliens and at the same time eats as many cherries as possible, and also special bonus targets when these suddenly become available. I went first and did rather well. Then it was Kanzi's turn: With a look of rapt attention he sat before the computer screen waiting for the game to start, his hand poised on the joystick (which itself was an arresting sight). As soon as the game started, he began gobbling up the cherries, always careful to keep his player inside the lines, which, of course, is one of the basic rules of the game. He did his best to avoid the hungry aliens, vocalizing excitedly as they got near his player. Not only did he absolutely delight in the game, but, as Sue had predicted, he did indeed do better than I did.

The possibility that bonobos in the wild engage in normative behavior is especially important when we try to extrapolate from primatology about the possible social behavior of Australopithecines. We now believe that the Australopithecines lived in groups of from sixty to seventy individuals, which is approximately the size of bonobo troops in the wild. As we can see from the above, the evidence shows that the nonhuman primates who live in populations of this size have developed advanced social, cognitive, and communicative capacities. One of the most extraordinary of all suggestions in recent years, however, is that they may even have developed primitive symbolic abilities. Once again, the bonobos at Wamba present an informative example.

When up to seventy individuals live in a troop, a feeding site is quickly stripped of all its fruit. So bonobos constantly have to travel to new feeding sites, just to meet their basic needs. But because traveling through the forest as a group can be a highly perilous endeavor, they split up into small subgroups and leave several hundred yards between them. This way, if there is an attack by predators or a rival troop, the risk of the entire group's being destroyed is lessened. But now consider the logistics involved: All these subgroups start from the same point and arrive at the same point at the end of the day after traveling through dense forest as silently as possible. How are they able to coordinate such a difficult task? Evidence reported by Savage-Rumbaugh and colleagues suggests that they accomplish this feat by leaving markers for each other along the way.[34] For example, at a fork in a path, they might position a log or trample on the vegetation in a way that indicates which path to take or whether to climb a nearby tree. There are even indications that they leave signs for each other about a rich feeding site, perhaps where insects can be found at such-and-such a depth.

The richer our understanding of chimpanzees and bonobos becomes, the more interesting it is to view these great apes as a potential resource for modeling the capacities of our early hominid ancestors.[35] When paleoanthropologists made their first great discoveries of the Australopithecines, they were primarily driven by the desire to find fossil evidence of the "missing link." But in recent years, we have begun to think seriously about the sorts of cognitive, communicative, and social capacities that the Australopithecines might have inherited from earlier hominins.[36] The picture that is emerging is one of creatures that were able to maintain ever more complex forms of social organization.

As we remarked earlier, for early hominids to live in relatively large and stable social groups, they had to acquire the ability to co-regulate emotional interactions by using facial expressions, vocalizations, and gestures. When emotional signals can be rapidly exchanged, two or more parties can reciprocally communicate their intentions and negotiate, much as in the example above of the juvenile bonobo who is caught trying to feed before the others. The great apes and humans can co-regulate emotions and intentions to a much greater degree than other animals. As we can see in the harmonious social existence that bonobos enjoy, the process of negotiation and co-regulation enables individuals to tame catastrophic emotions, such as fear and aggression, which tend to compel immediate action (see Chapter 1). Through emotional communication, two parties can regulate each other's feelings and behaviors to reduce fear, create a sense of safety, gradually modulate each other's aggression, and build cooperation.[37]

Co-regulated emotional interaction, which marks our third stage of f/e development, enabled Australopithecines to communicate very rapidly with one another, which was vital for them to live in groups of sixty or seventy individuals. This progression to the third stage of f/e development may, in fact, be intimately connected to the emergence of bipedality. For over the course of hominid evolution, living together in groups increased efficiency and partially freed up individuals from full-time devotion to meeting basic needs (food, shelter), which, in turn, created opportunities for them to be more creative. Using nonverbal means of communication for purposes other than short-term survival permitted the development of ever more advanced forms of thought and language. The key question, then, is this: What were the evolutionary forces at work that enhanced this capacity to engage in co-regulated emotional interaction?

THE ROLE OF CAREGIVING PRACTICES IN THE EVOLUTION OF EMOTIONAL INTERACTION

In the evolutionary scenario that we are sketching here, if, as many paleoanthropologists currently speculate, the emergence of bipedalism resulted in a gradual increase in the brain size of adult hominids and correspondingly longer periods of what scientists refer to as the "neotony" and "secondary altriciality" of their infants, then the emergence of bipedalism would have played a key role in the development of ever more sophisticated caregiving practices.[38] ("Neotony" refers to the retention of juvenile features in the adult of a species and the slowing of development; "secondary altriciality" refers to the complete helplessness of a human infant.) Some have even suggested that our infants can be said to be born approximately nine months prematurely to accommodate our large brains with female bipedality. The more immature the newborn the more essential caregiving practices are for the development of its brain. This is because the more pronounced the neotony and secondary altriciality of a species, the more variable are its newborn's basic biological capacities, and thus, the more critical species-typical caregiving practices are for the development of species-typical capacities.

We saw in Chapter 2 how these basic biological abilities can vary. Babies come into the world equipped with a wide range of sensory, information-processing, and motor-control capacities. Caregivers have to learn which rhythms, touches, and sounds calm an infant effectively; and, conversely, they must discover the sorts of animated voices, facial expressions, body movements, and gestures that will stimulate an infant's desire to interact socially or to explore her world. Furthermore, we have to bear in mind the extraordinary spurt in brain growth that the infant is undergoing at this time. The more she can develop her senses, the more the connections in the cortex that integrate this information will be forged, thereby further enhancing the child's ability to explore her senses and her environment. The importance of this intricate dance can be seen in infants who do not receive the kind of soothing they need to become calm and focused; indeed, they quickly become listless and self-absorbed, and they even lose muscle tone.

We would speculate that the challenges contemporary caregivers must deal with in this respect have much in common with the challenges that early hominid caregivers were faced with, and that the caregiving practices we enjoy are the product of caregiving practices that were developed over the long course of our evolutionary history. Perhaps even some of the most striking anatomical features of hominid evolution, such as the loss of facial hair and the development of a complex facial musculature, are intimately bound up with the growing helplessness and dependency of the hominid infant. For the more complex the emotional signaling, the more complex the facial muscles and the more subtle the expressions required to support this. But, as we saw in Chapter 2, although the capacity to develop species-typical facial expressions of affect rests on certain basic biological templates, the actual development of those facial expressions is contingent on species-typical experiences of emotional gesturing. The patterns of behavior that provide these experiences are culturally, not biologically, transmitted from one generation to the next. Furthermore, the skills required to establish the shared rhythms and gestures that calm an infant and create attachment are culturally transmitted. No doubt many hominid groups were unable to maintain these cultural practices and died out, but others were able not only to transmit these skills but also to develop new practices that raised their infants to a higher level of emotional development, and thence, a higher level of cognitive, communicative, and social development.

—4—

Problem-Solving Collaborations: Chimpanzees and Early Humans

IN THIS CHAPTER, WE will explore the hypothesis that early humans were becoming established at our fourth stage of functional/emotional development. In this fourth stage, long chains of problem-solving interactions start to emerge. Based on certain evidence from the fossil record, which we will discuss below, it appears that these early humans would have been capable of engaging in a minimum of fifty to sixty exchanges in a row. As we will see, the development of these basic psychological skills leads to more complex group behavior.

PATTERN RECOGNITION AND PROBLEM SOLVING

The fourth stage of f/e development holds the key to one of the major problems raised by the field of Artificial Intelligence in the latter half of the twentieth century. One of the central insights of AI is that pattern recognition is essential for the development of the human mind. AI scientists devoted considerable effort to exploring the processes whereby different sorts of pattern recognition skills—visual-spatial, auditory-vocal, individual-social—are slowly built up out of simple discrete units as an individual interacts with the world. The great stumbling block for AI, however, was to explain the origins of these pattern-recognition skills.

Because AI was so firmly entrenched in the mechanist tradition of psychological explanation,[1] it completely overlooked the role of emotions in the development of the mind. Instead, AI scientists sought to treat the development of pattern-recognition skills as a purely cognitive phenomenon: the result of certain strategies by which the brain processes stimuli. According to AI, these strategies are part of the brain's intrinsic architecture. But this argument simply replaced one mystery with another: Where did these built-in strategies come from? The solution, which continues to inform modern evolutionary psychology,[2] is that they must have been naturally selected at the "genetic dawn" of our species (whenever that might have been). But such a method of postulating the necessary explanation represents, as Russell once put it, "all the advantages of theft over honest toil."[3]

What we are arguing here is that the emergence of these pattern recognition skills was part of a long evolutionary process that goes back to the Australopithecines, and even beyond them to earlier hominid species, and, indeed, to the Common Ancestor. In the critical fourth stage of f/e development, infants begin to engage in long chains of co-regulated affective interactions, which enables them to recognize the various patterns involved in satisfying their emotional needs. Based on culturally transmitted caregiving practices, they learn what different kinds of gestures and facial expressions signify. They learn that certain kinds of facial expressions, tones of voice, and behavior are connected with an individual's mood or intentions, or with certain sounds and actions, and so on. Moreover, as we saw in Chapter 2, this ability to recognize patterns is essential if a child is to have and act upon expectations. Now the toddler knows when to expect different kinds of responses from her caregiver, or what love, anger, respect, and shame feel like. And just as a child is learning all these patterns in her second year, before she has developed significant language skills, so too, we would speculate, early humans, and even, to some extent, nonhuman primates and early hominids, were developing these pattern-recognition skills—and the sense of self that results—long before the explosion of symbolic thinking that is associated with anatomically modern human beings.

In Chapter 2, we divided this important stage into four levels according to how complex the emotional signaling and problem solving become. These levels include an *Action Level,* without much two-way emotional signaling; a *Fragmented Level,* with islands of intentional, emotional signaling and problem-solving behavior; a *Polarized Level,* with patterns of emotional signaling expressing only one or another feeling state; and an *Integrated Level,* with long chains of interactions involving different emotional patterns organized into problem-solving exchanges such as flirting, seeking closeness, and getting help.

It is intriguing to consider whether we can see a similar progression reflected in the fossil record between the cognitive and communicative abilities of *Homo habilis* and *Homo erectus.* That is, it is intriguing to consider whether there is evidence to support the hypothesis that early humans were gradually moving up through these four levels of the fourth stage of f/e development.

In Chapter 3, we inferred, on the basis of what we know about the great apes and the fact that early hominids would have enjoyed greater freedom of movement and the use of their hands, that these caregivers would have spent more time touching, massaging, and tickling their infants, playing with their arms and legs, and holding them gently while rocking back-and-forth. It goes without saying that the same practices would have carried over into early humans. But, as we will see now, the emergence of early humans, who existed between 2.5 and 0.25 million years ago, was marked by the introduction of an important new range of caregiving practices. Indeed, these new caregiving practices were instrumental in the transition from prehuman hominids to the genus *Homo.*

APE COLLABORATIONS

One morning, I showed up early at the lab to find Sue and Kanzi using a lexigram board and deeply engrossed in a conversation. When I say "conversation," I really do mean exactly that. As we will see in the following chapter, the apes at the Language Research Center have acquired quite remarkable symbolic skills by using a tool that Duane Rumbaugh developed in the 1970s, when he was working with the chimpanzee Lana. Duane designed the lexigram board to circumvent some of the limitations of earlier studies in ape language. In the beginning, researchers had tried to mold apes' vocalizations, but, as became clear from Virginia and Keith Hayes's research[4] with the chimpanzee Vicky, the structure of the vocal chords and nasal cavity and the manner of breathing make human sounds very difficult for chimps to produce. Researchers in ape language switched to working with a simplified form of sign language. The most famous of the latter studies was started by Beatrix and Allan Gardner with the chimpanzee Washoe, and then continued by Roger and Deborah Fouts over the past thirty years.[5] This study was far more successful, yet, because signing was still a fairly difficult task for chimps (given the musculature of their hands and faces), Rumbaugh developed the lexigram board as an alternative way to assess an ape's communicative capacities.[6]

Lexigrams are colorful iconic symbols arranged either on a computer keyboard or on portable laminated boards. By pressing the keys on the computer board in the proper sequence, Lana could turn on music, watch slides, open a window, dispense food and drinks, and invite people into her room to visit and play. To be sure, this technology is not without its own serious drawbacks. For one thing, it imposes a linear order on communication; and for another, it can be extremely difficult to memorize all 250 keys on the board. In fact, it often takes new workers at the lab up to a year to become proficient, which only makes Kanzi and Panbanisha's facility with the board that much more remarkable. But the lexigram board does provide us with a way of circumventing some of the limitations of the earlier studies that were due more to anatomical than to cognitive factors.

On the particular morning in question, what struck me most forcefully was not Kanzi's mastery of the lexigram board but the incredible gestural dance that Sue and Kanzi were engaged in, which was centered on the portable board but not confined to it. Both were busy touching symbols, mostly taking turns, but occasionally pressing a key before the other had finished. All the while, as they argued about the day's activity, they engaged in a constant flurry of looks, gestures, facial expressions, head nods, changes in posture, and arm and leg movements. Sue wanted us all to go for a cookout in the forest, an idea that required a considerable amount of consultation about what we should have for our lunch and where we should have it. Scattered throughout the forest are sixteen sites, each with a different name, and each stocked with a different kind of food in a cooler. Kanzi selects the location he wants by pointing either to a lexigram symbol or to a photograph of the location, or by hitting the lexigram key for the distinct kind of food kept at a particular site. On this particular occasion, Sue had wanted to go to Treehouse, but Kanzi was quite insistent that we go to A-Frame, and he kept pushing away the Treehouse photograph. He then started pointing towards the key for "ball," shaking his head, knitting his brows, and even frowning. All this left Sue slightly confused because there were, to the best of her knowledge, no toys at A-Frame. But when we arrived there, it turned out that Kanzi had indeed left a ball when he last visited A-Frame several days before with Mary, one of the scientists at the lab.

Without question, my favorite activity at the LRC is going on these cookouts with Kanzi or Panbanisha. They will choose the location that they want to visit and then lead the way, either by gesturing when we come to a fork in the trail or else by taking one of us by the hand and leading us in the right direction. Along the way, they are constantly gesturing: sometimes to request a game of chase, or to point to an animal in the bushes, or even to chide you for being so slow. When we reach the site, a ritual begins, Kanzi chipping in by breaking the sticks for Sue or helping to cook the food. In the NHK video *Kanzi: An Ape Genius,* one incredible scene shows Sue asking Kanzi whether he wants to light the fire for her, and telling him that he can fish the lighter out of her pocket. He then inserts his hand, extracts the lighter, and after a couple of tries he succeeds in getting a flame. Gingerly he touches the flame to the twigs and, when the fire starts up, he drops the lighter and stares wide-eyed at the fire, on his face a look of fear mixed with surprise and sheer wonder.

Although we have emphasized the way Kanzi's cognitive abilities match those of young children, Kanzi's abilities in some areas far exceed those you would find in a young child. For example, one can't help but be by struck by Kanzi's spatial memory on these walks. The forest is quite large and densely treed. But Kanzi always knows the fastest way to every location, no matter where he might be in the forest and regardless of whether he is on one of the trails or in the middle of the woods.

Not only is Kanzi's and Panbanisha's knowledge of the forest impressive, but their ability to communicate that knowledge is remarkable. Once when I was out with Panbanisha and Mary, Panbanisha took me by the hand and led me to a plant on the side of a tree and began shaking her head and making guttural sounds: She was telling me not to eat the fungus. When I responded that I would be careful never to eat the plant in question, Panbanisha led me to another plant and did the same thing. What was happening, Mary told me, was that, aware of my lack of familiarity with the forest, Panbanisha was teaching me which plants to avoid and which were safe to eat. Each time she imparted her information she would scan my face to make sure that I had understood the import of the lesson. Once I pretended that I hadn't understood the point: I asked her, when she was obviously telling me to avoid some particularly noxious looking weed, "Are you telling me that this is really good for you? That I should try to eat this every single day?" Her vocalizations became agitated and she actually stamped on the plant. Clearly she was determined to get her message across! At this point, she reminded me so much of a small child leading her father by the hand to show him something in the garden, and then stamping in frustration as daddy persists in looking in the wrong place.

These sorts of long problem-solving interactions are an everyday experience with Kanzi and Panbanisha, but what about apes in the wild? Once again, Jane Goodall's research is invaluable. One of the first great discoveries that Goodall made when she arrived at Gombe was that chimpanzees hunt for meat. Her report shook the primatological world, for there was a tendency at the time to regard chimpanzees as akin to the "noble savages" depicted in eighteenth-century writings about the state of nature; that is, chimpanzees were thought to be gentle creatures living in a state of idyllic bliss, quietly spending their days eating nuts and fruits and grooming one another. This picture was shattered when Goodall observed a group of chimps hunt down a red colobus monkey, kill it, and eat it. She watched them carefully block off all of the possible escape routes. Then one of the chimps stealthily crept up to a monkey and successfully captured and killed it. Had he failed, he would have driven it into the waiting arms of the other members of the hunting party. After the kill, there was a sort of Bacchanalian frenzy in which the successful hunter doled out portions of his kill to the other males.

At first, Goodall's critics objected that this must have been some sort of aberrant behavior, confined to the chimpanzees at Gombe. But over the past thirty years, hunting parties have been observed in most of the sites where chimpanzees are studied in Central Africa. Typically, these hunting parties, primarily composed of males, range in size from one lone chimp to a group of thirty-five. Solitary hunters are successful only about a third of the time, however, whereas large hunting groups are almost always successful.

Craig Stanford has spent the past ten years studying these hunting practices in the chimpanzees at Gombe.[7] He has concluded that meat is important to chimpanzees, not simply for the fat and protein, but also because chimps use meat to forge or cement political bonds, and even to solicit sex from receptive females. Interestingly, Sabater Pi has shown[8] that hunger alone cannot be the driving force behind these hunts, for the chimps usually form a hunting party only after they have just had a long feed. Rather, there seems to be an urge to engage in a group activity driven by strong social forces as opposed to biological forces.

Stanford has speculated that *Ardepithecus ramidus,* which lived around 4.4 million years ago, likely engaged in similar hunting practices. These early hominids lived in the same sort of forested environment as the chimpanzees at Gombe, in which there would have been colobus monkeys, small antelopes, and other ground-dwelling vertebrates. To be sure, it might have been a little more difficult for *A. ramidus,* which was bipedal but still climbing trees, to have caught these monkeys; but even so, they could easily have organized into hunting parties similar to those of chimpanzees. Indeed, Stanford even speculates that these hunting practices may have played a critical role in the evolution of patriarchal human societies.

For our purposes, what is particularly striking about these chimpanzee hunting parties is the amount of complex emotional communication that is involved. In *The New Chimpanzees,* a *National Geographic* film produced by Cynthia Moses,[9] there is some remarkable footage of a hunt. Through gestures, body movements, head nods, and facial expressions, the hunters carefully position themselves around a troop of colobus monkeys and signal to one another about who is to climb which tree, and in what order. Their movements are carefully monitored and, at just the right moment, the other members of the party start whooping and crashing branches, which sends the monkeys into a full-flight panic, many of them blindly crashing into the waiting arms of the other hunters. The hunt lasts several minutes. At first, the wild celebration that occurs immediately afterwards seems to be utter bedlam. But it soon becomes clear that the assembled members of the troop are adopting various food-begging positions; they are then awarded pieces of the kill in what appears to be a deliberate fashion, apparently determined by social status.

Christophe Boesch and Hedwige Boesch-Achermann have added considerably to our understanding of the sorts of sophisticated planning and collaboration that go on in these hunting parties. As they explain in *The Chimpanzees of the Taï Forest,* hunting is a complex activity that takes place in conditions of low visibility. To succeed, the hunters have to develop a fairly sophisticated understanding of how the other hunters will behave in any given situation, as well as how the different species of prey will behave. Not surprisingly, it takes approximately twenty years to learn how to hunt. Youngsters carefully observe the most skilled hunters and model their own hunting techniques on theirs. What distinguishes the most skilled hunters is their ability to anticipate the actions of their intended prey; and, still more complex, to anticipate the effect that the actions of the other hunters will have on the behavior of the intended prey. This "double anticipation" is especially impressive because the likely behavior of the intended prey varies according to the species. Only the two oldest and most experienced of the hunters at Taï were capable of such double anticipations. Boesch and Boesch-Achermann describe watching one of them, Brutus, "suddenly run away from a hunt in a direction where we could see no monkey. It was only by following him that we realized that he was simply thinking further ahead, predicting precisely the further movements of the prey, estimating the time he would need to climb and very cautiously so as not to move any branches that could betray him, position himself high enough, and dart up in a perfect surprise attack when the prey entered his tree."[10]

HOMO HABILIS AND *HOMO ERECTUS*

Early humans started to appear approximately 2.5 million years ago. So striking are the anatomical and behavioral changes seen in early humans that paleoanthropologists draw a fundamental distinction between prehuman hominids and the genus *Homo.* The most important of the anatomical changes involved is a sudden jump in EQ: the ratio of brain size to body size. For example, *Australopithecus africanus* was from 1.1 to 1.4 meters tall and its brain was from 400 to 500 milliliters, whereas *H. habilis* was around 1.5 meters tall and its brain was from 600 to 850 milliliters. Along with this jump in brain size, *H. habilis* is thought to have acquired a new range of cognitive abilities, which, as we will see below, is reflected in tool-making skills.

Homo erectus existed between 2 million and 400,000 years ago. The fossil record for *H. erectus* demonstrates striking anatomic changes. Apart from its skull, which still retained many of the ape-like features of early hominids (although it is fascinating to see how, over the course of its existence, the face of *H. erectus* became increasingly flattened and its cranium increasingly rounded), the body of *H. erectus* was very similar to that of modern humans. As the name implies, *H. erectus* was fully upright, had similar body proportions to ours, a waist like ours, and even a stature similar to ours (adults were between 1.5 and 2 meters tall). Their brains were the largest yet seen in hominids, ranging in size from 900 to 1,200 cubic centimeters: twice the size of a chimp's brain, though still somewhat smaller than ours. One of the most extraordinary of all finds was that of the Nariokotome boy, 1.5 million years old, found on the banks of Lake Turkana, who may have been around fifteen at the time of his death. His nearly complete skeleton indicates that *H. erectus* had the same slender, graceful frame as that of modern humans who live in the region; he also had little body hair, which suggests his skin had developed dark pigmentation to protect him from the sun.[11]

Paleoanthropologists have speculated that these early humans lived in small social groups and engaged in cooperative activities. Some have even speculated that pair bonding might have begun to emerge in these early humans. Much of the reasoning for such a hypothesis is highly speculative. There is, however, one extraordinary discovery that has stimulated a great deal of controversy around the early pair-bonding hypothesis.

Andrew Hill, a professor of paleontology at Yale University, was working at Mary Leakey's Laetoli site in Tanzania in the 1970s. As Hill tells the story, one day he and one of his colleagues were horsing around, throwing dried-up pieces of elephant dung at each other. Diving out of the way to escape one particularly accurate projectile, Hill quite literally stumbled upon a set of fossilized footprints, 24.6 meters long, left by two hominids walking side-by-side, close enough to have been touching. It seems that a volcano had erupted and scattered a fine layer of ash over the landscape approximately 3.6 million years ago. This had been followed by a rainfall that turned the ash into a sort of cement, followed by another volcanic eruption that spread another layer of ash; thus, many bird and mammal footprints were preserved, as well as the pair of hominid footprints that Hill discovered. According to some, the reason these footprints may provide evidence of early pair bonding is that one hominid was clearly larger than the other, and the smaller of the two bears signs that whoever left the prints was burdened on one side; this evidence has ever led to speculation that perhaps it was a female carrying an infant on her hip.[12]

ORIGINS OF TOOL USE

Assuming—and it remains an assumption—that these early humans were starting to hunt large animals, they needed some sort of tool to butcher their kills—if there were indeed kills, for in one popular scenario, our early human ancestors were not the bold hunters it was once supposed but rather somewhat cowardly scavengers who had to develop tools to break open the heavy bones of animals that had been killed by larger predators and then picked clean. But whatever the truth, the one thing we do know for certain is that the habilines were creating stone tools and passing on the methods for creating these tools for thousands and thousands of years.

Indeed, *H. habilis* ("man the handy man") is so-called precisely because of his tool-making abilities. These stone tools, which are known as Oldowan, date back to approximately 2.5 million years ago. They were usually produced by striking one rock against another to create a sharp-edged flake that was used to butcher animals and to strip tough plants, or as a digger to unearth tubers and insects. But as we already know from Nick Toth's studies of stone knapping, these apparently crude stone tools not only required a significant amount of manual dexterity but also reflect a significant advance both in memory and in planning abilities insofar as these early humans had to learn which stones were best suited for knapping and where they were located.[13]

How did these creatures acquire the skills required to create a stone tool? We have one intriguing clue about the origins of these skills in the extensive research that has been done on the chimpanzee nut-cracking site at the Taï National Park, in the rain forest of the Ivory Coast. In a groundbreaking study recently published in *Science,*[14] Christophe Boesch, Julio Mercader, and Melissa Pranger reported that the chimps collect rocks and panda nuts from up to 300 meters away and bring them to wooden "anvils," where they bash them with stone "hammers" to get at the nutritious flesh inside. The stones are fairly large (weighing up to 15 kg) and have a distinctive shape. The chimps have to hit the nut with just the right amount of force: enough to crack the shell without smashing the kernel. It can take up to seven years to master the technique. Young chimps first begin to bang on the nuts when they are still by their mother's side, and some have been observed hitting nuts with smaller stones, perhaps practicing the motions they have observed in their parents.

Boesch has described in earlier work how chimpanzee mothers demonstrate the technique by performing the necessary actions in slow motion and carefully monitoring their infants' attempts. They have even been observed to adjust the angle of the log for their infants. All this amounts to what Boesch describes as a form of teaching.[15] Perhaps even more important, for our concerns here, is that the skills involved have been culturally transmitted over many generations, indicating a strong and stable social organization.

Boesch, Mercader, and Pranger are quite insistent about describing this as a *cultural* phenomenon. For one thing, the nuts the chimps crack are available throughout tropical Africa, yet nut-cracking behavior has been documented only in the chimpanzees from Western Ivory Coast, Liberia, and Southern Guinea-Conakry. Hence they conclude that this nut cracking is a cultural behavior that distinguishes one population of chimpanzees from another. Furthermore, judging from the piles of shells and stone flakes found around some of the tree stumps, the same site has been used to crack nuts for over a century. Indeed, Boesch, Mercader, and Pranger believe that some of these stones may have been used over and over for many generations.

Even more fascinating is the authors' report that some of the flakes they found resemble those found at hominid sites in East Africa. In fact, they speculate that early hominids probably engaged in nut-cracking practices just like those of the chimps, and that over time, early humans began to manufacture more complex tools of different shapes and sizes. Such a hypothesis draws support from the recent discovery of a 2.34-million-year-old hominid "tool-making factory," possibly belong to *H. habilis,* in the Rift Valley in Kenya.[16] Hélène Roche and her colleagues of the Centre Nationale de la Recherche Scientifique (CNRS) discovered three thousand stone tool fragments scattered over an area of 17 square meters. They were able to determine not only that the tools were made on the site but also that these early human toolmakers carefully tested new rocks to see whether they produced the right kind of sharp flake and discarded them if they didn't. Meticulous research also revealed the order in which the flakes were struck from a core and the direction in which they were rapidly struck, the implication being that the toolmakers possessed a knowledge of the material and considerable manual dexterity. Interestingly, the tools were found with the bones of fish and mammals and therefore may have been used to cut up meat.

The technological advances seen in *H. erectus* are even more striking. Paleoanthropologists have documented how, with *H. erectus,* we see the first evidence of the use and control of fire; the first record of the novelty-seeking behavior that is characteristic of modern humans; signs of a more modern human social organization; and the first appearance of stone tools that demonstrate a conscious complex design.[17] These so-called Acheulean tools were formed by chipping the stone from both sides to produce a symmetrical cutting edge. To make such a tool required strength and skill because large shards were struck from big rocks or boulders, fashioned into bifaces, and then refined at the edges into distinctive shapes. Acheulean hand-axes can be extremely large and may have been used not just to butcher animals but also to cut up branches into firewood. Some have speculated that they may even have had a ceremonial and a monetary use.[18]

Clearly, we need to explain the cognitive factors that would have enabled early humans to progress beyond the sorts of crude tool-making skills seen in the chimps at Taï, which the Australopithecines were probably also practicing; but, even more important, we also need to explain the social processes whereby these skills were passed on from one generation to the next for 2 million years. That is, we are concerned here not simply with the sorts of changes that may have been occurring in, for example, their processing capacities; we are also interested in the changes occurring in their social organization that would have led to the sort of stable social structure necessary to maintain and transmit the design and manufacturing criteria of the Acheulean tool industry for over a million years.

CAREGIVING PRACTICES AND THE ORIGIN OF EARLY MAN

One of the most striking aspects of the Oldowan and Acheulean tool industries is how long they each lasted. Genetic determinists have seized on this point, maintaining that because we see so little technological innovation in *H. habilis* and *H. erectus,* their genomes must have remained unchanged. But there is a very different way to view the endurance of Oldowan and Acheulean tools. The consistency can be seen as evidence that the skills for manufacturing these tools were handed down from one generation to the next for extraordinarily long periods of time, and that there were no significant social or environmental pressures compelling them to acquire a new suite of tool-making abilities. Seen in this way, the transmission of such knowledge could have been possible only in long-enduring and continuous human societies, and the major question raised by the Oldowan and Acheulean tool industries then becomes: What was the source of the social stability that made possible the transmission of the necessary skills in *H. habilis* and *H. erectus*?

The answer, as we see it, lies in the introduction of caregiving practices that enabled these early humans to establish deeper feelings of engagement: to engage in longer exchanges, to understand one another better, and to develop more complex relationships. Such deeper emotional bonds are essential for the sort of continuous social cohesiveness that is required for stable social organization. This then raises the question of what would have enabled them to develop these deeper feelings of engagement.

One highly significant anatomical development that occurred at this time might have been the loss of bodily hair. Current evidence suggests that the early hominids lived in wooded areas and that it wasn't until *H. habilis* that they began living in arid environments, at which time they probably acquired hairless skin. Naturally enough, paleoanthropologists assume that the loss of hair was connected to the need to cool the body. But this hypothesis has been challenged in recent years by physiological studies that question the overall benefit of hairless skin as a cooling mechanism.

In particular, the loss of facial hair would have done little as a cooling mechanism; but it would have had an enormous impact on the role of facial expressions in the development of closer emotional bonds. So perhaps the growing importance of attachment between caregivers and their infants was the reason that we lost some of our facial hair. For the more visible the face, especially the area around the eyes, the more prominent facial expressions become in emotional signaling. As Darwin was the first to note, humans have a much more developed facial musculature than the great apes, enabling a wide range of subtly nuanced facial expressions.

To appreciate just how facial expression of emotion could have led to more sustained relationships in human evolution, one need only consider a baby's responses to animated facial gestures. She reacts with full open-mouthed smiles that reach up to the corners of her eyes; she makes happy vocalizing sounds; she imitates facial expressions; she moves her limbs and body to the rhythms of her caregiver. On the other side of the equation, given the variability of babies' sensory systems, information processing, and motor control, caregivers have to interpret a baby's expression of particular likes and dislikes, her particular needs and aversions.

Human caregivers also used facial expressions to entice their babies into an awareness of the world around them. A baby's first exploratory behaviors are to reach and touch the things that most interest and delight her: invariably, the caregiver's face, nose, hair, and mouth. What the baby most craves in the beginning is the caregiver's smile and attention. So strong is this desire that the baby begins to develop her motor control system so that she can turn and see her caregiver. Her need to engage with her caregiver drives her to strain her muscles to their very limits. Perhaps even more important, her need to engage with her caregiver drives her to strain her ability to attend to its very limit.

A baby's ability to attend for longer periods of time is very much a function of her emotional state; that is, her interest in what the caregiver is saying or doing. This attention span is critical for the child's developing ability to solve problems. To solve a problem as simple as Piaget's classic example of an eight-month-old who eventually stops pulling on a string that causes a bell to ring if the bell is detached, the baby must first discover the causal relationship between her smile and her caregiver's response.[19] Indeed, it is extraordinary to see how quickly a baby will stop smiling, and even become agitated, if a caregiver does not return her smiles.[20]

Another significant aspect of focused attention is the ability to ignore nonsalient or distracting information.[21] Researchers have now established that controlled attention and inhibition develop significantly in the first five years of life.[22] But the origins of this ability lie in the child's focus on the emotional content of the caregiver's message and ignore the misleading elements. For example, a caregiver may make all the right noises when she holds the baby, but if she is anxious and her expression is tense, the baby will in turn become tense and agitated. These observations suggest that, through the more finely tuned emotional expressions and gesturing that hominids were developing, their infants were acquiring much greater attentional and inhibitory control: basic skills that would have played a significant role in the ever-widening cognitive gap between the great apes and early humans.

Moreover, herein lies one of the most critical of all aspects of the break between early humans and prehuman hominids. The importance of this shift becomes vivid when we compare the attention spans of human babies with those of nonhuman primates. To be sure, we see some fairly striking attentional abilities amongst the great apes. For example, the chimpanzee infants at Taï have been observed to attend closely to their caregivers when they are cracking open nuts with stone or wood tools. Even more striking is the attention span of language-enculturated apes. But great apes in the wild seldom appear to engage in interactions for more than a few seconds at a time. Given the kinds of attention necessary to master the intricate skills involved in Oldowan and Acheulean toolmaking, it seems clear that early human infants were developing considerably more pronounced attentional abilities.

This crucial psychological development was not, we would argue, the result of some random genetic change. Rather, on the hypothesis we are canvassing here, it would have been the result of ever more sustained bouts of emotional exchanges in which infants were engaging with their caregivers. Furthermore, we now know that the development of a child's attentional capacity is intimately connected with the growth of the prefrontal cortex. And, as we mentioned in Chapter 1 and will clarify in more detail in Chapter 10, the prefrontal cortex does not maturate according to a fixed genetic timetable; rather, the development of the prefrontal cortical structures that enable a child to develop her basic cognitive abilities is fundamentally bound up with the nature and quality of her interactions with her primary caregivers. Thus, we would speculate that the origins of the genus *Homo*—mental, physical, emotional, social, and communicative—lay in the development and transmission of these critical caregiving practices. These early humans were learning how to entice their infants with different facial expressions and gestures and were engaging them in playful interactions. Perhaps they were beginning to vocalize in warm, enticing motherese.[23] These more elaborate interactions laid the foundations for the acquisition of the cognitive skills manifested in Oldowan and Acheulean tools, as well as for the next stage of f/e development.

—5—

Symbols, Words, and Ideas: *Archaic Homo sapiens* and Early Moderns

A GRADUAL TRANSITION

We saw in the opening chapter how at the fifth stage of functional/emotional development a child starts to use words in a meaningful manner. This exciting event depends upon a wealth of prior emotionally relevant experiences on which he draws. Without a lot of warm exchanges with caregivers, the child won't be capable of developing meaningful language, even though he might be neurologically capable of speaking. When he is so emotionally deprived, the child is still able to memorize words or to label objects, but he can't explain what a story means or why the characters behave the way they do. It is only through emotional exchanges that words and images acquire meaning for a child. Only then can he use words to express his feelings, rather than acting them out. Only then can he think symbolically and solve problems creatively.

The processes involved here develop gradually, through the following steps:

- Ideas are acted out in action (the child raises his arms to be picked up), but words may also be used to signify the action (while raising his arms, the child says, "Up"); words are not yet used instead of actions.
- Words referring to the body are used to convey feeling states ("Head hurts").
- Action words are used instead of actions to convey intent ("Hit Mommy").

- Feelings are conveyed through words but are treated as real rather than as signals ("Need a hug").
- Words are used to convey feelings, but they are mostly global feeling states ("I feel awful").
- Words convey polarized feeling states ("I feel sad").
- Words begin to be used to convey more differentiated, abstract feeling states ("I'm lonely").

These steps are particularly important for the argument that we will present in this chapter in regard to human evolution. A standard evolutionary premise holds that there is a sharp distinction between our earliest modern ancestors, *Archaic Homo sapiens,* and our own species, *Homo sapiens sapiens.* Initially, paleoanthropologists believed that our more distant ancestors emerged some time around 0.5 million years ago and survived for several hundreds of thousands of years; and then, for reasons that are unknown and will likely never be well understood, our own species emerged, relatively suddenly, some time around 200,000 years ago. Such an evolutionary scenario lent plausibility to the idea that the birth of our species was marked by some radical genetic mutation, frequently said to have been the selection of a language gene, or perhaps a mindreading gene.

Instead, we suggest that the transition from *Archaic H. sapiens* to *H. sapiens sapiens* was a gradual process and not a sudden cataclysmic event. Admittedly, evidence in the archaeological record leads one to postulate a sharp break. But consider how children begin to look like totally different beings when they start talking, typically between the ages of eighteen and twenty-four months. Once a child gets beyond the stage of using just a few words and phrases and starts to speak in sentences, he really seems to take off. Within a few months he is talking, thinking, and imagining to an enormous extent. However, as we saw in Chapter 2, when we look at the process more closely, we can actually break the child's development down into gradual steps. His understanding of actions and agency and objects is gradually acquired through preverbal experience, especially the exchanges involved in solving problems and forming a sense of self.

When the ability to speak and use symbols reaches the point where children can describe their experience, this looks like a new event; but, although in part this is true, it is also simply one more in a continuous series of developmental steps. Like a leaky faucet that we do not notice is dripping until the water finally overflows and we see it coming through the ceiling, so too the child's symbolic development is the result of many gradual steps leading up to the "language explosion" that typically occurs between eighteen and twenty-four months. Similarly, we are suggesting that anatomically modern *H. sapiens* evolved very gradually from early humans and needed to master various elements of presymbolic communication before using symbols to a large extent. Although the evidence for a "symbolic explosion" between 60,000 and 40,000 years ago is clear and seems momentous—and indeed it was—we should not forget that it was part of a very gradual process of f/e development.

Another reason why the appearance of symbolic abilities appears to be so sudden is that symbolic communication leaves a more visible record than presymbolic communication patterns. Now we have cave drawings, ceremonial beads, and more refined tools. But then, just as it has taken thousands of years of modern history to figure out that babies don't simply become people at age three or four when they more closely resemble adult humans in their ability to speak or understand human behavior, so too *H. sapiens sapiens* did not suddenly become thinking, problem-solving creatures when they began drawing on cave walls. With human babies, we are discovering the road map of monumental developmental achievements in the first two years; so too we are starting to discover the road map of achievements in the life of *Archaic H. sapiens* leading up to the appearance of modern humans. As science so often reveals, the processes of development are far more complex than they initially appeared, and once we look at them under a microscope, the pathways involved often have far more steps and are more gradual than was originally assumed.

BONOBO COMMUNICATION

My first week at the Language Research Center (LRC) was inspiring. I remember thinking about what it must have been like for some ancient mariner who, convinced that the earth was flat, had set out to see for himself the edge of the world. His doubts and uncertainties must have grown stronger with each passing day as the horizon remained just as far away as when he had set out. For someone trained in the dogmas of twentieth-century linguistic theory, the idea that an ape could understand spoken English was every bit as preposterous as supposing that the earth is round. Only humans, we had been taught, had the neurological capacity to understand spoken language.

Each day at the LRC brought a new challenge to this belief system. It might have been Kanzi performing at the same level as a three-year-old child on a battery of language tests. Or Kanzi fetching Sue's keys from another room when he heard her shouting to me to ask whether I had seen them. Or when he responded to Sue's suggestion that they "go to the trailer and make a water balloon" by going to the trailer, getting a balloon out of the backpack, and then holding it under the water faucet. But it wasn't just a matter of coming to terms with Kanzi's extraordinary comprehension skills. Every day seemed to bring a new surprise in regards to Kanzi's and Panbanisha's cognitive development. For example, one day I found Panbanisha drawing a picture of a clump of grapes on the wall of her bedroom; another day I watched her don a gorilla mask and chase the other apes around the cages in a fit of gleeful pretend play.

Each of these sorts of events is so common at the LRC that after a while one scarcely sees them as at all unusual. But when you start to think about them carefully, it is remarkable to see how much complex communication is taking place. Take, for example, the day we spent testing Panbanisha and Heather, Sue's three-year-old niece. My friend, Talbot Taylor, who is the G. T. Cooley Professor of English and Linguistics at the College of William & Mary, had put together a very clever list of sentences that would enable us to ascertain Panbanisha's linguistic proficiency. Tolly's idea was that we would ask one to perform an action and then the other to perform a similar action, but using slightly different words. So, for example, we might say, "Panbanisha, can you put the doll on the table?" and then, "Heather, can you put the bowl on the box?" This strategy would enable us to compare their comprehension skills in regards to various simple syntactical constructions as well as to vocabulary. After an hour of this testing, it was clear that there was very little difference between the two of them as far as their understanding of the batch of sentences Tolly had put together. But that was really the least significant part of the experiment. What we found truly mesmerizing was how the two of them behaved with each other.

Sue had arranged for the two of them to sit side-by-side on two plastic chairs. Watching them was like watching any two small children playing a game together. They were constantly exchanging glances, laughing with each other, and teasing one another. To keep them interested in the test, Sue rewarded them with a prize after each answer, a sweet or a small toy. At one point, Heather had just answered her question and Sue was about to give her a balloon when Panbanisha, with a mischievous grin on her face, leaped forward and grabbed it out of Sue's hand. Heather was not about to stand for this. We watched in amazement as she leaped out of her chair and started pounding on Panbanisha's back, insisting she give it back. Like an older brother who wanted nothing more than to torment her younger sister, Panbanisha responded by chortling over the balloon as if it were the most wonderful of all the rewards they had been given that morning. At this point, Sue, who was concerned that the session not end in chaos, intervened and insisted in a firm voice that Panbanisha return Heather's balloon that instant. With a chastened look on her face, Panbanisha turned towards Heather and ever so slowly extended the balloon. Grabbing it with a look of triumph, Heather promptly returned to her seat. To maintain the peace, Sue then gently gave Panbanisha her own balloon. The two of them then began to play with their balloons in such a way as to indicate, through loud vocalizations, that theirs was by far the better of the two. And every so often each would glance surreptitiously at the other to make sure that the other's was not in fact the better balloon.

The day that Panbanisha put on the gorilla mask is just as illustrative of the rich variety of child-like emotions that the bonobos display. The morning session had started off peacefully enough as Kanzi quietly groomed Tamuli while Panbanisha lolled around on some blankets with what can only be described as a look of boredom on her face. Suddenly Panbanisha spied the gorilla mask that Sue sometimes wears with a gorilla suit when she wants to keep the bonobos away from certain areas of the LRC. In a trice, Panbanisha had donned the mask. Even more incredible, her behavior was transformed the instant she had the mask on: She started lumbering around the room in a remarkably convincing imitation of a gorilla, swinging her arms in a ponderous manner to suggest that they had suddenly become incredibly heavy, and moving in a stiff-legged manner from the hips. Kanzi had a look of intense annoyance on his face and kept shooing Panbanisha away with his arm. But Tamuli was another matter.

Tamuli was one of the controls at the LRC. She had never been exposed to the lexigram board and was kept in another part of the compound whenever Sue was doing language testing. Poor Tamuli's response to Panbanisha was of a completely different order: She was absolutely terrified! Unlike Kanzi, she couldn't grasp that all this was pretend play. Rather, she started scrambling around the room, desperately trying to elude this frightening creature who had somehow managed to invade their sanctuary. Her terror had the unfortunate consequence of making

Panbanisha double her efforts to be the most convincing gorilla ever. The two of them dashed around the room, Panbanisha making grunting noises and Tamuli squealing; all the while, Kanzi sat frowning in the corner, expressing his distaste for such juvenile shenanigans.

I am often asked: Was there some critical point during that first visit to the LRC when I realized that I must either return to Toronto and somehow explain away everything I had seen or else press forward with my work with Sue and come to terms with the behavior I had witnessed? In retrospect, I suspect that my turning point occurred the first time we all went for a walk together in the woods. We had stopped in a clearing to pick wild raspberries when Kanzi decided it was time for a game of chase. Not that he was about to exert himself. No, Kanzi's idea of a fun game was to delegate Sue to do the chasing and one of us to be the chasee. For the latter, he slowly looked at me and then, to ensure that he had been understood, he hit the lexigram symbols for "visitor" and "man." As I was considerably taller than anyone else on the walk there was no mistaking his intentions. Unlike the other researchers, I had no "name" key on the board. So Kanzi had improvised a solution by combining two different kinds of terms to pick me out. But there was something even more arresting than this creative use of symbols. In much the same way that a small child warily eyes a stranger, Kanzi had been glancing at me surreptitiously all morning. By choosing me to be the chasee, Kanzi was actually testing me out to see how I would respond. He was also deliberately drawing me into the group, both physically and psychologically. And, of course, he was ordering me around, which effectively brought me down to size.

What followed was exhausting and exhilarating. Through facial expressions and gestures, Kanzi expressed his delight and kept changing the players in the game to keep things interesting. Next I was the chaser and Tolly the chasee. Then Tolly became the chaser and Mary the chasee. This went on for some time, Kanzi all the while imperiously directing the proceedings as if he were some Herbert von Karajan, slightly dizzy with his power, when all of a sudden he abruptly stopped and stared intently into the upper branches of a nearby tree. A look of concern came over his face and the hairs on his back started to bristle. When Sue asked what was wrong, he pointed at the branches and vocalized softly. We all stared intently and, after several minutes of this, with no one moving or speaking, there was ever so slight a rustling in the leaves. Almost immediately Kanzi relaxed and nodded at us, signaling that we could now resume our walk. There was so much to absorb here. Obviously, there was the question of what he had seen that had caused him such alarm. Perhaps he had seen one of the mountain cats that are indigenous to the region; perhaps it was something completely different. But whatever the animal, Kanzi clearly regarded it as a threat to our safety. What I found even more interesting was how quickly he had communicated his warning with his facial expressions and gestures and how quickly the mood of the group had changed from one extreme to another. One could not help but be struck by the role that Kanzi had assumed on this walk. Inside the lab, there was never any question but that Sue was in charge. But the forest was Kanzi's domain, and he had stepped effortlessly into the role of the leader by watching over our safety and organizing our activities.

At first, Kanzi was dismissed by the scientific community as some sort of aberration: the bonobo version of a genius who had fortuitously landed in Sue's lab. But when Panbanisha demonstrated the same cognitive and communicative skills, it became clear that the explanation for their development lay in the language-enriched environment in which they were raised. At this stage of the research, it seems reasonable to conclude that the advances we see in the bonobos' modes of communication led to a confluence of events. Indeed, as is the case with infants, so too with Kanzi and Panbanisha: Not only have their interactive vocalizations, facial expressions, and other means of signaling all improved but so has their ability to plan and execute motor actions. It seems likely, then, that new central nervous system pathways were stimulated to deal with the challenges of their unique environment at the Language Research Center.

Having said that, it is important to note that the more we have learned about nonhuman primate behavior the more we can recognize islands of comparable behaviors in great apes in the wild. As we saw in the foregoing chapters, the more a species can engage in long chains of emotional exchanges, as bonobos in the wild can, the more complex becomes their group behavior. Some have suggested that bonobos are able to maintain a state of fairly pronounced social harmony by establishing a much more matriarchal type of social structure and also by engaging in various kinds of sexual behavior that serve to defuse tensions.[1] Although there may, of course, be some truth in both of these hypotheses, we would speculate that there is a much more basic social process at work here. For the more advanced becomes the emotional signaling of a species, the more it can separate perception from action. This enables individuals to think before they act. It is this latter capacity that is critical for establishing and maintaining harmony in a more complex group, and to begin to communicate symbolically. We looked at some of the evidence that apes in the wild can communicate symbolically in Chapter 3.

This possibility is further corroborated by Christophe Boesch and Hedwige Boesch-Achermann's account, in *The Chimpanzees of the Taï Forest,* of the role that drumming plays when the chimps are moving through the forest in much the same conditions as the bonobos described in Chapter 3.[2] The males appear to drum on buttressed trees to indicate their position and the direction in which they are moving. But the scientists noticed that when Brutus, the alpha male, drummed, the group as a whole abruptly changed its travel direction. In fact, Boesch and Boesch-Achermann report that "on some occasions, Brutus' drumming sequence appeared to transmit a specific message" either instructing the group about the direction in which it was to proceed or telling them they were to rest for a specific duration.[3]

The Chimpanzees of the Taï Forest provides us with a rich catalogue of the social and cognitive capacities of chimpanzees in the wild. We will restrict ourselves here to just two of the remarkable anecdotes reported, which provide further evidence that chimpanzees in the wild may be capable of symbolic communication. The book is framed with two different accounts of how the chimpanzees react when one of the members of their group is seriously wounded. In the first episode, Mognié, a six-year-old female, has been the victim of an attack, probably from a leopard, and has suffered severe wounds in her chest and belly; a piece of intestine is sticking out from one of the wounds and her respiration is audible through the hole in her chest. Her mother, Mystère, rushes over and licks some of the blood off the chest wound. Mystère remains with Mognié for two days, spending large amounts of time in a nest, although the researchers could not see what she was doing. Mognié then remained on her own for seventeen days. When she rejoined her family: "Mystère ran towards her and they embraced screaming briefly. Mystère lifted Mognié's chin and looked at her wounds that looked fully healed except for a small pink spot in the largest wound."[4]

Now contrast this with the episode reported at the end of the book. Tina, a ten-year-old female, had just been attacked by a leopard, and when the scientists arrived they found six males and six females sitting silently around the body, as if keeping vigil over it. A group of high-ranking females arrived and each sniffed but did not lick Tina's wounds. A low-ranking female and some infants tried to approach the body, but they were chased away. Three of the males then groomed the body, which is unusual insofar as none of them had ever been seen to groom Tina while she was alive. Later in the day, these males softly tapped Tina's chin and shook her hands and legs while looking at her face, as if testing to see whether there was any kind of reaction. Two and a half hours after Tina's death, her five-year-old brother, Tarzan, was allowed to "smell gently over different parts of the body and he inspected her genitals. He was the only infant allowed to do this. Then, Tarzan groomed her for a few seconds and pulled her hand gently many times, looking at her."[5] After this, the chimpanzees slowly started to drift away from the body.

Goodall has reported several sightings of chimpanzees mourning over a dead family member,[6] but here we have a vivid portrayal of a group's collective response to the death of one its members. Unlike the first episode, where there is a deliberate attempt to heal the injured animal's wounds, no such attempt is made this time; on the contrary, the animals are clearly aware of the gravity of her condition, and, perhaps, of her imminent death. They may, of course, merely be waiting to ascertain whether or not she revives. But there appears to be something far more profound going on here: almost a sort of funereal rite complete with social conventions governing who is allowed to mourn over the body and who is prohibited from approaching the body; and then a special role assigned to Tina's closest relative, Tarzan, who would otherwise have been prohibited from touching the body (because of his age); he performs a ritual that seems to mark the closure of what almost strikes one as some sort of ceremony.

THE ARRIVAL OF *HOMO SAPIENS*

The appearance of anatomically modern human beings has long presented evolutionary theory with one of its greatest puzzles. Part of the problem concerns the question of where these creatures came from, and when. Even more perplexing is the extraordinary technological ascent of this new species. Prior to the emergence of modern human beings, the pace of hominid evolution—both morphological and behavioral—was exceptionally slow. The changing anatomy of modern humans has conformed to this pattern, which suggests very stable conditions in the developmental matrix that is so critical for the expression of genes. But over the past 10,000 years, the pace of cultural change has been dizzying. What is it about this new creature that has enabled it, and it alone, to transform its environment—indeed, to transform the world—in what, in evolutionary terms, is the mere blink of an eye?

For the latter half of the twentieth century, the prevailing explanation of this phenomenon was that some sort of "monster genetic mutation" must have transformed the manner in which the evolving human brain processes information. There are many reasons for why this Big Bang theory exercised the appeal that it did, but one of the chief was simply that there was no competing developmental explanation. The theory presented in this book, namely, that caregiving practices that were culturally transmitted and thus learned anew from one generation to the next and, indeed, across species promoted the f/e development of prehuman hominids and then early and ultimately modern humans, provides a compelling answer to this question that is in close harmony with recent critical developments in genetic theory itself and the growing number of anthropologists who are incorporating aspects of sociality (in addition to natural selection) into their theories.[7]

Originally, it was thought that *H. sapiens sapiens* evolved from *H. erectus* independently in several different parts of the world, that there was genetic exchange between these different populations, and that modern racial differences are the product of local environmental conditions. Then paleoanthropologists speculated, from computer analysis of mitochondrial DNA, that *H. sapiens sapiens* may have evolved in Africa about 200,000 years ago and then migrated around the globe. But recent evidence suggests a more intermediary position between the Multiregional and the Out of Africa models.

Scientists were able to extract some of the mitochondrial DNA from a *H. sapiens sapiens* skeleton found on the shores of Lake Mungo in southeastern Australia. By comparing this DNA with that of nine other ancient Australian skeletons, two Neanderthals, and over three thousand people from around the world today, it was discovered that so-called Mungo Man had a unique genetic marker and indicates that a lost genetic line of *H. sapiens sapiens* was living in Australia prior to the arrival of Australian Aborigines. Lending further support to this intermediary model is Alan Templeton's recent genetic analysis indicating that there has been interbreeding between hominids living in Asia, Europe, and Africa for as much as 600,000 years, which suggests that successive waves of early humans expanded out of Africa over and over again and interbred with indigenous populations.[8]

Such paleoanthropological evidence lends important support to our hypothesis that the transition from *Archaic* to modern *H. sapiens* has been a slow process. If, as we argue, the underlying cause of the evolutionary changes manifest in the fossil record—morphological, behavioral, cognitive, and social—has been in part due to a series of advances in caregiving practices built on the practices of preceding generations, the emergence of *H. sapiens* can only have been gradual. This culturally mediated process goes back to the Common Ancestor and beyond. To be sure, some sort of "representational explosion" occurred in modern *H. sapiens,* but even here, the process whereby humans reached this critical threshold was gradual and not the result of some cataclysmic genetic event.

Paleoanthropologists speculate that *Archaic H. sapiens* began to appear sometime between 600,000 and 100,000 years ago. The skulls of these creatures were more rounded than those of *H. erectus,* and they also had a less pronounced brow ridge; the forehead was relatively high and the chin increasingly pointed. They had a cranial capacity at the low end of the modern range (from 1,150 to 1,250 cubic centimeters), an expanded parietal region, and changes in the basicranium that some have tentatively linked to an elaboration of the larynx. They were starting not just to look and possibly to sound more like us but even to act in ways that we would recognize as distinctly human.

To some extent, the category of *Archaic H. sapiens* is a grab-all for fossils whose cranial vaulting and bone thickness suggests that there was an intermediary species between *H. erectus* and *H. sapiens sapiens.* Our f/e model suggests that there must, in fact, have been intermediary steps between *H. erectus,* who were entrenched at the fourth stage of f/e development, and *H. sapiens sapiens,* who were firmly established at the fifth stage and starting to enter the sixth. Hence, we will adopt this convention and assume that there was an archaic species of *H. sapiens* who started to move through the steps outlined above up to the striking symbolic skills displayed by *H. sapiens sapiens.*

One might even speculate whether the emotional and cognitive advances that *Archaic H. sapiens* were undergoing played a role in the anatomical changes exhibited in the fossil record for *Archaic H. sapiens.*[9] A baby born 100,000 years ago would have had vastly more enriched emotional signaling and symbol-formation opportunities than his counterpart

600,000 years ago. Could this phenomenon also explain a change in the physical structure of the face and skull? After all, there is evidence that environment affects skin pigmentation, height, stature, fat content, motor capacities, and, of course, the manner in which the brain is wired. Is it not possible that there is also a connection between interactive learning experiences and facial muscles, and thus some of the differences observed in the shape of the skull?

There are indications in the fossil record to support the view that *Archaic H. sapiens* was slowly advancing through the early steps of the fifth stage. In the beginning they were still producing Acheulian tools, but we begin to see subtle refinements leading up to the next major revolution in stone tool technology: the Mousterian tool industry, which appeared around 200,000 years ago and lasted until 40,000 years ago. Mousterian toolmaking marks a significant technological advance. The whole process of manufacturing tools was standardized into well-defined stages and became much more labor intensive, and far more time was devoted to reshaping and sharpening tools. Various shapes are found in the Mousterian tool kit: pointed (used for hunting); burins (used for cutting grooves); borers (used for making holes in soft material); drills (used with much the same twisting motion as modern drills); scrapers (used for tanning hides); and hand axes and backed knives. These different shapes provide us with important indications of the cultural changes that were taking place, not just in hunting practices but also in the construction of shelters, the use of fire, the production of clothing, and, possibly, as we will see in the following chapter, the fashioning of ornamentation.

As one would expect, there are also signs of important social developments in the fossil record. For example, there is evidence that not only was *Archaic H. sapiens* using caves as shelters, but it is likely that these caves had inlaid stone floors. Furthermore, there is evidence that they were beginning to construct open-air structures. The remains of a 400,000-year-old hut were found in Terra Amata, near Nice in France. The hut appears to have had a hearth in it, no doubt used for heat, but possibly also to cook food. Near the entrance of the Lazeret Cave, also near Nice, are the 20,000-year-old remains of a 36-by-11-foot structure that had two hearths in it, providing us with a possible glimpse of a cohabiting extended family.

Even more intriguing is the evidence that, around 300,000 years ago, *Archaic H. sapiens* was starting to engage in burial rites, which suggests the beginnings of a more abstract belief system associated with the phenomenon of death. At the site of Atapuerca in Spain—the so-called *Sima de los Huesos*—the bones of thirty-two dead humans were found. Lending credence to the possibility that this was some sort of burial ground are the numerous Neanderthal burial sites that have been discovered (although some have warned that what appear to be tombs may simply be the consequence of accidental roof falls). We see a consistent pattern in these Neanderthal burial sites: Males and females were placed in small shallow graves with their knees bent, food and tools were placed beside them, and wildflowers were strewn about. Some have even suggested that we can see a pattern in the kind of objects that were buried with males versus females, which would suggest an early form of social organization based on sex, possibly akin to the models of early hunting-gathering societies thought to have emerged in the Pleistocene.

Early moderns in Africa (between 100,000 and 60,000 years ago) already had the same morphological features as modern humans, but they likely resembled Neanderthals in their behavior. We say this in part because paleoanthropologists now believe that Neanderthals had more sophisticated planning abilities than was previously thought,[10] and because of the recent evidence that Neanderthals were manufacturing beads during the Châtelperronian Period (from 32,000 to 30,000 years ago) at Arcy-sur-Cure, and that they transported fire to the cave at Bruniquel, France, where they built a stone structure that contained a burnt bear bone.[11] Early moderns were collecting iron and manganese compounds to use as pigments, most likely for body decoration; they built fires at will, buried their dead, and hunted large animals.[12] Also, like Neanderthals, they produced a relatively small range of stone artifacts that were stable over a long period. Most of the raw materials they used came from local sources, suggesting a small home range, a simple social network, and a sparse population. They rarely if ever used bone, ivory, or shell, and there is no evidence of symbolic representation.

Recent archaeological discoveries also show that the representational capacities of modern humans were developing over hundreds of thousands of years and began before the appearance of anatomically modern humans. In May 2000, Lawrence Barham reported the discovery, in a cave at Twin Rivers, in Zambia, of more than three hundred fragments of pigment, believed to be between 350,000 and 400,000 years old. These pigments had been ground up into a powder and ranged in color from brown to red, yellow, purple, blue, and pink. Barham believes they were used for ritual body painting and, perhaps, for cave painting.

Indeed, Barham points out: "If we link pigments with group activities such as ritual, then language is part of the equation. Rituals are grounded in a shared understanding of the significance of the events and the rules that govern the performance. . . . Perhaps the development of ritual and language are linked, one expressed through the other with pigment left as the only material trace."[13] As controversial as this last hypothesis will no doubt remain for some time, it is clearly time that we rethink the Big Bang hypothesis.

THE DESCENT OF THE LARYNX

One of the great appeals of the Out of Africa model was that it seemed to provide a simple solution to the puzzle of modern cultural development; for if a single species swept away all other living hominids, it must have been because they enjoyed some special advantage, either physical or mental. And since there is nothing particularly striking about the morphology of *H. sapiens sapiens,* then, according to this line of reasoning, they must have possessed some special mental capacity; that is, there must be something unique about the manner in which the brain of *H. sapiens sapiens* processes information, and if so, this accounts for the dizzying pace of modern human technology.[14]

One significant problem for this Big Bang hypothesis is that, insofar as the difference between nonhuman primate and human cognitive capacities seems to be more a matter of degree than of kind, one would expect the same to hold true for the difference between the cognitive abilities of anatomically modern and earlier humans. To be sure, all our so-called executive functions—e.g., our attention span, our ability to inhibit nonsalient information, and our short- and long-term memory—are more advanced than those of nonhuman primates, but the more we learn about the nonhuman primate mind the more we can discern fundamental continuities between nonhuman primate cognitive capacities and our own.[15] All, that is, except for one: the capacity to speak. For only humans, according to many influential linguists and psycholinguists, possess the capacity to acquire language. Hence, what distinguishes anatomically modern humans, according to the Big Bang hypothesis, is that they were the first hominid to speak.

Herein lies one of the reasons why there has been so much heated debate over the question of whether great apes are capable of developing primitive linguistic skills. For according to the Big Bang theory of human evolution, modern human beings—and modern human beings alone—suddenly acquired the capacity to speak sometime in the Pleistocene, for reasons that will remain forever unknown.[16] On this popular line of argument, the explanation for our remarkable development does not lie in enhanced cognitive abilities or an increased EQ; it lies in the unique properties that language provides for the communication of thoughts and the transmission of knowledge.[17]

Such a hypothesis has a long philosophical pedigree to back it up, as Chomsky recognized early on.[18] It was first articulated by the father of modern philosophy, René Descartes, and has been repeated by discontinuity theorists ever since.[19] The argument was given further support by modern linguistic analyses of the unique structural features of language, which, according to Chomsky's influential theory of Generative Grammar, accounts for what Chomsky described as the uniquely *creative* aspects of human language; that is, the capacity of humans to understand and produce infinitely many novel sentences.[20] In Chomsky's view, a child's brain automatically begins to process the complex linguistic data to which it is exposed at birth and spontaneously generates a grammar. Not surprisingly, Chomsky's hypothesis received considerable support from the computer revolution, for given the exciting advances being realized in the development of natural language programs, it seemed plausible to view the human brain as a form of computer, biologically equipped with certain basic rules—what AI scientists refer to as "heuristics"—that enable it to acquire the grammar of the natural language to which a child is exposed at birth.[21]

The Big Bang hypothesis received further support from archaeologists, who argued that the colonization of Australia 60,000 years ago would have required the construction of a boat, which takes considerable planning that is different in kind from the planning abilities seen in earlier hominids or apes.[22] How else, according to the Big Bang theorists, can we account for the sudden appearance of this cognitive capacity unless by postulating a genetic mutation? And how else, according to the archaeological argument for the Big Bang, can we explain the sudden appearance of representational art and the appearance of ornamentation and representational artifacts, such as jewelry and decorative clothing, unless the creatures in question had acquired representational capacities?

Influential support for the Big Bang hypothesis came from Lieberman and Laitman's argument that the descent of the larynx in human beings, which occurred at roughly the same time as the appearance of representational art, serves as a sign of the recent evolution of language.[23] A descended larynx enables humans to produce a far greater range of distinct sounds, but it exposes us to the serious risk of choking if we attempt to eat and talk at the same time. Lieberman argued that such a potential risk would only have been selected if it was associated with some significant adaptational advantage; that is, just the sort of benefit that would be bestowed by the advent of speech.[24] Thus, according to Lieberman and Laitman's hypothesis, we may conclude, based on the recent descent of the larynx, that anatomically modern humans were the first and only hominids to speak. In fact, according to Lieberman and Laitman, the Neanderthals might have been capable of making some verbal sounds, but because their larynx was not nearly as low as ours, these sounds would have been highly nasalized and indistinct and would have rendered spoken language virtually impossible.[25] To back up their claim that only anatomically modern humans could speak, they point out that Neanderthals became extinct even though their brains were slightly larger than ours and their stature more robust. What could be a more plausible explanation for their demise than their lack of language, which provided this new intruder with an insuperable advantage?

The obvious problem for this hypothesis is that the soft tissue of the larynx cannot be preserved in the fossil record. However, Lieberman and Laitman argue that one can infer the location of the larynx by studying the shape of the skull and the roof of the mouth. When Lieberman and Laitman first presented this argument, there was no evidence in the fossil record that *Archaic H. sapiens* had the same shape as is seen in anatomically modern humans, but recent discoveries have provided such evidence. The Lieberman and Laitman hypothesis has been further challenged by the discovery in the Kebara Cave in Israel of an adult male Neanderthal, called Moshe, who apparently had been deliberately buried there (he was placed on his back in a shallow pit, his right arm folded across his chest and his left arm across his abdomen). This evidence of burial rites is a significant indicator of cultural practices; but the finding that has really sparked all the controversy was the discovery of his hyoid bone: an extraordinary find, given the fragility of this small horseshoe-shaped bone that rests freely in the throat. According to Arensberg and colleagues,[26] the hyoid is proof that Neanderthals could speak. Indeed, they argue that not just the Neanderthals but possibly even *H. erectus* had the capacity to speak.[27]

To complicate matters still further, recent studies have shown that the widespread belief that apes are anatomically incapable of speech—because of the shape of their mouth, lips, and tongue and the location of their larynx—may not be the straightforward issue that was once assumed. Close analysis of the vocalizations that Kanzi makes when he uses the lexigram board reveals that his high-pitched sounds, which can be difficult for the human ear to distinguish, are similar to the English words that he is communicating when using the lexigram board.[28] This is something that I had become aware of almost from the beginning of my work at the LRC. I was sitting one day with Tolly and watching as Kanzi chose something to eat from the variety of foods that Sue had laid out for him. Kanzi was making what, to me, simply sounded like inarticulate vocalizations. But to Tolly's highly trained ear, it was clear that Kanzi was saying the English name of the food that he was about to eat before he chose it. Once Tolly had pointed this out to me, I could hear it too. For example, Kanzi would say "onion" before picking up the onion; but, because of the high pitch of his vocalizations, this came out as "oñynn." Equally interesting was our observation that he was engaging in this private speech in much the same way as a small child would.

Also very important here is the work of David Armstrong, Bill Stokoe, and Sherman Wilcox, and, more recently, Michael Corballis, on the gestural origins of language.[29] Armstrong, Stokoe, and Wilcox point out that the greatest problem with Lieberman and Laitman's hypothesis is that it conflates *speech* with *oral speech.* Moving the larynx and lips is really just one type of gesture, and signing (moving the hands, arms, facial muscles, head, and torso) is another. To be sure, there are certain advantages to oral speech over signing (it is easier to communicate in the dark, or in heavily forested areas, or when one is using one's hands); but then, there are advantages to signing over oral speech because one can communicate simultaneously through facial expressions, posture, and movement as well as with signs, and one can also communicate easily in noisy environments.

According to Armstrong, Stokoe, and Wilcox, whatever the reason hominids shifted from gesture to oral speech, the origins of language lie in the former. Quite recently, Pär Segerdahl has taken an intriguing step towards explaining why hominids may have shifted from gestural to oral speech.[30] Scientists have long been accustomed to think of gesturing as constituting a vestige of humans' earliest communicative behaviors. How else are we to explain why we gesture unconsciously while speaking, and, indeed, why we even gesture while speaking on the telephone? Or why so many aphasics are still able to communicate effectively through gesture? Or why a baby's first communicative acts are gestural?

In the standard view—which goes back to the eighteenth-century philosophies—hominids first communicated nonverbally, by gesture, and then they acquired language, which by definition was spoken.[31] In the modern version of this argument, we continue to gesture while we speak because the newer parts of the brain that process language were superimposed on the older parts of the brain that control nonverbal communication. Gesturing is thus seen as a redundant element in linguistic communication; in much the same way, communications engineers build in redundancies that enable listeners to understand a radio or television broadcast given that various sounds or pixels will always be lost in a transmission because of signal degradation or interference. In fact, the prevailing view of linguistic communication today amongst linguists and psycholinguists is that it involves several modalities, each controlled by a different center in the brain: linguistic (spoken or signed); paralinguistic, which involves any part of vocal behavior that affects the communicative content of an utterance but is not part of the linguistic system (intonation, stress, rate, pauses, pitch, and rhythm); and nonlinguistic, which involves any part of bodily behavior that affects the communicative content of an utterance but is not part of the linguistic system (gestures, posture, facial expression, gaze, head and body movement, and proxemics).

Segerdahl's hypothesis casts a different light on the role that gesturing played in the evolution of language. Segerdahl suggests that gesturing with their hands may have enabled hominids to develop language skills very early on by helping them to clarify the meaning of their (presumably indistinct) vocalizations. Consider, for example, how a caregiver can understand her infant while no one else around her can; or how a therapist can learn to understand a child with severe facial-motor apraxia once she gets to know that child; or how an ethnographer begins to understand the language of a completely foreign culture by immersing herself in that culture. Segerhdahl's hypothesis thus draws on Tolly's groundbreaking work showing that what singles out language as a communication system is its reflexivity.[32] In Segerdahl's hypothesis, the evolutionary dynamic at work in the origins of language would have been the development of the capacity to engage in such *reflexive* acts.

We are taking this important hypothesis a step further here. So much of our mutual understanding depends on shared context, history, attitudes, and emotions. Through emotional signaling, we clarify what we mean, intend, desire, and feel. A child begins to understand and perform these acts long before he can begin to speak. A similar phenomenon may have occurred with Neanderthals, and, indeed, with far older hominids; for even though they probably lacked the capacity to articulate clearly, they might have been able to clarify their vocal intentions with those acts of pointing, head nods, facial expressions, and hand movements that we continue to use to clarify what we mean, want, intend, and so on. With time, the location of the larynx may have shifted so that vocal modulations could take over the reflexive role originally played by hand gestures, thereby freeing up the hands for such activities as making more sophisticated tools, painting on caves, and fashioning jewelry and pottery.

It is speculative, admittedly, but worth considering, that the increased use of vocalizations was stirred by the increasing complexity of co-regulated affective communication. In other words, as we became more complex in our affective interactions, we had to include more and more subtle affective variations. The tongue and the oral motor cavity provided the motor capacity for humans to express this needed extra variation, that is, in greater flexibility and capacity for variation than possible in the hands or fingers. Therefore, it is possible that as we used this system more for this new purpose, it was further developed in the sense that function does affect neural pathways and that somehow its increased use may have been associated with the changing structure.

The descent of the larynx may well be connected with a significant advance in human culture, therefore, including a sharp increase in language skills; but not because it marks the sudden appearance of language, fully formed, as it were, in the way that Athena is said to have emerged from the head of Zeus. For language skills would have been developing slowly, over millennia, and would have been passed down from one generation to the next through the sorts of caregiving practices that are now referred to as motherese.[33] If there was a sudden representational explosion that occurred roughly contemporaneously with the descent of the larynx, perhaps it was in part because their greater articulatory capacities enabled humans to engage in other types of activities while they were speaking: something that would have been far more difficult when gesturing was critical for mutual understanding.

THE COHESION OF THE GROUP

We will return to this discussion of the origins of language in Chapter 9. For the moment, we wish to emphasize how a gradual improvement in emotional signaling leads to a confluence of events. The parts of the brain responsible for vocalization, facial expressions, and motor planning alter their wiring in a favorable direction. In human evolution, each generation would have showed the next, through infant caregiving, how to signal emotions and thus how to master preverbal and, ultimately, verbal communication. As each new generation learned to do this better through environmental challenges, as well as new central nervous system pathways that were spurred on by these challenges, it provided a relatively more culturally advanced or richer set of learning interactions for the next generation.

The importance of the skills that the child is acquiring at this stage for his ability to join the group into which he is born cannot be overstated. Long before a child acquires his first words he is acquiring the ability to understand the meaning of someone's facial expression, posture, movements, tone of voice, and direction of gaze. Indeed, as we will see in Chapter 7, it is only on the basis of these skills that a child is able to take the momentous step into language proper. The child must also be able to recognize when the words of others are not consistent with the emotional signals that they are giving off nonverbally. To develop a truly shared sense of reality, the members of a group have to be able to understand what each other really feels as well as what each other says.

To achieve the sort of social stability that is reflected in the fossil record and artifacts, humans had to develop cultural practices that enhanced their infants' emotional, cognitive, communicative, and social capacities. By learning how to enhance their infants' ability to engage in emotional exchanges, and how to read and respond to their infants' emotional overtures as well as responses, they were enabling their infants to develop in a variety of ways. They became aware of themselves; they learned to attend; they became curious; they understood causality; they developed empathy and morality. In this way, babies became social agents: members of a group, albeit with their own distinct personality and sense of self.

—6—

Representation and the Beginning of Logic: *Homo sapiens sapiens*

WE SAW IN CHAPTER 2 how the sixth stage of functional/emotional development involves learning to connect symbols and form larger categories. The child who masters this stage can think causally and figure out how one idea leads to another. This ability sets the stage for advancements in thinking. The child becomes more reflective and forms a symbolic sense of "self" and "other" that provides the basis for reality testing.

A parallel progression can be glimpsed in the fossil trail left by our early modern human ancestors. The earliest known *Homo sapiens sapiens* fossil dates from around 130,000 years ago and was found at Omo in East Africa. By 40,000 years ago, *H. sapiens sapiens* had migrated into Europe, and between 20,000 and 10,000 years ago they migrated to the Americas. The Cro Magnon humans, named after the fossils found in the Cro Magnon cave near Les Eyzies in the Dordogne, France, lived between 40,000 and 10,000 years ago. They were different from coexisting Neanderthalers in various ways: They had a higher forehead, less prominent brow-ridges, smaller teeth, less robust bodies, and reduced sexual dimorphism (with males around 1.8 meters tall and females around 1.7 meters). Curiously, their average brain size was slightly smaller than that of the Neanderthalers (around 1,350 milliliters).

The Cro Magnon were able to adapt to the severe conditions of the Ice Age; they battled large carnivores and successfully hunted the formidable mammoths (which provided them with a source of clothing and fuel; even the bones were used for dwellings). In fact, the Cro Magnon were making advances across various domains. For example, they began to live in larger and larger groups, their social structures developing greater complexity; they cared for their sick and their elderly; they performed burial rites for the young as well as the old. Their technological advances are seen in the sealed dwellings that protected them against the elements; they were heating these dwellings and cooking their food; they were even using fuel as a source of illumination. They were developing complex tools and weapons for hunting, such as bows and arrows, spears, and animal traps; they wore clothing made from furs or hides that they sewed with bone needles; and there are signs that they were starting to trade. Jewelry and ornaments made from shells and bones are evidence of artistic representation. But the great expression of their culture is the remarkably sophisticated cave art created between 25,000 and 13,000 years ago. Typically, these drawings are found deep underground, far from the living quarters of the caves. The very inaccessibility of these drawings suggests that they had a sacred function. From the way that certain aspects of the animals are depicted, it seems clear that specific artistic conventions were being passed down from one generation to the next.

The levels of emotional development that we've laid out help us to understand the factors underlying these social, cultural, technological, and personal advances. They also give us a new way of looking at one of the great mysteries in paleoanthropology: the dramatic appearance of cave art. Because of the sophisticated skills and concepts demonstrated in these paintings, evolutionary theorists of a determinist bent have been tempted to postulate that the humans who created them must have experienced some sort of genetic mutation that suddenly endowed them with advanced representational abilities. We have already seen a similar Big Bang argument used to explain not only the origins of language but also the "language explosion" that children typically experience between the ages of eighteen and twenty-four months. The comparison between a child's language explosion and the symbolic explosion that occurred during the Aurignacian period is illuminating. As we saw in Chapter 5, and will see again in Chapter 7, the factors leading up to each so-called explosion are very similar. Each time, there were long, slow advances in emotional signaling.

THE ALTAMIRA CAVE PAINTINGS

In 1879, one of the greatest of all discoveries in the history of paleoanthropology was made, not by an archaeologist, or even an amateur scientist, but by a five-year-old girl. Maria Marcelino Sanza de Sautuola had gone with her grandfather to explore one of the caves at Altamira, in Spain. Wandering off on her own, she walked about 30 meters into the cave and then simply looked up. What she saw above her on the ceiling were some of the greatest Paleolithic works of art ever discovered.

These fifteen large bison (and other animals: deer, wild boar, a horse) were produced between 13,500 and 11,000 years ago, not all at once but over the span of several centuries. Drawn with ochre and zinc oxides, some of the remarkably detailed and realistic animals have as many as four colors. The artists also exploited the natural contours of the rock to create a three-dimensional effect.

Psychohistorians of art have compared these paintings to the sort of visual realism (though created with much greater skill) that we see in the drawings of children between the ages of eight and twelve. For we see in these paintings the same attention to detail, form, and relative size, and an exploration of new techniques for representing depth and linear perspective, that we see in the paintings of children in the later stages of concrete operational and early formal operational thinking.[1]

So developed are the skills manifest in these paintings that many evolutionary theorists were led to conclude that the humans who painted them must have crossed some great Cognitive Divide. The idea lent itself perfectly to the Big Bang hypothesis about the origins of language and the idea that language, and symbolic capacities in general, suddenly emerged in human beings as the result of a genetic mutation. In fact, the cave drawings soon became one of the most important of all the arguments in the discontinuity thesis that anatomically modern humans were the first—and only—humans to possess language capacities. Each year, however, has brought to light some new archaeological discovery that challenges this argument. The picture that is now emerging contradicts simple genetic explanations. For it appears that hominids were developing representational skills very slowly; not, as one might expect, over thousands, but over hundreds of thousands, and possibly even millions of years, and that these skills were to a great extent culturally transmitted.

REPRESENTING THE BODY: FROM THE MAKAPANSGAT COBBLE TO THE VENUS OF WILLENDORF

The possibility that representational skills originated with our earliest hominid ancestors was given a boost in 1998 when a dark red water-worn piece of ironstone, approximately 3 million years old, was discovered in the South African cave of Makapansgat.[2] Apparently, it had been found by an Australopithecine in a site that was at least 30 kilometers away and brought back to the cave. What is so striking about this stone is its resemblance to a human face. Was this the reason it was collected?

In Chapter 4 we argued that the Australopithecines had probably reached as high as our fourth stage of f/e development. Typically, a child of about eighteen months can recognize a family face in a photograph. Thus, the ability to recognize the stone as an image of a face, and furthermore to collect it for that reason, would certainly bear out placing the Australopithecines at this high a level of f/e development. But then, if we accept this argument, we must seriously begin to question the assumption of a sharp discontinuity between the genus *Australopithecus* and the genus *Homo*.[3] Perhaps the depiction at the American Museum of Natural History in New York of the pair of individuals who left the Laetoli footprints (see Chapter 4, note 12) is not so far-fetched after all.

Numerous natural objects collected and carried because of some striking visual property (so-called manuports) have been found, although none as early as the Makapansgat cobble. Two other finds indicating the early development of symbolic skills are the Acheulian figurines found in Berekhat Ram, Israel, and Tan-Tan, Morocco.[4] The former is a fragment of volcanic tuff measuring 35 millimeters that resembles a female figure. Perhaps most striking about this figurine are the grooves around its neck and arms. Alexander Marschack has shown that these markings were deliberately put there, as much as 800,000 years ago. Not only did the *Archaic* humans who discovered this rock likely recognize its resemblance to a human female, but they may even have tried to enhance this aspect.[5]

If, as these datings suggest, these manuports were collected by *Archaic H. sapiens,* this would certainly seem to support our hypothesis in Chapter 5 that these early humans had begun to acquire symbolic capacities;

for the figurines apparently present us with a deliberate attempt to enhance the mimetic features of a natural object. One cannot help but wonder whether there is a parallel here to the manner in which children turn corncobs into dolls, or sticks into guns.

Another intriguing sign of advancing symbolic skills in *Archaic H. sapiens* is the pieces of ochre, discovered in many parts of the world, that were carried into settlement sites in South Africa between 900,000 and 800,000 years ago. They may have been used for body painting, skin protection, or tanning hides. We have substantial evidence that the Neanderthalers used pigments for some sort of funereal rite. For example, a Neanderthal skeleton found at Le Moustier had been sprinkled with red powder; red pigment was found around the head of the skeleton of La Chapelle and near two skeletons at Qafzed, Israel, and many others.[6] And a small ochre pebble with oblique striations was found in an Acheulian layer at Hunsgi, southern India. This pebble is between 300,000 and 200,000 years old and seems to have been used as a crayon on a rock.

As we can see, the picture that is emerging as the result of recent archaeological discoveries is one of gradually developing representational abilities. The skills manifest in the cave art drawings did not come about through a sudden mutation but rather through a prolonged process that led up to the flowering of these skills between 33,000 and 20,000 years ago, at the beginning of the Aurignacian period.[7] One of the most striking manifestations of this development is the appearance of the Venus figurines.

These small representations of women (and a few men, too) were made of clay or carved in stone, ivory, and wood. Perhaps the most famous of these figurines is the Venus of Willendorf, which, like many of its genre, is faceless and has large breasts, stomach, and thighs. The most popular of the many theories about the function of the Venus figurines is that they were fertility objects and possibly symbols of an earth goddess cult. Bruce Dickson has even suggested that these artifacts may present us with the first evidence of institutional forms of religious belief and practice.[8] Whether or not we can go this far, the Venus figurines clearly represent a significant step forward in the capacity of *H. sapiens sapiens* to engage in abstract causal thinking. For to use a representation of the human body in some sort of rite indicates an ability to combine ideas in a way that has meaning for a group.

THE FIRST DRAWINGS

The first signs of drawing are two engraved ochre stones, between 100,000 and 70,000 years old, that were found in the Blombos Cave near Cape Town, South Africa. The Blombos Cave has pushed back our views about the cognitive abilities of early moderns in several key respects. Among the kinds of evidence found there are

- the oldest bone tools;
- the earliest evidence for fishing;
- sophisticated spearheads made with techniques comparable to those seen in Europe 20,000 years ago;
- purposeful use of space within the cave, including distinct hearth areas and discrete areas where bone and stone were worked to manufacture tools; and
- large pieces of ochre pigment ground into a fine powder.[9]

The two engraved pieces of ochre found at the Blombos cave were first scraped and ground to create flat surfaces and then marked with cross hatches and lines to create a consistent geometric motif. The resulting pattern is similar to the sorts of abstract patterns seen along with the animal drawings in the cave art of France and Spain (produced somewhere between 60,000 and 90,000 years later). Whether or not this engraving provides us with evidence that the persons who created it had fully syntactical language, as some have claimed, it seems likely that the intent behind these engravings was symbolic.[10]

Numerous bones and stones found in Europe are marked with similar engravings. Twelve horse bones from the French rock shelter of Abri Suard and a fragment of a bovine rib from Le Pech de l'Azé, around 300,000 years old, have what appear to be intentional engravings comprised of a series of connected double arcs. And several bones from Bilzingsleben, Germany, from around the same time have a series of parallel incisions.[11] These designs are particularly intriguing because they suggest that our early human ancestors were progressing through the same sort of representational development as children. Typically, a child's first drawings are scribbles. She then builds up a repertoire of twenty or so elementary units and begins to combine them into patterns.[12] The earliest identifiable drawings on bone or stone are composed entirely of incisions or cup marks. To begin with, the incisions are simply a single vertical line (seen on several of the bones from Bizlingsleben); but, dating from about 300,000 B.C., we see series of parallel incisions; and by 54,000 years ago, a fan-like engraved motif on the Siga Phalange that was found in a cave in the Crimea.

A child in the early stage of drawing does not have the sorts of fine motor control skills necessary to produce accurate mimetic representations. At the first stage of artistic development, the child-as-actor and the actual representation are undifferentiated; the child's bodily movements, the trail of the brush on the paper, and the movement of the imaginary object are all compressed into a visual and motor act.[13] Some scribbles are clearly just random physical acts, but some are clearly intended to be representational. It is really only from the context that we can discern whether a scribble was intended to be representational.[14]

The same point likely holds true for these first apparent drawings. Perhaps, like young children, these early humans also lacked the sorts of fine motor control skills necessary to produce accurate representations. As with a young child's scribbles, whether or not markings can be interpreted as symbolic depends on the context in which they were produced. For example, two bones, 110,000 years old, each with a hole bored through the top, were found in Bocksteinschmiede, Germany. These bones were probably used in ritual contexts, and, for that reason, the carving can be seen as representational.[15] A similar point applies to the first European example of cup marks, found at the French Mousterian site of La Ferrassie. A Neanderthal child was buried underneath a large limestone rock with eighteen small cup marks on the surface, mostly in pairs, and placed in such a way that the cup marks were facing the child's body. Whatever the significance of the cup marks, it seems clear that the intention behind them was symbolic.

Even more striking is the rock art discovered at Ubirr, in Australia. These paintings, which date between 60,000 and 40,000 years ago, are surprisingly advanced. They contain stick figures that, among other things, depict a running male figure carrying hunting gear. Many of the figures are arranged in groups that appear to depict ancient hunting practices, or, perhaps, ritual activities.

The paintings thus serve as evidence of important cultural advances taking place at this time insofar as they clearly served an important symbolic function. They may also hint at the growing ability of these early modern humans to reflect on their self-identity both as a group and, perhaps, as individuals. Indeed, as we will see in the following section, the representational advances displayed in these paintings were part of a striking development in the cognitive abilities of these early modern humans.

BEGINNING TO THINK LOGICALLY

Evidence that these early modern humans were beginning to build bridges between their ideas, and beginning to think logically, is provided by the first colonization of Australia, which required them to build some sort of crude transport to cross the sea about 60,000 years ago.[16] By the start of the Aurignacian period, we see evidence that they were developing the capacity to think logically and to form a clear picture of reality; indeed, the social advances that occurred at this time would have demanded a significant advance in the capacity of early modern humans to build bridges between ideas.

For example, the open-air settlements around Pavlov and Dolní Vestonice in southern Moravia in the Czech Republic date from this period (around 27,000 years ago). In the Dolní Vestonice settlement, which housed between 100 and 125 inhabitants, pit houses have been discovered in which the floor had been sunk a meter into the ground to facilitate sealing between the roof and floor for protection against winter storms. The walls, built of wooden posts, were covered with animal hides. The planning and knowledge required to build such complex structures certainly suggest a capacity to think logically. There is even evidence that they were using fuel derived from mammoths to heat their structures.[17]

The discoveries made at the settlements of Sungir, on the outskirts of Moscow, provide similar evidence. The remains of surface dwellings reveal hearths and pits of various sizes, work areas where sophisticated stone and bone implements were manufactured, and four tombs between 25,000 and 20,000 years old filled with tools, jewelry, weapons, and clothing. These tombs contained the remains of two men, a boy, and a girl. It appears that the males were clad in shirts, long pants with attached footwear, over-the-knee boots, short outer cloaks, and fur hats; the girl wore a hood. One of the men's clothing was decorated with

3,000 ivory beads and his hat was circled with perforated arctic fox teeth. Five thousand beads were sewn into the boy's clothing and he wore a belt made of more than 250 drilled arctic fox canines.[18]

The full significance of these finds remains to be established, but it seems likely that the Sungir tombs present us with evidence of a complex social hierarchy, possibly hereditary, given the boy's clothing;[19] complex funereal rites associated with social status; sophisticated millinery skills; and even advanced hunting techniques.[20] Thus, from a f/e perspective, the archaeological evidence of dwellings and tombs suggests that these early modern humans were able to perceive connections between natural events and plan their actions accordingly, and to develop cultural and religious practices that would have fostered a shared sense of reality amongst the members of the group.

REPRESENTATION AND LOGICAL THINKING IN NONHUMAN PRIMATES

One of the more fascinating questions to surface in comparative research over the past decade is whether great apes are capable of intentionally creating artistic representations. There has been a fair amount of controversy over the paintings done by Koko and Michael, the gorillas studied by Penny Patterson. These paintings are all highly abstract and easily dismissed as the random creations that result when a gorilla is armed with a paint brush. Yet the paintings display several noteworthy features: the striking and harmonious use of color; the control shown in the strokes; and, most important of all, the presence of mimetic features. For example, in the painting *Bird,* which Koko made of a bluebird that used to visit her, there is a definite sense of a bird in flight; and in the painting *Toy Dinosaur,* made by Michael, one can discern what looks like a deliberate attempt to represent the scales on the dinosaur that was one of Michael's favorite toys.

I have seen Kanzi do something very similar myself. To begin with, his attempts at drawing were very much like those of a young child, that is, scribbles that were probably just random physical acts. But soon after this, the scribbles started to become more purposive and to be combined into very simple patterns. When I returned to the lab a year later, I could see that an important advance had taken place in Kanzi's artistic abilities. I watched as he carefully tried to make a painting of one of the lexigram symbols. He first looked carefully at the lexigram and then produced very slow and controlled strokes on his paper. At one point, his line became a little wavy and I watched as he tried to rub the line out with his hand and then redrew it.

The resulting picture did indeed bear a striking resemblance to the lexigram. My most memorable experience in this regard, however, was the day that Mary took me in to look at Panbanisha's bedroom. There on the wall were several paintings; intention was unmistakable.

Are these representational skills an aberration, an island of expertise, as it were, in an animal that is incapable of thinking logically? To be sure, there are many cases of so-called autistic savants who are significantly cognitively impaired yet capable of executing the most remarkably sophisticated drawings. One might thus be tempted to conclude that something similar is applicable here: that Kanzi and Panbanisha possess remarkable representational skills, far in advance of their mental abilities. But, in fact, there is also evidence that they have the ability to build bridges between their ideas.

At the beginning of the twentieth century, a number of psychologists set out to prove that animals are incapable of such a feat. Perhaps the most famous of these studies are Thorndike's experiments on the "learning curve."[21] Thorndike designed a puzzle box to measure the number of times a cat placed inside would randomly pull on chains and levers in order to escape. He found that when practice days were plotted against the amount of time required to free itself, a learning curve emerged; it fell rapidly at first, and then climbed gradually until it approached a horizontal line that signified the point at which the cat had "mastered" the task. According to Thorndike, his results showed how animal "learning" takes place in a series of brute repetitions that gradually "stamp" the correct response into the animal's behavior. If one should happen to see the cat for the first time at the very end of this process, one might naturally assume that the cat was exhibiting some form of insight in its ability to free itself so quickly from the box. And so, too, with any of the other known examples of so-called animal intelligence: We are tempted to credit an animal with the capacity to reason only because we have not witnessed the "stamping in" process.

Thorndike's experiments received their first serious challenge in the 1920s when Wolfgang Köhler set out to ascertain whether chimpanzees are capable of insightful problem solving.[22] Kohler's famous studies of the chimpanzees Chica, Grande, Konsul, and Sultan on the island of Teneriffe played a pivotal role in the development of ape research. Köhler would place a chimp in an enclosed area in which a bunch of bananas had been hung out of reach. To reach the bananas, the chimp had to use an object as a tool. Scattered around the area were various wooden boxes and sticks of different lengths. The chimpanzees were quite adept at using the sticks as rakes to pull in bananas that had been placed out of reach; they would drag the boxes to a point underneath the bananas, stack them one upon another, and climb up to reach the bunch. But the star of Köhler's experiments was Sultan. In one experiment, Sultan was given two sticks, neither of which was long enough to reach the bananas. After several frantic attempts to knock the bananas down by leaping with one of the sticks in his hands, Sultan sat down and could be observed reflecting quietly on the problem. Suddenly, he leaped up and pushed the thinner stick into the hollow of the thicker one, thereby creating a pole that was long enough to pull in the bananas.

Our understanding of chimpanzee tool-use has grown by leaps and bounds over the past few decades. We now know that different chimpanzee cultures have different toolmaking practices, depending on the materials at hand and the problems that need to be solved. The chimpanzees of the Taï forest have been observed to select their hammers according to the type of nut that needs to be opened, and to collect and carry the appropriate type of hammer to a site where those nuts are found. They also select the twigs they use to extract food from various types of nest according to the appropriate thickness, and chew the twig to the appropriate length for a given site. Indeed, as Boesch and Boesch-Achermann point out, it is hardly surprising that they are able to engage in this sort of causal reasoning given the sort of complex social life they experience.

In my work with Kanzi, I have often wondered whether he hasn't reached as far as the seventh stage of multicausal thinking. For example, one morning Sue showed up at the lab having slept only a couple of hours. The day before had been insane, one person after another assailing her with some crisis that urgently needed to be handled. When I left her at 1:30 in the morning, she was heading back to her office to answer the day's e-mails and then check that everything at the center was fine. She was looking a little worse for wear when she showed up at 6:30 the next morning. When he saw her, Kanzi came bounding over as he usually did, but Sue simply didn't have the energy to play. In fact, she was so tired that, for the very first time in my experience, she went to fix herself a cup of coffee. This seemed to alarm Kanzi as well. In a flash, his mood changed completely; he rushed over to Sue and, very gently, began to poke and prod her to make sure that she was alright. After she had passed this physical, he began to explore other possible problems. He pointed at the lexigram for bad and then stared intently in her eyes. He appeared to be asking Sue whether she was angry with him, and certainly Sue interpreted it in this way because she quickly assured him that she wasn't. Then Kanzi hit the key for Panbanisha: Was he asking whether Sue was angry with one of the other bonobos? This was how Sue interpreted it, for she quickly assured him that she wasn't upset with anybody. But Kanzi still wasn't satisfied. He had been eating cereal when we came in, which he now fetched and offered to her. When Sue declined this, he went and fetched his toy doggie, which Sue cursorily patted. It was incredible to see how concerned he was about her; but even more extraordinary was how systematically he was trying to figure out what the problem was.[23]

I have never seen Kanzi try to solve a problem by trial-and-error. Rather, he always studies a problem before he acts, and he always assesses and adjusts his actions. Maybe most interesting of all, he comes up with innovative solutions to problems. For example, he once wanted to ask Mary where I was, but there is no lexigram for my name on the board. I had been carrying a rucksack around all day, which I'd left in the corner of the group room. Seeing this, Kanzi rushed over and pointed to it, and then looked around. When Mary asked him whether he was looking for "Stuart," he immediately vocalized "uh huh."

We saw in Chapter 3 how Kanzi became quite adept at playing Pac-Man. It is really not surprising that Kanzi should perform so well on this sort of visual-spatial task. After all, consider the chimpanzee hunting practices that we looked at in Chapter 4. We saw there how Brutus, one of the chimpanzees that Boesch and Boesch-Achermann have studied in the Taï forest, is able to anticipate the direction in which a prey will flee and even project how long it will take it to arrive at a certain point. When one thinks about the sorts of visual-spatial skills involved in these complex problem-solving interactions, it makes sense to place the chimps at still a higher level of f/e development. Indeed, one might argue that, at least in some domains, chimps have reached our sixth stage of f/e development.

The real lesson here is not that great apes should be ranked higher on the sort of Piagetian stage-model scale that we looked at in Chapter 3; it is that we should adopt another of Piaget's important concepts for our assessment of great ape intelligence, the concept of *décalage.* In Piaget's structuralist theory, once a child has acquired a construct—say, the seriation scheme—it should apply to all domains; but in practice, a child may master seriation in one domain (e.g., length) a couple of years before she does so in another domain (e.g., weight). This is what Piaget meant by *horizontal décalage:* You cannot infer from a child's ability to solve seriation problems in one domain that she can solve seriation problems in all domains.

A similar phenomenon applies to the great apes. The capacity of great apes in the wild to solve visual-spatial problems is as high as Level 7 of our f/e framework. This is a consequence, given their environmental contingencies, of their visual-spatial system having developed first. They had to become sophisticated visual-spatial thinkers in order to deal with the problems they faced involving space and movement. Visual-spatial problem solving reaches Level 7 when it involves planned actions that can flexibly accommodate or be altered to the challenge at hand. With Kanzi we see a transition from visual-spatial to verbal logical thinking; but this transition was only possible because of his latent capacity for advanced visual-spatial reasoning: a capacity that the lexigram board actually capitalized on.

The same point must apply to our earliest human ancestors. As we have seen in the foregoing chapters, these early humans were making tools and perhaps even building shelters long before they had a fully developed language. These creatures could not have survived for the length of time that they did unless they were capable of some sort of nonverbal logical thinking. As we will see in Chapter 7, this nonverbal building of bridges between ideas may then have led the way to early humans' ability to engage in verbal bridge building between ideas and, indeed, to form larger and larger groups.

We have hypothesized that advances in affective communication not only fostered the ability to create and build bridges between symbols—it is affective communication, after all, that enables symbols to be used interactively and facilitates this logical bridge-building pattern—but also enabled early modern humans to communicate with one another very rapidly and to negotiate, almost implicitly, the basics of group survival (such as safety and danger, approval/disapproval, dominance/submission).

Living in larger and more stable collectives provided individuals with the opportunity to be more reflective and creative and may have been a critical factor in our extraordinary technological ascent. Because we have become so conditioned to think reductionistically, we assume that, if there was a turning point in human evolution sometime in the Pleistocene, it must have been produced by a genetic mutation. The real engine in our evolution, however, was affect signaling, which underpins the development of the shared sense of reality on which a large group structure depends. As we will show more fully in Chapter 12, "The Cohesion of the Group" may not be nearly as catchy a slogan as "The Survival of the Fittest," but it comes much closer to capturing the essence of the human evolutionary dynamic.

—7—

The Engine of Evolution

THE BASIC PRINCIPLE OF evolutionary logic, to which we alluded in the introduction to this book, is that, in Charles Darwin's famous words, there must be some "organisation" on which "the whole has progressed."[1] But, as McShea has documented,[2] there have been many fruitless attempts over the years to define this organizing principle, each of them failing to capture all the subtle nuances or anomalies in evolution. Time and again evolutionary theorists have attempted to reconfigure the Great Chain of Being, with some principle of complexity assuming the role that medievalists assigned to Divine Intention. But, as McShea shows, apart from a subtle tendency to define *complexity* tautologically—as that which is naturally selected—there are many examples where one might argue that *simplicity* is really the driving mechanism.

The failure to operationalize Darwin's nebulous concept of organization indicates that evolutionary theorists may have to consider distinct "organizing principles" governing the progression of different genera. Such a point, we have argued, is particularly important regarding the evolution of humans; for our central hypothesis is that, in the higher primates and hominids, there is a second fundamental mechanism that is not hard-wired but culturally transmitted.

In other words, we are challenging the continuum picture that lies at the heart of any attempt to find a universal construct or unitary mechanism underlying evolution. Furthermore, we are challenging the principle that the basic mechanism of evolution is simply the survival of an organism's genes. As we have seen over the past few years, when this argument is applied to human nature, it turns out that, no matter what the behavior, one can always find some way of explaining why it would have been selected for its role in enhancing reproductive success.

To be sure, an organism's ability to promote the survival of its genes is a necessary condition of evolution, but it is not a sufficient explanation for the evolution of the higher-level reflective thinking skills and complex social behaviors seen in human beings. We have proposed an alternative hypothesis for the evolution of these capacities that not only fits with the existing paleoanthropological and biological data but also accords with our current understanding of the factors involved in the development of these higher capacities. In our alternative hypothesis, the growth of greater ability in co-regulated emotional communication along with changes in the developmental manifold are the primary organizing principles of evolution in higher primates. Indeed, these two primary factors are inextricably connected.

When an organism is involved in a continuous flow of back-and-forth communication, it is constantly sampling subtle variations in its environment. As we saw in Chapter 2, in humans, affects mobilize the ability to focus on and process these sensory variations. Each time the organism responds, it also changes that environment ever so slightly. As it changes the environment, the organism creates variation that it must then process. In higher-level biological systems, that is, the primate and human brain, this continuing exposure to greater variation challenges the organism to discriminate more subtly among these variations. Each more discriminated perception leads to a slightly altered response that in turn alters the environment further, resulting in a cycle of increasing differentiation. For this to occur, however, the organism must be involved in a continuous flow of co-regulated communication. Only a continuous flow permits a continuous sampling of subtle variations.

Such an expanding range of environmental variations would likely have occurred in the rearing environment of hominid infants as gradually more complex caregiver-infant interactions took place. This process may have occurred in the following manner: As the infant gestured or vocalized to its caregiver, the caregiver's response changed ever so slightly to deal with the infant's affective overture. In turn, the infant now had an opportunity to process an ever so slightly novel affective response from the caregiver. As this process repeated itself, their co-regulated communication became more and more differentiated. Therefore, once an infant begins two-way communication with affective signaling, there is a tendency (with an adaptive caregiver) to become part of an increasingly more differentiated interactive system.

The functional capacity for co-regulated emotional communication ultimately led to the ability to operate in triads, extended families, and small groups, as we described in Chapters 2 and 4. Thus, as these social patterns were forming, there was an intricate interaction between culture and individual and social differentiation, all at the same time. That is, culture itself was an integral part of the organism's changing sensory-affective world as it emanated from and, in turn, enriched intra- and interpersonal development.

In all likelihood, as we shall see in Chapter 10, this type of learning interaction also facilitates new neural pathways in the brain. Interestingly, the pathways that have to do with subtle sensory affective discrimination, pattern recognition, and ever more complex subtle response patterns are part of important functions (and related central nervous system structures, such as the prefrontal cortex) that in part distinguish modern humans from nonhuman primates and earlier human species. We can see, therefore, that the pathway towards human intelligence includes gradually increasing functional capacities in co-regulated communication that enables problem solving, social organization, and the like. Simple reproductive success does not appear, therefore, to be an adequate explanation of higher levels of intelligence and high levels of social organization.

To be sure, one might still want to insist that higher levels of intelligence enable organisms to influence their environments and thereby perpetuate their genes; but a more reasonable organizing principle for the evolution of humans' higher reflective abilities is to think in terms of greater functional/emotional capacity, which we define as *the ability of the organism to engage in co-regulated affective communication, where this communication becomes more and more differentiated over time.* This definition of complexity is different from other proposed definitions of complexity, including general information-processing models,[3] and avoids the sorts of methodological problems that plague these other definitions.[4]

Co-regulated emotional communication that gets ever more subtly differentiated enables the organism to respond to and, in the process, alter its environment, and then respond to these new contingencies. As we saw in Chapter 2, this capacity to respond to and alter one's environment can be seen as a general definition of intelligent behavior. Therefore, our hypothesis holds that the evolutionary line leading to human intelligence may have been characterized by a gradually increasing functional capacity for co-regulated communication, culminating in a nervous system with the capacity for affective reciprocity. As we discussed in Chapter 2, it is the affective system that brings physiology and emotional and social functioning together. In a newborn baby, physiologic reactivity is the foundation for the development of the affective signaling system, which emerges as physiological reactivity takes on affective qualities through infant-caregiver interactions. As we saw in Chapter 2, it is this enhanced signaling system that enables the infant to construct patterns that eventually take on meaning and leads to different levels of emotional and intellectual organization.

—— PART THREE ——

The Development of Language and Intelligence

Introduction

IN THIS SECTION WE look at the development of language and intelligence. Significantly, for all the attention that the evolution of language and intelligence has received, the role of emotions in their development has received relatively little consideration. As we will see, there are interesting historical reasons for this glaring omission; these have been perpetuated in the Western view of language and intelligence as autonomous faculties. By looking at the evolution and development of language and intelligence through the lens of our functional/emotional framework, we introduce a critical new dimension into the important body of work examining the relationship between emotional, communicative, cognitive, and language development. Through this lens, we are better able to understand the complex interrelationships between the different aspects of a subject's development, including early emotional interactions and the formation of neuronal pathways in the brain and the development of new cognitive capacities. We are also able to frame a new hypothesis about the origins of speech. In the final chapter of this section, we present our most recent work on understanding the pathways to and from autism and explore how the remarkable progress of a group of children has helped to reveal the relationship between emotions, language, and cognition.

—8—

The Origins of Language

IN THE PRECEDING CHAPTERS, we concentrated on the development of intelligence and social skills and paid relatively little attention to what many would regard as the most important of our higher human attributes: language. Indeed, for a large contingent of evolutionary theorists, it was the acquisition of language that made us human, and not our larger brains or our ability to walk on two legs. Their argument is that, because of its unique structure, language enables human beings to develop complex and novel thoughts that can be communicated at a speed and with an economy unrivalled by any nonhuman communication system.[1] This line of thinking adds that the power unleashed by language accounts for the explosion of representational artifacts 30,000 years ago; our species' extraordinary technological advance ever since; our domination of all other species and our conquest of *Homo neanderthalis;* the emergence of agriculture around 10,000 years ago; and the development since that time of civilizations and complex social institutions. Meanwhile, the great apes are said to have remained in the same ecological niche that they inhabited 3.5 million years ago, and in much the same state. The only explanation for this overwhelming disparity, according to this discontinuity theory, must be that we possess language, whereas the other primates are constrained by their linear forms of gestural communication.[2]

As it turns out, this argument has quite a long history in Western thought. In the eighteenth century, it was widely argued that what sets humans apart from animals is that we are the only species that possesses speech. Speech was said to enable us to exercise voluntary control over the ideas coursing through our minds. According to the influential writings of the Abbé de Condillac, speech forces us, because of its unique structure, to analyze our thoughts: to break complex ideas down into their component parts, which can then be named, recognized, recalled, reflected upon, recombined, and, of course, communicated. At some point in our prehistory, primitive humans could communicate only nonlinguistically and thus were incapable of thinking. What remained an utter mystery for Condillac, however, and for eighteenth-century philosophers in general, was how primitive humans took that first prodigious step from what was called "natural language"—communicating by gestures and facial expressions—into the realm of speech, which provided early humans with the mental capacities necessary to form societies.

This mystery has remained with us ever since. Noam Chomsky made famous the so-called Big Bang hypothesis, according to which language must have emerged from a monster genetic mutation sometime during the Pleistocene. Given the preeminent cognitive, social, and communicative advantages bestowed by language, such a "language gene" would have been naturally selected. Chomsky has been assailed from many quarters for advocating such a miraculous thesis. But really, his argument simply implies that the origins of language are fated to remain a mystery forever. All that one can safely say is that language could not have evolved from prelinguistic communication systems, given the pronounced structural differences between language and those systems; any scenario one might sketch would fall victim to the cardinal sin in evolutionary theory of "Just-So" reasoning. Hence, for all intents and purposes, one might just as well settle for an evolutionary myth such as the Big Bang hypothesis.[3]

THE DEVELOPMENTAL VIEW

In the preceding chapters we have presented an evolutionary scenario that shows how language, along with our higher cognitive and social capacities, could have emerged through the interaction of biological and cultural processes over millions of years. As we have seen, the advances taking place in emotional signaling were building over the millennia through the consolidation and extension of caregiving practices. These signaling systems employ not only gestures and movements but also vocalizations, body movements, and facial expressions of emotion which, judging from the increasing complexity of hominids' facial musculature over the millennia, played a critical role in co-regulated emotional signaling. Moreover, just as complex as facial expressions are the intonations we can make with the voice, enabling us to produce a range of different sounds that can convey endless varieties of intent.

This gradual improvement in the range and complexity of signaling led to a confluence of events. In Chapter 1, we discussed how emotional signaling enabled the formation and manipulation of symbols: a key ingredient for the complexity and abstract use of language. But using symbols for thinking can occur through pictorial representations and complex gestures. Language as we ordinarily think about it also involves speech and rapid interactive abstract communication. To look at how full language as we know it evolved and developed, we need to look at how other factors came together with symbol formation to create this wondrous human ability.

As interactive vocalizations, facial expressions, and other means of emotional signaling improved, the abilities to plan, to attend to salient information and inhibit nonsalient information, and to execute motor actions all improved. We see the full significance of this point when we compare children who are developing typically with those who struggle with various biological challenges. As children develop, they usually begin to use more detailed facial expressions and to use their arms, hands, and fingers in a more purposeful manner, for example, to point out things when they're walking, or to lead you places and show you what they want. It is these advances in emotional signaling that lead to the improvements that we see in attention, planning, and motor sequences. When because of some biological challenge children do not develop the capacity to enter into these sorts of long co-regulated interactions, we do not see these other systems developing. Thus it is the emotional signaling that drives these other systems to develop. And so, too, with our hominid ancestors: The parts of their brains as well as the musculature responsible for vocalization and facial expression of emotions were altered and enhanced as their signaling capacities advanced, and other parts of the brain—the parts of the prefrontal cortex that enable us to develop our executive functions—likewise began to develop.

How did language emerge from this process? To answer this question, we need to look at how, in early infancy, emotional signaling occurs through touching, holding, smelling, and rhythmic rocking. As the signaling becomes more complicated and children begin to enter into co-regulated emotional interactions with a caregiver, they start to achieve the same sense of closeness and feelings of nurturing through distal communication, such as vocalizing across space or reading each other's facial expressions, body posture, and movements. In other words, through a still preverbal dialogue, the need for dependency and nurturing closeness can now be achieved in a new way.[4] As Harlow and Suomi showed us in monkeys,[5] or Savage-Rumbaugh and King in great apes,[6] or Spitz and Bowlby in humans,[7] this need for dependency and nurturing closeness is one of the most basic of all needs of higher primates.

Once this basic nurturing can be satisfied across larger spatial terrains, the child starts to develop the ability to communicate her desires and intentions, and to engage in communication for its own sake. This pleasurable aspect of distal communication frees vocal patterns from their use of meeting basic needs, such as the needs for food and protection. Just as emotional interactions enable the separation of perception from action, now they enable the formation of symbols that are freed from concrete utilitarian functions and can be used for showing things to others, sharing observations, communicating ideas and insights, and gossiping. This more advanced use of symbols has its roots in the pleasure and security provided by basic nurturing social interactions.

In summary, a confluence of events enabled language (as well as speech; see later in this chapter) to form. All these events had common roots in the complex co-regulated emotional signaling as described in Chapter 1. These critical language-forming events include the improved motor skills needed to produce more complex and subtle facial expressions, gestures, and vocalizations; symbol formation itself; complex problem-solving and group behavior; and the evolving ability to use communication as a vehicle for closeness and dependency (pleasure in communication in its own right, not tied to meeting basic needs). Once these were met, the opportunity arose for more interesting and creative communications. As we showed in Chapters 3 to 6, we can trace this chain of events over millions of years through the fossil record of different epochs in our prehistory, as well as in observations of nonhuman primates today. We demonstrated how each stage of f/e development corresponded to observable epochs in the evolution of our human capacities. This process didn't stop with the early forms of speech and language but advanced, and in all likelihood is continuing to advance, to higher and higher levels of reflective thinking.

This developmental model of language represents a striking departure from one of the main themes in Western views about the origins and development of language, namely, the idea that language is a form of telementational system.[8] The telementational view of language, which dates back to Descartes, is the idea that language is a rule-governed system for transmitting ideas from one mind to another; in the modern version of this view, a system *for encoding and decoding thoughts.* What is so striking about this view of language, apart from its failure to acknowledge the role of language in the day-to-day affairs of human beings,[9] is that it excludes any consideration of the role of emotion in the origins—or development!—of language skills. Rather, language is treated as a mechanical process for transmitting thoughts, which is the reason we see modern theorists likening linguistic communication to two computers, or even two faxes, "communicating" with each other.[10]

Closely connected to this point is the bias towards competence in theories about the origins of language—that is, the focus on what adult speakers know and do when they speak a language. With this as one's starting point, it is no wonder that so many contemporary theorists are drawn to genetic explanations to account for how adults could possibly have acquired this ability. For just as emotion is absent from this scenario, so is development.

Our theory of the origins of language addresses both these omissions. Advances in emotional signaling were culturally transmitted from one generation to the next, each generation learning from the previous one through infant/caregiver interactions how to convey emotions, ideas, and intentions, and thence, how to master preverbal and ultimately verbal systems of communication. Such striking phenomena as the development of the representational skills manifest in the cave art drawings, or the technological advances seen in the rise of agriculture and civilizations, were not the result of a genetic mutation but rather were developed through a series of gradual increments in complex emotional gesturing leading to representational thinking.

THE CONTINUITY/DISCONTINUITY DEBATE

Since the 1990s, there has been an explosion of articles and books on the origins of language.[11] This literature has seen the emergence of a debate between two opposing camps: the so-called *discontinuity* and *continuity* theories. The former maintains that language must have evolved relatively quickly sometime during the Pleistocene as the result of a monster genetic mutation, or a fortuitous series of genetic mutations,[12] whereas the latter insists that language must have evolved very slowly, over millions of years, as the result of behavioral modifications in response to changing environmental contingencies.[13] The debate between these two theories has reached an impasse, neither side willing to embrace the opposite's view of language itself, much less of the forces that brought about the emergence of language. Our approach introduces a new dimension into this debate in a manner that enables the continuity approach to address some of the most pressing concerns raised by the discontinuity theorists.

There are several reasons why this debate over the origins of language has received so much attention. One is simply because the "generativist"[14] school of linguistics founded by Chomsky seized on this issue as a way of supporting the discontinuity view that a child acquires language automatically sometime between the ages of eighteen and twenty-four months as a result of genetically controlled factors.[15] Another reason is because the large contingency of discontinuity theorists, known as "nativists," who regard our species-typical abilities as innate, saw in Chomsky's Big Bang hypothesis a paradigm for the genotypic view of evolution. That is, if language, long regarded as the quintessential higher human capacity, was the result of natural selection, it seemed that much more compelling to argue that less complex human capacities could also be construed as the result of genetic mutations in our recent evolutionary past.[16]

On a deeper level, the generativist view of the origins of language tapped into the resurgence of mechanist thinking that took place during the middle of the twentieth century as a result of the birth of Artificial Intelligence. The so-called postcomputational mechanist revolution promised to disclose the hidden operations of the mind by mapping mental processes onto computer programs.[17] This new model of the cognitive unconscious—heuristic algorithms said to be embodied in the brain—represented a strong reaction against the Freudian view of the unconscious; for AI promised to replace Freud's vision of a powerful and mysterious force underlying human behavior with a computational model of neurological processes that lie "beneath the threshold of consciousness."[18] That is, in place of the challenging task of grappling with unconscious motives, wishes, desires, and intentions, scientists could analyze the operations of genetically selected "programs of the brain."[19] Hence, the computational revolution would bring the science of the mind into line with other biological sciences in terms of things that can be objectively measured and concretized.[20]

The generativist explanation is said to resolve the question of why human knowledge, technology, and society have made such extraordinary advances over the past 20,000 years, whereas apes have remained in the same state for millions of years. Because of a sudden genetic mutation, *Homo sapiens sapiens* migrated around the globe rapidly and subjugated the Neanderthals, who, on this line of argument, were no match for this new breed of humans because they could communicate only nonlinguistically. Some of the other factors cited to support this idea that language must have appeared quite suddenly, sometime between 200,000 and 40,000 years ago, are marked innovations in tool use and the appearance of representational artifacts (we examined these in Chapter 6), and the existence of categorical perception and the descent of the larynx, which we will look at in greater detail in the following section. But most important of all is the argument that, given its complexity (syntactic, morphological), its capacity to refer to abstract ideas and to spatio-temporal distal events, and its creativity, language differs categorically from all nonhuman primate communication systems.

Using an analysis of the "design features of language,"[21] however, continuity theorists argued that primatologists have discovered evidence of several key elements of human language in nonhuman primate communication.[22] Moreover, a great deal of primatological research refutes the discontinuity argument that monkeys and apes live in the immediate present and can respond only to immediate stimuli, or that they have no control over memory of events and are unable to weigh alternatives, make predictions, or plan for the future.[23] The big problem for continuity theory, however, remains the discontinuity objection that there is a fundamental distinction between language and "purely functional and stimulus-bound animal communication systems." This distinction is said to lie in the fact that "language, being free from control by identifiable external stimuli or internal physiological states, can serve as a general instrument of thought and self-expression rather than merely as a communicative device of report, request, or command."[24] Hence, primate research cannot shed light on the origins of language because no matter how subtle and advanced these gestural/vocalization systems might be or become, they cannot be compared to the complexity and creativity of the language system. And it is precisely here where our theory of the development of language steps in.

A NEW THEORY OF LANGUAGE DEVELOPMENT

We saw in Chapter 4 how early humans were learning, through amusing facial expressions, animated vocalizations, and rhythmic movements, to extend the brief encounters that characterize the early stages of development into longer and more complex exchanges. That is, they were learning how to create all sorts of opportunities to communicate: through structured routines, in which familiarity would have been combined with novelty; through highly responsive and modulated behaviors; and through action-response games that would capture children's attention and help develop all their senses. No doubt peek-a-boo, or some variant on this simplest of games, has an incredibly long ancestry. Early humans would have been developing the "formats" that Bruner describes, in which a caregiver and child engage together in a task, such as "peek-a-boo," getting dressed, bathing, or playing with toys.[25]

Formats, according to Bruner, are a means of entering language and culture simultaneously; indeed, the two cannot be meaningfully separated. They are simple game-like microcosmic versions of the everyday means by which competent members of a culture cooperate in integrating their vocalizations and actions for the purpose of achieving some shared goal. But the formats of childhood are not always so purposive. Some formats may have a particular purpose, such as bathing or dressing, but many formats are performed simply to amuse a child, or to occupy him, or just for the fun of it. Formats, for Bruner, serve as the nursery for language/cultural development. Formats are crucially adaptable to the child's developing skills; indeed, this adaptability is exploited by the caregiver as she encourages the child, step-by-step, to try more sophisticated communicational means of participating in their interactions.

To engage in these formats, early humans had to learn how to read their infants' signals; for as Lev Vygotsky said, the caregiver may be the more experienced partner in these interactions, but the infant plays just as important a role in setting the tone and rhythm of the interactions. Hence, the caregiver is the one who "raises the bar," but only to the extent that the infant is ready to have it raised. Moreover, not only did caregivers have to learn how to read their child's physical state, they also had to learn how to read his burgeoning emotions, intentions, and desires. One of the most important consequences of a child's developing communicative abilities is the effect this has on the emotions he is capable of experiencing and expressing, and the impact this has on his cognitive development. But then, caregivers had to learn how to respond to these new and different emotions.

Long before hominids had begun to speak, therefore, they would have been acquiring the communicative skills required to engage in these sorts of proto-conversations with their infants. Bateson[26] and Trevarthan[27] coined the term "proto-conversation" to refer to the early stages of dyadic interaction. They had in mind the idea that the members of a dyad are involved from the beginning in shared gaze and vocalization dialogues that maximize the optimal levels of arousal for the infant and minimize the infant's negative affects. Since the early 1970s, careful study has shown that language acquisition occurs through this prolonged process of preverbal communication that takes place in the first eighteen months of a baby's life.[28] Typically, a child's language development surges from eighteen to twenty-four months, an event no doubt tied to the maturation of prefrontal cortical structures. But this surge invariably occurs when the child reaches the fifth stage of functional emotional development, made possible by the child's having progressed through the preceding four stages. And the essence of this progression is that it is a dynamic, social process and not an endogenous, maturational phenomenon.

Interestingly, a similar surge can be seen in the early language research conducted by Sue Savage-Rumbaugh with the chimpanzees Sherman and Austin. Sherman and Austin suddenly began to acquire symbolic skills after years of very little progress when Savage-Rumbaugh shifted from a behavioral modification program and began instead to use lexigrams in the ordinary day-to-day activities that engaged the chimps' attention and interest. The more Sherman and Austin were able to engage in long chains of co-regulated affective gesturing, the more advanced became their receptive and productive skills.[29] One of us (S.I.G.) has observed a similar phenomenon countless times, using very similar techniques, in work with young children who have severe language deficits.[30]

As far as concerns the origins of language, what we would argue here is that, whereas a baby needs from eighteen to twenty-four months of affective gesturing to reach the point of being able to speak in sentences, prehistoric humans needed 2 million years to reach this point. The advances taking place in their affective signaling were continuously building over the millennia through the consolidation and extension of caregiving practices. These affect signaling systems employ not only hand gestures but vocalizations, body posture, and numerous subtle facial expressions, and offer an almost infinite variety of subtle communicative possibilities. Consider, for example, how many different facial expressions can convey the various degrees of pleasure, delight, scintillation, and amazement. And each one of these seemingly discrete affects has different textures and tones to it. Equally as complex as facial expressions are the intonations that can be made with different sounds that also convey an infinite variety of affective intent.[31]

Herein lies the source of the linguistic creativity that Chomsky sought to account for in mechanist terms. That is, linguistic creativity arises not from some sort of generative mechanism but rather from the continuous flow of back-and-forth emotional signaling that provides a constant source of new affects or emotions to stir the next sequence of ideas or words. As we saw in Chapter 4, the preverbal gestural problem-solving interactions create the context for mastering meaningful verbal symbols. Without this basic level of *knowing through doing,* words may be used instrumentally, but they have no other intrinsic meaning for the user. We often see this phenomenon in children with a severe language deficit who have undergone intensive behavioral therapy. Through constant repetition and operant conditioning, they can be brought to use a large number of words to obtain certain goals; but they have a great deal of trouble generalizing the use of these words to novel circumstances, let alone combining them with other words in creative utterances.

Co-regulated, emotional interactions are thus essential for many aspects of language development. The more emotionally charged their interactions, the more motivated children become to master language skills. The more they develop these skills—and children vary considerably in pace[32]—the more meaningful and socially appropriate their language becomes. Grammar development, including the alignment of verbs and nouns, also appears to depend upon the presence of specific experiences. But just like phonological or pragmatic skills, these specific experiences also grow from emotional interactions: Emotion fuels planning and sequencing abilities, and is eventually invested in words. For as the child's feelings become more complex, he searches for more complex ways to express those feelings. To be sure, this is very much a symbiotic process; language development itself comes to play a role in the child's emotional development. But the critical point here is simply how critically involved the child's emotional development is in her development of grammar. Thus, the thirty-month-old sharing his observations about the world isn't simply imparting information; he is articulating—and indeed, starting to work out—what he feels about people, places, and objects.

One reason generativists missed this point was simply because, in most families, the kind of interactions we are talking about occur ordinarily and routinely. Unless one is looking for them, they are easily overlooked. It was by studying children suffering from language problems, autism, and deprivation syndromes (children from orphanages or multirisk families) that we were able to see what happens when these processes are missing, either for biological reasons (as in autism) or for environmental reasons (when a family or institutional caregivers cannot supply them). For virtually all children who develop language problems, a biological or social factor is preventing them from undergoing the sorts of specific experiences that need to be present for language-learning to occur.

The same point applies to the complexity of language. Discontinuity theorists were deeply impressed by the young child's ability to process complex auditory and visual cues in the early stages of language acquisition. They were unaware—for good reason, as we are only just starting to understand these mechanisms—that there's a developmental process or pathway that enables the growing baby to process more and more complex perceptions, including sounds and sights. This developmental process involves the gradual perception of emotional variation in sounds and also in facial expressions and body movements, gestures, and the growing ability to enter into reciprocal interactions around these vocal and visual emotional signals. Eventually, this leads to ever more refined differentiation of perception and thence the use of words.

As we argued in the early chapters, emotional signaling brings together the different processing domains—including auditory and visual-spatial, as well as olfactory, tactile, and proprioceptive—and enables them to function as an integrated whole, as well as to become differentiated, which builds to higher and higher levels of complexity. The levels described in Chapter 2 characterize these progressively higher levels of perceptual integration that occur during emotional interactions. Part of the reason why it is so difficult to envision differentiation in human perception is because it occurs in almost infinite amounts, every split second of every waking moment every day of the growing baby's life. This process is so rapid and so ongoing that it's easy to take it for granted and notice it only when it reaches certain critical milestones or when it's been derailed.

What biology brings to the table is not this whole interactive dynamic process of emotional and perceptual differentiation but a nervous system that is capable of being differentiated. In other words, the potential for perceptual differentiation is embedded in genetically mediated neurological structures, but these dynamic interactions we have been describing between caregivers and their infants are necessary for this learning process to take place. Furthermore, as the infant and caregiver are involved in these emotional interactions, this leads to new neurological pathways being laid, as well as new skills that in turn support higher levels of differentiation and integration.[33]

For example, the ability to master the first few words, which we described in Chapter I, creates the opportunity for new levels of differentiation and integration, that is, new tools of learning. But, as we showed, this capacity for symbol formation was itself the result of earlier emotional interactions. Without this complex interactive process, no amount of neurological development would, in and of itself, create the ability to master the use of meaningful symbols. Rather, in circumstances where these interactions are missing, we tend to see patterns very similar to what we observe in many autistic children: limited sounds are tied to actions or are used repetitively or perseveratively rather than meaningfully. Therefore, our genes are part of a biological complex that in the appropriate social and physical context confers on us an initial capacity to learn and provides a neurological substrate for subsequent learning if, and only if, certain emotional experiences take place in a sequence that has evolved over millions of years.

THE FUNCTIONAL/EMOTIONAL ORIGINS OF SPEECH

So far we have talked about language in quite general terms; in this section we will take a closer look at speech, because speech has been even harder to explain than language in general. For hundreds of years, the title of this section would have been interpreted as a statement about the origins of language. Thanks to Armstrong, Stokoe, and Wilcox's *Gesture and the Nature of Language,*[34] few would make such an assumption today; for speech, it is now widely recognized, is just one modality of language, possibly a later form of language that was made possible only by earlier, gestural forms.[35] But that still leaves us with the major question of why humans began to speak and what significance this has for the evolution of the human mind; and intimately connected with these evolutionary questions, why children begin to speak and what significance this has for the development of their minds.

The first systematic response to these questions, mentioned earlier in this chapter, was articulated by the Abbé de Condillac in his *Essai sur l'origine des connaissances humaines.*[36] Condillac argued that the evolution of the human mind and the growth of knowledge depended on the invention (Condillac's term) of language/speech. (Like all eighteenth-century philosophers, Condillac treated these terms as interchangeable.) Speech, Condillac argued, isn't just a tool for the communication of ideas: Without speech, thinking as we know it would not have been possible. Therefore, to understand the origins of thought, we need to understand why and how humans started to speak.

Why did Condillac, and eighteenth-century thinkers in general, place so much emphasis on *speech?* Why were they so convinced that "without speech no reason, without reason no speech"? In large part, the reason was because the philosophes believed that gestures are a residue of our primitive linguistic origins and as such play only a tangential communicative role in spoken languages (e.g., to convey emotion). The reason why speech was accorded such preeminence in eighteenth-century thought was because of the view that speech, unlike gestures, enables human beings to acquire voluntary control over the ideas coursing through their minds.

The philosophes believed that the so-called natural language of gestures could be used to communicate thoughts but that such an iconic system was severely limited both in its capacity to represent abstractions and in the speed with which one could convey what one was thinking. The transition from this *natural gestural language* to *speech* was thus considered to be of the utmost importance for the emergence of the modern mind on the grounds that one can speak nearly as quickly as one can think as well as convey far more abstract concepts than is possible in (what they conceived of as) a purely mimetic system; and still more important, because they thought that speech, unlike the natural language of gesture, forces one to *analyze* one's mental processes: to break complex thoughts down into their component parts, which can then be named, recognized, recalled, reflected on, recombined, and expressed in novel ways.[37]

Thus Condillac argued that "a man who has only accidental and natural signs, has none at all at his command. . . . Hence we may conclude that brutes have no memory; and that they have only an imagination which they cannot command as they please."[38] In other words, without speech, humans, like animals, were thought to be incapable of analyzing and ordering their ideas, incapable of summoning up ideas at will and hence recombining ideas to form new thoughts or recall past events. Indeed, it was thought that, since they lacked speech, the deaf do not possess free will.[39] Apparent counter-instances in which deaf adults, or, for that matter, animals, behaved in what appeared to be an intentional manner were explained away as instincts or the association of ideas.

Condillac concluded that the mind of prelinguistic humans, like that of animals, was governed by involuntary, mechanical associations. But this created a formidable problem: Given that the essence of speech/language was said to be that it is a voluntary creation, how was it that the primitive human mind was able to make the transition from *natural language*—the language of gestures, body movements, and facial expressions—to the voluntary use of signs to refer to ideas? If language were an invention, didn't this require an act of free will? And if it is only speech that gives us free will, how were nonlinguistic humans able to take that first momentous step and invent language?

The philosophes felt that, although science could only speculate about how the transition from involuntary to voluntary behavior might have occurred in the distant mists of early history (compare Rousseau's *Second Discourse*), it could more reasonably answer this question by observing closely how an infant makes exactly the same transition. For everything that has been said here about the lack of free will in animals and prelinguistic humans must apply to infants, because, as the very word "infant" betokens, they too lack speech. The infant's transition from cries and gestures to verbal interaction was thought to mirror primitive humans' transition from *natural language* to *speech.* The philosophes thought that children first create meaningless sounds that their parents imitate, and then invent words that the parents incorporate into their own speech;[40] but because children learning how to speak are subject to the same sorts of social coercion as everyone else, they soon abandon their idiosyncratic words and increasingly conform to the verbal behavior of their society. What they never could explain, however, was how it was that society itself was able to acquire language.

It is striking to see how little science has been able to add to this basic story over the past three centuries. To be sure, paleoanthropologists have come up with various reasons as to why speech may have come to supplant the older, gestural system; for example, speech, it is argued, freed up the hands and allowed humans to carry on other activities while communicating, such as hunting or cooking or making tools and clothes. Or to communicate over longer distances in impenetrable situations, such as the forest. Or to communicate at night.[41] But whatever the precise reason, it has seemed clear to many that speech must have been genetically selected sometime between 100,000 and 40,000 years ago (see also the discussion of the descent of the larynx in Chapter 5). For how else can we explain how humans acquired the ability to speak? The fact that speech must have been genetically selected and came to supplant gesture is said to be reflected in the maturational timetable that just about every child goes through: the cooing beginning around the age of three months, babbling at around seven months, and first word around the age of twelve months.[42]

Part of the problem with this argument is that it completely ignores the sociality involved in speech; that is, it treats the biological factors that underpin the ability to speak as canalized, or somehow isolated from the social factors that are intimately, and indeed, inextricably bound up with a child's development of speech. After all, while children are going through these stages of vocal development, they are simultaneously going through important stages of gestural development: From six to nine months, a child will start to shake his head to indicate no; around nine months, start to gesture to request things; and by twelve months, intentionally point at things. One cannot talk about the reasons children start to favor speech over gesture without considering the enormous influence of motherese on the child's linguistic development.[43] And as we can see from Ellen Groce's fascinating book, *Everybody Here Spoke Sign Language,* a history of the sign/spoken English bilingual community that existed for such a long time on Martha's Vineyard (because of the unusually large number of deaf in the community), social conditions can render communicating by sign just as natural as communicating by speech.[44]

At still a deeper level, however, the problem with the genetic determinist explanation of the evolution and ontogeny of speech is that it rests on a tautology; for ultimately the evidence that speech must have been genetically selected is simply that virtually every child speaks. But the interesting question is why speech should have been so favored, and why it continues to be. Is there something about the activity of speaking that may explain its universality? Before one thinks about the origins of speech, however, one needs first to understand the evolutionary forces that brought about a vocal system that was sufficiently complex and finely regulated to support speech. The initial question that needs to be addressed, therefore, is why a more complex vocal system would have conferred an evolutionary advantage. Not only does our hypothesis about the role of emotional signaling in the human evolutionary trajectory enable us to understand the evolutionary advantages of a more subtly nuanced vocal system, but further, it shows how these same caregiving practices would have played a critical role in the subsequent evolution of speech.

Once again, it is useful to start by considering the ontogeny of speech in a young child. As we saw above, children typically start to speak around the time of their first birthday: first in single words, and then, a few months later, in two-word combinations. During the ensuing language explosion, they master as many as ten new words every day. This can be a constant source of wonderment and amusement as they begin using some word that their parents can barely remember uttering. From this point on, their language abilities seem to explode and, generally by the age of twenty-four months, they are beginning to speak in sentences.

The tendency to view the child's development of speech in genetic terms is certainly understandable, therefore, given the regularity with which so many children progress through these milestones. But to begin mastering the use of spoken words, children must go through a great deal of development before and during their progression through the milestones. We are not simply referring to his advances in the gestural and receptive skills that, as many theorists have noted, provide the substratum for the child's subsequent ability to speak.[45] Rather, we are referring here to the child's developing ability to separate perception from action. For, as we saw in Chapter 2, mastering symbols and the capacity to think require that the child move beyond fixed perceptual motor patterns in which a stimulus evokes a response. By separating the perception from the action, the child can develop a "freestanding" perception or image that can become associated with experiences and form the basis for later use of symbols. As we have pointed out, this capacity to separate perception from action occurs only as a result of engaging in emotional signaling with caregivers. Only when infants can use emotional signals to communicate and negotiate are they no longer at the mercy of fixed perceptual motor patterns. More important, they are also no longer at the mercy of "catastrophic" emotions, such as rage, fear, or overwhelming neediness, which tend to lead to fixed reaction patterns.

Children cannot begin to speak, however, until they are capable of controlling the vocal system, which occurs only when they are capable of engaging in long chains of co-regulated emotional signaling involving all communicative modalities (the vocal system, facial expressions, gestures, body movements). Our hypothesis is that our hominid ancestors had to go through the same stages of development to attain the same mastery over the vocal system before they could begin to speak. Thus, rather than viewing speech as evolving—suddenly or otherwise—as an alternative to the gestural system of linguistic communication, we need to consider the reasons why the vocal system developed *alongside* the gestural system, indeed, as an integral part of it, and eventually came to serve as the primary mode in linguistic communication. That is, we need to understand the factors that promoted not simply the transition from nonverbal to spoken language but more fundamentally the capacity of nonhuman primates, the Australopithecines, and early humans to produce and discern an ever wider and more modulated range of sounds. What were the factors promoting the advancement of vocalizing to the point where it became complex enough to support a transition from nonverbal to spoken language?

The first point that we need to bear in mind here is simply that the voice introduced a critical added dimension to emotional signaling. The more complex emotional interactions became, the more nonhuman primates and early hominids needed to augment their existing communication modes. A more subtly nuanced vocal system would have been recruited as part of this dynamic, and would have conferred a further evolutionary advantage as a natural consequence of the growing complexity of the communication it made possible. When we look at bonobos, for example, we can easily see how their vocalizing is every bit as important an element in their emotional signaling as their gestures, facial expressions, and body movements. To be sure, we are a long way off from understanding the intricacies of their vocal exchanges. Yet one thing that is clear is how vocalizing allows for many more signals to be exchanged, and in wider circumstances, thereby enhancing the capacity of the group's members to negotiate relationships of power and submission, to tame catastrophic emotions and aggression, to deal with safety and threats, and, of course, to build stronger bonds of intimacy and group cohesion. Another thing that seems evident when observing bonobos is that they experience great pleasure in the very act of vocalizing. This is an important point, for any theory of the evolution of speech should proceed from the idea, mentioned earlier, that the very act of speaking is pleasurable: not only for the speaker, but for the receiver as well. The staggering success of the telecom industry is an indication of just how much we love to speak.

The hedonic principle operating here, it would appear, is that the more differentiated the emotions and the more subtle the variations, the greater the pleasure one experiences in communicating these variations through vocalizations and other gestures. And the vocal system does indeed enable one to convey different and more subtle qualities of emotion than the purely visual system. Through the voice we can convey infinite variations in warmth and closeness, distance and anger, curiosity and disinterest. If you watch a silent movie you certainly feel some warmth, but it is not nearly the same as hearing the actors' voices. Or consider those telephone ads showing families separated for the holidays, or grandparents separated from their grandchildren: The "characters" are able to experience great feelings of closeness solely with their voices. These ads wouldn't be nearly as effective if the vocal system were not such a rich source of emotional signaling, such a powerful means of conveying warmth and intimacy.

Thus, to understand the processes that are at work here, we need to situate the evolution of the vocal system in the context of the reciprocal emotional gesturing that had been developing for millions of years. As the ability to engage in interactive problem solving advanced, hominids became increasingly capable of taming catastrophic affects (see Chapter 1) based on flight-fight and similar reactions, which are governed by relatively more primitive parts of the brain.

Significantly, some of the same neurological structures and pathways that are involved in emotional signaling also regulate the facial muscles, the muscles of the inner ear, and the pharynx and larynx (the throat and the voice box). The importance of this point cannot be overlooked, for it would appear that Mother Nature was operating with her usual zest for efficiency. Of the many neurological structures and pathways involved in these complex processes, one that has been particularly well studied is the vagus nerve. It has been shown that the myelinated components of the vagus (evolutionarily more advanced components than the unmyelinated component of the vagus, which deals with vegetative functions) regulate or control the facial muscles, the inner ear, and the pharynx and larynx. Furthermore, one of us (S.I.G.) collaborated in research that showed the same pathway is involved in the interactive regulation of behavior.[46] It would thus appear that this important pathway is simultaneously influencing parts of the nervous system that have to do with the part of the middle ear used to decipher and discriminate sounds, the organs required for producing sounds, and the muscles used for facial expressions.

If emotional interaction fostered the development of these neurological pathways, it would simultaneously foster the ability of hominids to discriminate sounds (regulating the muscles of the middle ear) and to create sounds through new nerve pathways regulating the pharynx and larynx. Thus, what may have happened is that co-regulated emotional signaling fostered the ability of hominids to communicate in a variety of ways, ranging from facial expressions, with all the complexities inherent in that, to the greater and greater use of vocalizations from the receptive and expressive perspectives. In other words, a range of communicative systems may have been promoted efficiently by just a few pathways. That these pathways tend to regulate several of these interrelated systems also lends support to our hypothesis on the central importance of co-regulated emotional signaling for symbol formation, language, and intelligence. Here we have evidence that the same types of interaction that in all likelihood made symbol formation possible also enhanced the likelihood of more complex facial signaling, more finely nuanced vocalizations, and the development of spoken language.

By no means do we wish to minimize the importance of the instrumental arguments that others have emphasized for the transition from gestural to spoken language (see above). Freeing the hands for other activities would indeed have bestowed a considerable advantage for the sorts of manual activities that paleoanthropologists have envisaged. By the same token, once we consider the importance of the hands for emotional signaling, we can see how these same manual activities would have put further pressure on the vocal system to develop; for if the hands were occupied and one was staring at the task at hand, the vocal system not only would have had to take over basic communication but also would have had to pick up for the finer textures and nuances that would have been supplied by gestures and facial expressions.

SPEECH AS AN EXPRESSION OF NURTURANCE

Initially, humans vocalized to convey hunger, anger, fear. How was vocalizing elevated beyond communicating basic needs to the pleasure of gossiping, relating some interesting discovery, or sharing one's amusement at something funny? In the development of each baby and child, we have observed that this transition occurs when the toddler discovers that vocalizations and/or words can be as much a source of closeness and shared pleasure as a warm hug. Toddlers begin experiencing greater closeness and the meeting of their basic needs for closeness and nurturance through the act of vocalizing itself.

In other words, in healthy human development, the need for nurturance is as basic as the need for food, but it is a basic need that comes to be filled in an ever more differentiated manner. The baby moves from hugs and caresses to shared vocalizations and communication through emotional signaling as a way of satisfying this fundamental need for nurturance. It is remarkable how, as adults, humans can feel the same nurturing warmth over the telephone with people thousands of miles away simply by hearing the warmth in their voices and basking in the meaning of their words. We've often wondered, also, whether the underlying satisfaction for seemingly dull small talk isn't at least in part that it provides an excuse to hear each other's sounds, words, and simply to operate together as a group. In this way, a group may come to provide aspects of the nurturing that was originally provided in a more concrete way by caregivers.

Another way to look at this point is that the critical feature in human evolution was—and is—the importance of nurturance as a basic need that is as pressing in many respects as the needs for food and protection. But the basic need for nurturance can undergo many developmental transformations towards higher and higher forms of expression, negotiation, and satisfaction. In fact, the transformations we have described in co-regulated emotional interactions (the sixteen stages described in Chapter 2) can be viewed as transformations in the way we nurture one another. We appear to try to nurture each other at all the levels at the same time, up to the highest levels of which we are capable.

The question we are left with, then, is this: Why did early humans make the important transition from using communication to satisfy basic needs to valuing communication for its own sake? It would be reasonable to hypothesize that here, too, as emotional signaling enabled basic nurturing interactions to be more and more finely negotiated between adults, and between infants and adults, communication itself progressed beyond simple nurturing satisfaction. At that point, humans began to love speaking for its own sake. Gossiping and chatting, even arguing, can all be understood as providing different means of nurturance.

In addition, there are all sorts of social conditions that would have enabled more opportunities for such forms of communication; for example, when the members of a group spend large amounts of time together engaged in joint cooperative tasks, the opportunity is provided for such an advance to take place. And perhaps the anatomical changes in the vocal tract that occurred sometime between 100,000 and 40,000 years ago (see Chapter 5) are closely associated with these changes. As we said earlier, such a phenomenon would hardly have been the result of a random genetic mutation that miraculously and fortuitously enabled modern humans to begin speaking. Rather, as emotional signaling was progressing, it was fostering pleasure in vocalization for its own sake. This, too, could have fostered favorable changes in the nervous system for more finely regulated motor control as well as the changes in the larynx, pharynx, and middle ear discussed earlier. What we are proposing, therefore, is that co-regulated emotional signaling fostered the development of a dynamic system that came together and created a basis not only for language but also for speaking.

—9—

The Role of Emotions in Language Development

THE TELEMENTATIONAL VIEW OF language, dating back to Descartes, as an abstract system for transmitting thoughts from the mind of one speaker to another (see Chapter 7) led, in the twentieth century, to different mechanistic views about how a child's brain must be programmed, either by experience or by genes, to process speech "input." Conspicuously missing in all these accounts is the role of emotion in the development of language. The explanation for this lacuna also likely goes back to Descartes, who assumed that language and the emotions are housed in different parts of the brain. Thus, emotional factors may impinge on language but have no intrinsic role in its development. In generation after generation over the past three centuries we see philosophers, linguists, and psychologists basing their theories of language development on this fundamental assumption. But recent evidence has shown that emotion lies at the very heart of language. In this chapter we will show how this perspective helps us understand the critical processes involved in the way we develop and use language.

THEORIES OF LANGUAGE DEVELOPMENT

The psychology of language is currently dominated by three main schools of thought: generativism, the theory developed by Noam Chomsky;[1] cognitive linguistics, sparked by the work of Lakoff and Langacker;[2] and interactionism, inspired by the work of Austin and Bruner.[3] Generativism looks at language acquisition as a biological phenomenon: a maturational process in which a child automatically acquires the grammar of the language spoken in her culture. Language acquisition, in this outlook, is a mechanical process: a child acquires grammar as a result of information that is encoded in her "language gene," which controls how her brain processes speech. As its name suggests, cognitive linguistics looks at the role of cognition in language development. In this outlook, cognitive development precedes and enables linguistic development. Thus, instead of treating language and cognition as separate processes, cognitive linguistics seeks to explain the acquisition of the complex features of grammar that generativists have identified in terms of the concepts that a child constructs as she interacts with objects and people in her environment. And finally, interactionism looks at language development as a social phenomenon. In particular, it studies the types of enculturated skills that a child acquires when she learns how to talk. In this outlook, verbal behavior emerges from and along with nonverbal communication to provide a new way of solving problems together with others. The child is "learning how to do things with words"; and the "things" that she is learning how to do are socially constructed acts, possibly unique to her culture.

Each of these theories has illuminated different aspects of language acquisition. Generativists, for example, have been able to chart interesting patterns in the syntactic constructions that children acquire.[4] Cognitive linguists have been to explain the cognitive factors underlying the acquisition of many of these constructions.[5] And interactionists have shed light on the linguistic/cultural skills that a child acquires.[6] What all three of these theories have in common, however, is that none of them considers the role of emotion in language development. A rigorous literature search on the relationship between emotional and linguistic development over the past century turns up little, apart from the research of those working in the area of language disorders. The tacit assumption amongst psycholinguists appears to be that emotional development is at best *extrinsically* related to language development; that is, it enters the discussion only as a motivational factor that may enhance or impede the linguistic, cognitive, or social-communicative processes at work. Or else emotional development is subsumed under language development as a separate domain of category terms that a child masters in her march towards acquiring a theory of mind.[7]

In this chapter we present a radical departure in Western thinking about the nature of language development, which dates back to the view of language instituted by Descartes,[8] by placing emotions at the very heart of the language development process. That is, we argue that emotions serve not just as a motivational factor but as the critical architect of language development. A child's first words, her early word combinations, and her first steps towards mastering grammar are not just guided by emotional content, but, indeed, are imbued with it. A child's capacities to speak fluently and creatively, to become a competent member of her sociolinguistic community, to use her burgeoning language skills to master more complex aspects of language, and to use language to enter other domains of knowledge are all the consequence of intrinsically emotional processes.

Language, in this view, is not acquired mechanically as some sort of abstract system for transmitting one's private thoughts; nor is language development the result of mapping words onto concepts that a child has constructed in the privacy of her mind. Language comes first from *lived experience* and is much more complex than (but fundamentally similar to) a child's smiles and frowns, gestures and head nods, cries of joy and shouts of anger, the meaning of which she learns through shared emotional experiences with her caregivers. Likewise, the meaning of the verbal sounds or signs that a child masters are grounded in the long chains of co-regulated emotional interactions that, as we saw in Chapter 1, underpin the growth of the child's mind. In general, the more constricted a child's emotional development the more delayed and problematic her language development. Indeed, as we shall see, the view of language acquisition as a mechanical process championed by generativist theorists is most aptly suited to children suffering from profound biological challenges that have seriously impaired their emotional interactions with caregivers.

Chomsky's generativist theory proceeds from a famous argument, known as the "poverty of the stimulus," which sought to establish that language acquisition would not be possible unless a child were endowed with knowledge of the possible forms that grammar can take. That is, a child must come into the world possessing a set of "super-rules" that make language acquisition possible; only thus can we explain how "children's grammar explodes into adultlike complexity in so short a time. They are not acquiring dozens or hundreds of rules; they are just setting a few mental switches."[9]

From a psychological point of view, the most important implication of this argument is that, strictly speaking, language—or at least, grammar—is not something that can be *learned*. In Chomsky's words, "[K]nowledge of grammar, hence of language, develops in the child through the interplay of genetically determined principles and a course of experience. Informally, we speak of this process as 'language learning.' It makes sense to ask whether we misdescribe the process when we call it 'learning.'... I would like to suggest that in certain fundamental respects we do not really learn language; rather, grammar grows in the mind."[10] Thus, the guiding principle in Chomsky's view of language development, as described by Steven Pinker in his book *The Language Instinct,* is that language "develops in the child spontaneously, without conscious effort or formal instruction, is deployed without awareness of its underlying logic, is qualitatively the same in every individual, and is distinct from more general abilities to process information or behave intelligently."[11]

We will not look here at the various arguments with which generativist psycholinguists have sought to back up this argument, such as their attempt to document language "universals," or to seize on various language disorders as proof that language is an autonomous faculty.[12] Nor will we consider the legion of criticisms that this theory has attracted,[13] or the impoverished view of language which it adopts.[14] Rather, our main concern here is to consider what's wrong with looking at language development as a mechanical or biological process. When we return to our discussion of Chomsky's theory in a later part of this chapter, it will be to try to understand the reasons why Chomsky, and indeed, his generation, could have been so drawn to a biological model that they completely overlooked the role of emotion in language development.

To be sure, biology does play a critical role in early language development, for a child needs a brain that has the capacity to recognize and organize affectively meaningful experience into patterns. However, although the basic ability to construct patterns is present at birth, to some degree the ability to create more complex patterns and higher and higher levels of organization depends on favorable emotional interactions with the world, leading to new sequencing and pattern-constructing abilities. Therefore, what we observe is a biologically endowed nervous system with lots of potential. But at each step this potential requires highly specific kinds of emotionally engaging interactions, which in turn lead to new abilities and potentials. The human nervous system is thus continuously being modified as the child interacts in an emotionally meaningful way with her environment. We now need to consider the kinds of emotional experiences a child must undergo to develop language.

IMPLICIT OR PROCEDURAL KNOWLEDGE

The earliest stages of language learning lie in presymbolic processes that are outside conscious awareness. Such presymbolic, or subsymbolic, processes—for example, a toddler learning how to relate to a particular caregiver or learning the difference between acceptable or unacceptable behavior—are sometimes referred to as "implicit" or "procedural" knowledge. It is knowledge of the world that is organized, for the most part, without the use of symbols. Most of us, for example, have a sense of what's dangerous or safe without necessarily thinking about it.

Implicit or procedural knowledge, however, has not been sufficiently conceptualized or systematized. What are the different types? Are there different levels of such knowledge?

The four presymbolic levels of emotional interchange we have described in earlier chapters may constitute a useful framework for conceptualizing and systematizing procedural or implicit knowledge. These include shared attention and regulation, engagement, two-way purposeful communication, and complex, co-regulated, emotional problem-solving interactions and the formation of a sense of self.

Procedural knowledge can, therefore, become quite complex. Before being able to form symbols, a toddler develops a sense of self, forms expectations, and experiences emotional polarities. Well before symbols dominate the horizon, toddlers and preschoolers have developed an understanding of the basic themes of life, the nature of closeness and dependency, the scope of acceptable assertiveness and aggression, the behavioral patterns that lead to approval versus disapproval, and the boundaries of safety and danger. We propose, therefore, that the procedural or implicit knowledge that results from complex emotional interchanges between infants, toddlers, and caregivers leads to the stages of development that enable a child to separate perception from action and to form symbols. This in turn prepares the child for her entry into language proper.

THE EXPERIENTIAL NATURE OF EARLY LANGUAGE LEARNING

Preverbal gestural, social, and problem-solving experience precede the creation of meaningful verbal symbols. Without this basic level of *knowing through doing,* the child has no experience to draw on to render the use of words truly meaningful, when this becomes phonologically possible. We see the full force of this point in children with significant language impairments, who, through intensive behavioral therapy, can be trained to say a certain number of words to obtain a selected object or goal, and even to generalize these words to some extent. But the meaning of these words is fundamentally tied to their instrumental uses; the words have no connection with any lived emotional experience. For a child developing typically, however, a word such as "peach" is bound up with the sweet delicious fruit whose juice dribbles down her chin. Similarly, the child knows what "love" means through hugs, cuddles, and flirtatious glances; the meaning of the word is inextricably bound up with her loving experiences with her caregiver. So too with the words "open," "up," or "door," which are all grounded in their experiential contexts.

Consider the following "representative list of early words" for children learning English:[15]

juice	mama	all gone
cookie	dada	more
baby	doggie	no
bye-bye	kitty	up
ball	that	shoe
nose	hat	

If one looks at this list in isolation from the context in which these words are acquired, one might find it possible to treat this monumental step that the child is taking into language either as a mechanical process (related to the child's evolving phonological skills) or as a strictly cognitive process (reflected in the grammatical classification of the child's acquisition of the fifteen concepts involved here). But consider how these first words are acquired. For example, consider the child who lights up with glee when her caregiver says, "Would you like to go outside?" and is then pulled from the door handle with "First we have to put your shoes on, sweetheart"; "You know you can't go outside until you have your shoes on"; "Now we have to put your other shoe on," and so on. The child learns how to say "shoe" in the context of her excitement at being able to go outside and explore a whole new world. Otherwise, the incentive to learn the word is slight. Over time, the child may learn to dissociate "shoe" from this vivid emotional context and treat the word as what linguists call a "category term": the name of a class of objects that are defined by such-and-such features. (Or perhaps not? Perhaps this is one of the reasons why some adults continue to experience such a feeling of exhilaration when they shop for new shoes?) But to begin with, the *meaning* of "shoe" cannot be divorced from this or similar vivid interpersonal emotional experiences.

Every one of the words on the above list will have similar stories associated with it. Think about something as simple as a child's saying the word "hi" to a person she meets. How do we teach that? Do we say, "We say hello to first-degree relatives and everyone who lives within a quarter mile of the house"? Or does the child associate the word "hi" with a warm, friendly feeling in her gut? Then with people who invoke that feeling and with whom they feel comfortable, she waves and says, "Hi." When a relative comes over who's a little cold, the child hides behind mother's legs. Is it the emotional cue or the memorized script that children rely on when they learn how to use the word "hi"? Clearly, it's the cue.

The same point applies to the child's early word combinations and growing mastery of grammar. That is certainly not to say that we cannot identify important semantic-syntactic rules or significant phonological patterns in these early word combinations.[16] But underlying the learning of these possible rules and patterns is the child's advance through the first four stages of functional/emotional development.

We came to see the full significance of this point in our work with children with autism who had not progressed beyond the fourth stage of f/e development. For example, they would say, "Door, door, door" or "Open, open, open . . ." while standing before a door, indicating that they wanted to go out. There was no connection between the nouns and the verbs. We hypothesized that to link the nouns and verbs they needed to connect a strong emotion they were feeling with the action they wanted to perform. When a child said, "Open, open, open . . ." we got stuck behind the door. Now the children had to get rid of us to continue their activity. The children pushed us away and we, in turn, would try to establish some back-and-forth communication. We'd go away a step and come back, saying, "Go, go?" and the child would give some purposeful gestures as if to say, "Go! Get out of here! I want to open and close that door." Eventually, after some time, we would say, "Go away? Go away?" and the child would make gestures to that effect. Once the child began purposeful gestures instead of random and repetitive ones, she was more intentional and emotions could direct his behavior. We helped her make this connection by being a pain in the neck (but a gentle, playful pain in the neck).

Once children had this emotion and began connecting it to the word "go," guess what they did? They began using that verb and noun properly. They said, "Go away!" "Mommy, go!" and "Dad, stop!" Eventually, it was "Leave me alone." Within months, we would have meaningful sentences from children who had never before used language meaningfully. In fact, using this technique with the kids with echolalia (they already could say the words) led to a rapid use of meaningful language. It takes longer with children who can't form words because of motor-planning problems. The kids who can already say the words do well quickly. We have to get the affect system hooked up. Neurologically, they can speak. The words are simply not hooked up to the emotions; but that connection, from our experience, is not very difficult to establish once the child has begun to use words, especially in young children.

Our current theory of autism is that its basis lies in the biological problems that interfere with the children's ability to make this connection; that is, naturally hooking up the affect system to their sequencing system (motor planning), as well as emerging symbols. This is why the child who has some motor skills or has some ability to repeat words she hears begins rote and repetitive actions or words. These children line up their toys endlessly or repeat words because their motor skills or words don't become invested with meaningful emotional experience.

Once kids can connect their own emotions to their sequencing ability and, eventually, to their symbols, they become motivated to use language meaningfully. Gradually (children vary in pace) their language becomes more meaningful and socially appropriate. They tend to decrease or stop echoing, lining up their toys, and generally being perseverative. Depending on the individual children and the environmental and neurological problems, they may or may not retain some autistic traits. However, they tend to communicate more meaningfully and flexibly than would be possible by using the older methods, which do not involve emotion.

Thus, our work with autistic children constitutes further important evidence that contradicts a prewired language module. This work shows how specific experiences need to be present for language development, including how different parts of speech are aligned.

The meanings and the uses of words—the semantic and pragmatic aspects—are embedded in the gestural interactions that are used to explore and know the world. The ability to form a word is linked to what is already partially known through this exploration. A known object takes on additional meaning through context and further emotional experience with it. Therefore, the meanings of words emerge from emotional exchanges that provide the foundation and context for their use.

One might still argue that a child's ability to combine symbols in a systematic manner is genetically determined. However, as we can see in the example above, the ability to form grammatical connections also stems from emotional experience. That is, it was only when we created powerful emotional experiences (such as getting stuck behind the door) that the child began to say such things as "Go away from door." A child's growing ability to speak in sentences has to do with her expanding experience. *Various levels of intensity in emotional experience help a child define and categorize her thoughts and actions with the myriad emotional nuances that accompany them* (e.g., "happy mommy" or "yummy dessert"). Once we recognized the significance of this point, we began to develop a very different set of "language milestones" from what one finds in standard mechanistic theories of language development.

A NEW SET OF "LANGUAGE MILESTONES"

Generativists stress how nearly all children develop the same language skills at virtually the same time in the same invariant sequence. Thus, children are said to "acquire language like clockwork. Whether a baby is born in Stockholm, Tokyo, Zimbabwe or Seattle, at 3 months of age, a typically developing infant will coo. At about 7 months the baby will babble. By their first birthday, infants will have produced their first words, and by 18 months, 2-word combinations. Children of all cultures know enough about language to carry on an intricate conversation by 3 years of age."[17] The reason we see this regularity is said to be because the development of language is "under maturational control"; that is, neurobiologically determined by "unitary timing constraints."[18]

This argument presents a striking example of how, by narrowing one's vision, one can find data to validate one's theory. One thing that is missing here is the considerable variability that children display in the development of their linguistic skills: not simply in the age at which they master the various milestones listed above, but also in regards to which of the above steps they actually go through.[19] It is not uncommon for a child to go directly from babbling to multiword utterances, or for a child to "lose" a word or a construction that she seemed to have mastered. Also, in a significant number of cases, a child who had appeared to be developing normal linguistic skills may suddenly, between the ages of two and three, revert to a nonverbal state (which is often the parents' first indication that their child is suffering from a pervasive developmental disorder).

The real problem with this argument, however, is that it overlooks the preliminary stages of development that are vital to a child's language development. As we said earlier, before a child can develop language skills, she must first attend to the external world, and indeed, want to attend to the external world (first stage). She then has to become more interested in human interactions, and to become emotionally invested in interactions with her primary caregivers (second stage). She then has to start engaging in emotional interactions (third stage). Once the child can open and close small circles of communication, she can now advance to much longer and more complex problem-solving interactions (fourth stage). It is only when she has mastered these first four stages of f/e development that the child can separate action from perception and begin to master the use of symbols and words.

THE RHYTHMS OF LANGUAGE

Vocal and motor interactions between caregiver and toddler exist in various rhythms, rapid exchanges, slow exchanges, simple rhythms, novel rhythms, and so forth. We all get bored and tune out when we hear a slow, even, monotone, but we perk up and listen to a rapid, novel, changing rhythm, particularly when these are in tune with the meanings of the words being expressed (speeding up or slowing down to emphasize an emotional point). As infants and toddlers are identifying and using vocal patterns, there is an intimate relationship between what can be perceived (processed auditorially) and what can be articulated (a child hears sound patterns and tries to imitate them). A toddler or child who has a hard time planning and sequencing actions, and therefore with reciprocal emotional cueing, may have a relatively more difficult time engaging in and recognizing the rhythmic patterns while listening to or using expressive language.[20]

These rhythmic, interactive patterns are a vital dimension of language. Research on the temporal aspects of auditory processing and language has produced such intervention strategies as slowing the presentation of sound sequences according to temporal dimensions.[21] Another vital component of the temporal system, however, is the rhythmicity of sound sequences. Different perceiving patterns are likely. Too slow a presentation, for example, may make pattern recognition more difficult. Research in progress on the Interactive Metronome is revealing an optimal rhythm for sound perception and its relationship to rhythmic motor activity.[22] This research is also identifying differences in the degree to which individuals with different processing challenges can perceive these patterns.

Interest in rhythmicity and timing goes back centuries to the philosophical arguments of Aquinas, Newton, Leibnitz, and Kant about the nature of time and how humans come to perceive it.[23] Interestingly, fetuses and newborns alike perceive time and estimate the duration of events.[24] Newborns are able to discriminate the speech rhythms of different cultures.[25]

Our own research shows that there is an intimate relationship between timing, rhythmicity, and synchrony on the one hand and the ability to plan and sequence actions and solve problems on the other. Through reciprocal emotional interactions with her caregivers, the child continues to be involved in timed and rhythmic communication. Back-and-forth communications, which build on the capacity for timing and rhythmicity, in turn, give rise to these higher-level planning and sequencing capacities, ultimately involving complex reflective thought and verbal communication.

There are various steps to the process by which rhythmicity and timing influence higher-level cognitive and academic abilities. First, timing and rhythmicity are necessary for learning to sequence and plan actions (to be purposeful with one's environment). This ability for motor planning and sequencing is a foundation skill in human development that begins in early infancy. The infant actually begins to learn to interact with her environment in a rhythmic and timed way in utero, when we can observe her responding to sounds and to types of touch and movement. She is also hearing her mother's heartbeat and sensing her breathing rhythms. Almost immediately after birth, newborns will respond to sights and sounds and begin patterns of looking and listening. These early interactions with the world are organized in time and rhythmic sequences, as are the infant's cycles of alertness, sleep, and wakefulness, as well as other biological rhythms (which are partially endogenous and partially influenced by the environment).

The infant uses her timed and rhythmic interactions with the world to begin two distinct patterns. One is to establish her synchrony with her caregivers, as we observe when babies in the early months of life are involved in shared looking, smiling, rhythmic motor activity, and a variety of emotional expressions with their caregivers. This capacity for synchrony, based on coordinated rhythmic activity, keeps on developing through life.

The second pattern, which builds on the first, is the growing ability to plan and sequence actions. Initially, this involves purposeful, back-and-forth exchanges with caregivers through simple smiles, looks, expressions, and other motor behaviors, such as reaching, taking, returning. Over time, this ability to plan and sequence is used to establish social relationships, and the infant and toddler learns complex problem solving, for example, taking a parent by the hand and showing her the desired food. Later on, toward the middle part of the second year, the toddler uses this ability to plan and sequence to investigate and solve problems in her environment on her own. She may search for a hidden toy or figure out where her mother is when visiting a new house. Eventually, this same ability to plan and sequence becomes the basis for forming symbols and sequencing ideas (saying things such as "I want juice now" or "Where is Daddy?"). It then becomes the basis for thinking (answering questions such as "Why do you want to go outside?" with "Because I want to play."). Subsequently, it is the basis for sequencing ideas together into a complex logical pattern, such as writing an essay or performing other higher-level cognitive and academic tasks. In just about all advanced thinking and problem solving, the ability to plan and sequence behavior and thought is at the foundation.[26] Throughout this process, there is an intimate relationship between timing, rhythmicity, synchrony and motor planning, and sequencing.

Recent research supports this model. For example, the rhythmicity and timing of infant-caregiver interactions correlates with early cognitive capacities.[27] School-age children with better timing and rhythmicity capacities have higher academic and cognitive capacities than those with weaker timing and rhythmicity capacities.[28] Problems with timing and rhythmicity are associated with difficulties with modulating attention, behavior, and motor performance, and improving this capacity appears to improve attention and academic skills such as reading.[29]

ENTERING THE FLOW OF CONVERSATION

As an illustration of how co-regulated emotional gesturing orchestrates the higher levels of the mind, consider the seemingly simple yet puzzling everyday phenomena of how we decide what to say. In a back-and-forth conversation with a close friend most of us can talk a blue streak. We're using words interactively at a very rapid rate. Surprisingly, during this process, if we're fortunate, we're being both creative and logical.

How do we figure out what to say? Do we think of all possible words, phrases, or sentences and then decide which ones to use? In a rapid, spontaneous conversation with a close friend, there seems to be very little of this type of prior thinking. The words just seem to flow. We posed this question to a student who was attempting to master the theory presented in this chapter. The student thought long and hard and then explained, "You just have a sense or feeling about what you want to say and the words sorta flow." We responded, "Precisely!" For most of us, if we're not in a highly self-conscious state or having a conversation in which we are consciously anticipating our arguments, the words just "sorta flow."

The question is, what enables the words to flow? We've noticed that children who are unable to engage in back-and-forth emotional gesturing are often unable to keep up a conversation. For example, children who can exchange only three or four emotional signals in a row tend to have short bursts or islands of verbal exchanges and then either beat to their own rhythms, become self-absorbed, or idiosyncratically switch to another topic. They have a hard time with long conversations, whether it's a pretend-play conversation, a discussion about school or peers, or their favorite foods.

Conversely, children who can easily engage in a continuous flow of emotional gesturing and signaling seem to have an easier time maintaining long conversations. In fact, we tend to see certain parallels between the themes or emotional areas where the emotional gesturing is the strongest and where the verbal dialogue is the strongest. For some children, this may be an area of particular interest, such as playing house, and for others, it may be certain types of feelings, such as intimacy and pleasure.

How does emotional gesturing help a child maintain the flow of verbal interaction? How do her emotions influence the words she uses? Most of us have had the experience of trying to come up with an idea, be it for a present to buy a friend or a topic for an essay. As we're trying to think, we may ask ourselves rhetorical questions such as "What does my friend like or enjoy?"; but there's another component that we can sometimes bring to consciousness, even though it operates largely at a preconscious or subconscious level. This involves what the student we referred to earlier called "a sense or feeling about what you want to say." For this discussion, consider how the emotional signaling going on between two people would support their spontaneous, creative, ongoing conversations.

As one party initiates the conversation with an emotional gesture (tone of voice, facial expression), the other party experiences both the emotional gesture and the verbal content. The two together stir an emotional reaction. This in turn leads to an emotional/verbal response. If they can maintain a continuous flow of such emotional gesturing through their vocal tones, body postures, and facial expressions, this rhythm can provide the substrate for the words each uses. In this way, two individuals remain tuned in to one another, responsive to one another, and stir each other to remain creative and logical in their conversations. The continuous flow of emotional signaling organizes and maintains the seemingly higher-level symbolic exchanges.

This model of language and communication has important implications, not only for understanding the emotional basis of higher-level symbolic and language skills, but also for helping children who face challenges. For example, children or adults who have difficulties with "word retrieval" (finding the word they want to say) can be helped by creating highly emotional states that are motivating but not overwhelming. These states, we have found, enable the individual to find the word more easily. Over time, using heightened affect states enables the individual to retrieve words more readily, even in more typical exchanges.

Another important aspect of conversation that needs to be emphasized is the role of emotional gesturing in simultaneously supporting the creative and the logical and reality-based aspects of language. The continuous flow of emotional gestures provides a constant source of new emotions that can stir the next sequence of ideas or words. In this way, a conversation with a good friend is a shared creative enterprise. That's what makes it so much fun to just hang out and chit-chat. Because some individuals are less comfortable or gifted at expressing a full range of emotional gestures, their conversations often appear more formal and planned. Again this dynamic has been useful in therapeutic work. With children suffering from Asperger's syndrome, for example, we engage them in pleasurable emotional interactions containing more and more novelty and surprise, gradually accentuated, so that they feel secure while experimenting with new emotional exchanges. As we help a child laugh and giggle, and experience a wider range of emotions from coyness and flirtation to curiosity and mild annoyance and assertiveness, we often observe that their spontaneous verbal exchanges become much more creative and humorous. As they become more creative, they also become more capable of making inferences and engaging in higher-level abstract thinking because these also depend on generating new ideas. The key in this process is a caregiver or clinician who can challenge a child gently and gradually to experiment with a continuous flow of a broader range of emotional signals. We have found that direct work with words and concepts divorced from the world of emotional gesturing does not work nearly as well, and sometimes it's counterproductive because it leads the child more into scripts than into spontaneous and creative exchanges.

Just as emotional communication facilitates the creative use of language, it also facilitates the logical and reality-based use of language. Reality, for a child, is established through being able to relate to someone outside of one's self in a meaningful way. It's the ability to communicate across the frontier of one's own psyche to someone else's psyche that establishes a psychological boundary between one's self and another. This enables individuals to determine what's subjective and inside themselves and what's objective and outside themselves.

This process, however, is not a one-time event. The child must constantly sample and communicate with what's outside herself. She does this through a continuous flow of back-and-forth emotional signaling that uses all the sensory and motor capacities at her disposal to communicate a full range of emotional themes, from dependency and love to anger and aggression. As the child does this, she is also constantly sampling the reality of another person. As she responds to that person's gestures with gestures of her own, she is also building bridges between their emotions and her emotions, and between their associated words and her words. The dialogue is now not only creative, but logical and reality-based because it involves communicating with someone outside herself.

Reciprocal gesturing and motor planning and sequencing, particularly its rhythmic dimensions, may also be especially important for oral-motor sequencing capacities and overcoming oral-motor dyspraxia (children who find it hard to learn to speak), as well as stuttering and lack of expressive intonation. Improving reciprocal affective gesturing and the related capacities for motor planning and sequencing may, therefore, contribute in various ways to the processes that support language development. These may range from the basics of auditory perception and imitative production of sounds and words to symbol formation and meaning.

THE APPEAL OF MECHANISTIC VIEWS OF LANGUAGE ACQUISITION

To understand why Chomsky's biological theory of language acquisition had the appeal that it did in the 1960s and '70s, we first need to understand the reasons Chomsky himself was so drawn to a mechanistic model. In this respect, it is important to see how much his ideas were influenced by the climate that gave birth to Artificial Intelligence,[30] and, perhaps even more important, the influence of his teacher at the University of Pennsylvania, Zelig Harris.

In the late 1940s, Harris was one of the leading figures in the school of linguistics called Descriptivism. Harris insisted that linguistics has two primary tasks:

1. To establish the hierarchical structure of sentences
2. To sort the units of this hierarchy into classes with equivalent distributions

According to Harris, the study of linguistic form—that is, grammar—should be kept completely separate from the study of semantics, or meaning. Linguistics, according to Harris, should be concerned only with the discovery of formal patterns.

Language, in this view, can be treated as a formal system governed by a set of implicit rules that the linguist strives to discover by studying the speech patterns of native speakers. This view of linguistics was strongly influenced by the striking developments that were occurring in a branch of mathematics called "recursive function theory." In 1937, Alan Turing, the great English mathematician, showed that a "formal system" can be defined as any mechanical procedure for producing formulas. Turing's definition of a "mechanical procedure" was to become one of the driving forces in the creation of computers and the science of Artificial Intelligence.[31] The impact that these developments had on Harris is that he saw the linguist as trying, like the mathematician, to develop a formal deductive system complete with axiomatically defined initial elements and theorems concerning the relations among them.

In Harris's highly mathematized vision of linguistics, the two tests that a system must meet are, first, it should be able to generate utterances that a native speaker of the language will regard as well formed; and second, the system must be able to analyze novel utterances. To do this, the linguist studies distributional regularities in speech. Morphemes—the smallest units of meaning in a language—are classified into groups that resemble one another with respect to their distribution. For example, by studying speakers' reactions to utterances, the linguist establishes that *cat, dog, book, chair* can each occur in the frame

The _____ is on the mat.

Or, to take another example, *good, bad, nice, clean* can each occur in the frame

The _____ boy.

These two examples present us with the members of two different "form-classes." Because the first form-class seems to be approximately the same as what we call nouns, this can be symbolized with *N*; and as the second seems to correspond to adjectives, this can be symbolized with *A*. The next thing we discover is that *A N* phrases, such as *good boy,* have the same distribution for the frame

The _____ is on the mat.

This fact is recoded in the equation A N = N. But Harris made it clear that he was not trying to defend the traditional parts of speech (e.g., "noun," "verb"), which typically appealed to the logical analysis of the meaning of words to identify parts of speech. Even though "form-classes" might to a considerable extent coincide with traditional parts of speech, the two are not identical.

Chomsky's initial work was highly influenced by Harris. He adopted Harris's maxim that *meaning must be completely separated from grammar,* and that the goal of linguistics is "to formulate a general theory of linguistic structure in which such dry and abstract notions as 'phoneme in L,' 'phrase in L,' 'transformation in L' are defined for an arbitrary language L in terms of the physical and distributional properties of utterances of L and form properties of grammars of L."[32] Like Harris, Chomsky approached the task of constructing a grammar mathematically; namely, just as an equation generates a class of infinitely many well-formed formulae (wff), so too, according to Chomsky, we should treat a grammar as a set of rules generating all the wff sentences of a language. Meaning is not supposed to enter here at all. Hence the importance of Chomsky's famous example: *"Colourless green ideas sleep furiously"*; this was supposed to demonstrate that speakers will recognize this sentence as well formed, even though it makes no sense. Conversely, the grammar must not generate sentences—for example, "Furiously sleep ideas green colourless"—that speakers will recognize as ill-formed. If a grammar generates ill-formed sentences, it must be wrong.

In Chomsky's initial treatment of what he called "Phrase Structure Grammar," S stands for sentence, NP for Noun Phrase, VP for Verb Phrase, D for Determiner, Aux for Auxiliary verb, A for Adjective, N for noun, and V for verb. Chomsky introduced the "rewrite" sign "⟶" in place of Harris's "=" so that we would see the following formulae as rules for constructing sentences:

S	⟶	NP + VP
VP	⟶	Verb + NP
NP	⟶	Det + N
Verb	⟶	Aux + V
Det	⟶	*the, a, some . . .*
N	⟶	*man, dog, ball . . .*
Aux	⟶	*will, can . . .*
V	⟶	*hit, sit, catch . . .*

FIGURE 9.1 Chomsky's Phrase-Structure Grammar

This "phrase-structure grammar" generates a sentence such as *"The man will hit the ball,"* and it provides a structural description for every sentence that it generates. This structural description can be written either as a tree, or with brackets. The language generated by such a system is the set of all sequences that can be reached from the symbol S by following the rules and making a choice whenever one is offered.

Chomsky's next move in *Syntactic Structures* was the crucial one: He introduced the theme of *syntactic universals*. Chomsky's idea was that the infinite class of all possible grammars of the Harris/Chomsky type is well defined. We can define it as containing any finite set of rules, each of which is of the form "A." A set of rules that conforms to this definition is known as a "context-free phrase-structure grammar." Then Chomsky proved mathematically that there are well-defined classes of morpheme-sequences that can't be generated by any "context-free phrase-structure grammar." In other words, the class of "context-free phrase-structure grammars" is a subset of the class of all possible grammars or all possible languages. If one assumes, therefore, that "context-free phrase-structure grammar" is the appropriate tool for describing the syntax of human languages, then one is assuming that all human languages belong syntactically to a limited subset of the class of all possible grammars. In other words, there exist certain syntactic universals of human language; that is, language is NOT, as the Descriptivists had assumed, infinitely variable. Indeed, Chomsky argued that the Descriptivists were tacitly committed to this conclusion. Despite their overt commitment to unlimited linguistic diversity, they were tacitly committed to the existence of syntactic universals; that is, to the idea that every human language can be represented using the above kinds of rules for constructing a sentence.

The next thing Chomsky showed is that "context-free phrase-structure grammar" is actually inadequate to generate human languages. The problem is, there are certain sentences, called "constructional homynyms," that "context-free phrase-structure grammar" cannot explain without becoming hopelessly complex *("The shooting of the hunters").* According to Chomsky, this will have two different transformational derivations, either from the structure of *"The hunters shoot"* or *"They shoot the hunters."* Chomsky concluded that "context-free phrase-structure grammar" thus needs to be supplemented with "transformational grammar": transformational rules on phrase structures that we use to transform, for example, assertions into questions, or active sentences into passive sentences. In the above example, the underlying phrase *"the shooting of the hunters"* has two transformational derivations, one active, with "hunters" as the subject *("The hunters shoot")* and the other passive, with "hunters" as object *("They shoot the hunters").* In other words, we have to distinguish between the surface structure of the sentence, *"The shooting of the hunters,"* and its deep structure (namely, *"The hunters shoot"* and *"They shoot the hunters"*).

The point of formalization, according to Chomsky, is that once the linguist has discovered a grammar, he then needs to make it explicit and precise enough that it can be tested mechanically. And herein lies the key to the influence that Chomsky's early theories exerted on psycholinguistics: According to Chomsky, *someone learning a language is in exactly the same position as the linguist;* that is, the language learner has to "predict distributional regularities." Moreover, a language learner has to be able to produce and understand new utterances. So once the language learner has formulated a grammar, that language learner can be viewed as essentially a machine (à la Harris) that can "generate the sentences of her language." That is, the process that a child goes through in learning a language can, according to Chomsky, be likened to the analytic process that the linguist goes through when constructing a grammar. Thus, when a child produces a novel utterance, "this is of course a kind of prediction."[33]

In retrospect, it seems clear that the success of this argument lay in its tapping into the computational revolution that was transforming psychology at the time. Chomsky capitalized on computational ideas in his attacks on behaviorism, especially when he argued that "in principle it may be possible to study the problem of determining what the built-in structure of an information-processing (hypothesis-forming) system must be to enable it to arrive at the grammar of a language from the available date in the available time."[34] This so-called "steady state" argument cast the behaviorist in the role of a neophyte programmer who was struggling to model one of the most complex of all human behaviors by using a relatively crude "Markov chain" technique. Thus, it is hardly surprising that the new breed of computationalists, all of whom were familiar with Shannon's proof that a "brute force" approach could not be employed to construct even the most basic of chess programs,[35] should so readily have accepted the idea that the brain must be equipped with certain "super rules" that enable a child to acquire the myriad "surface" rules of the natural language to which she is exposed at birth.

As far as contemporary behaviorists were concerned, what Chomsky meant by "knowledge of language" was similar to what geneticists understood by "canalization." And, as Pinker[36] has shown, what Chomsky understood by "rule-following" was similar to what ethologists understood by "instinctive behavior." Thus, from a behaviorist perspective, Chomsky's early theory represented the convergence of the two primary advances taking place in behaviorism in the 1950s; namely, Artificial Intelligence and genetic determinism. From AI, Chomsky took the idea that the brain of the child must be preequipped with a basic language program for it to arrive at a grammar of a natural language from the limited (and degenerate) data that is presented to a child. And from genetic determinism he took the idea that a gene or genes contain the "blueprint" for the construction of this Language Acquisition Device. By marrying these themes, Chomsky presented an argument that played a crucial role in the transition from precomputational mechanism, which eschewed all talk of the mind, to the postcomputational view that higher mental processes, as well as reflexes and conditioned behaviors, could be explained in mechanist terms.

The essence of Chomsky's contribution to this transition from classical behaviorism to cognitive science was his idea that language acquisition is a maturational process, as this had been defined by Gesell; namely, one in which a child's development is directed by internal factors (genes) and always unfolds in a fixed sequence.[37] Hence, we find the constant refrain in Chomsky's and subsequent generativist writings that a child acquires language in a fixed sequence that is under genetic control. In this maturational view of language acquisition, gene-environment interaction amounts to a form of potentiation: "Language acquisition is a matter of growth and maturation of relatively fixed capacities, under appropriate external conditions. The form of the language that is acquired is largely determined by internal factors."[38] That is, the child must be exposed to the "right" kind of environment (whatever that is) if all the information stored in the "language gene(s)" is to be activated (or, as Chomsky would later put it, for the "parameters" of any particular natural language to be set). Herein lies the reason why generativism so swiftly assimilated the idea that "the functioning of the language capacity is . . . optimal at a certain 'critical period' on intellectual development."[39] For on the mechanistic view of language acquisition, it seemed straightforward to assume that the linguistic "information" that is "encoded" in the genes can be released only at specific junctures in the maturational process, the timing of which is itself directed by the genes.

THE FUNCTIONAL/EMOTIONAL VIEW OF INTERACTIONISM

Chomsky's biological model of language acquisition was countered by the interactionist thinkers mentioned at the beginning of this chapter. Jerome Bruner in particular showed that his model overlooks the role of caregiver-infant interactions of a child's language development.[40] Interactionism highlights how exchanges between infants and caregivers are initially carried out with sounds and gaze to establish and maintain joint attention. These are then supplemented by natural gestures and then sounds to initiate or coordinate routines between the two. The infant starts to look where her caregiver wants and starts to attend to objects and situations. Soon after this, the infant begins to use gaze and gestures to direct her caregiver's attention, and then to use gestures and conventionalized sounds to initiate exchanges. The child begins to use words in place of vocalizations and/or gestures. Remember that the child is trying to get her ball, to stand up, to be fed, to attract attention, to go to point B; and she is exploring means or soliciting help for attaining these ends. The gradual development of language skills is integral to the child's growing ability to satisfy her needs or expectations and to express her desires or intentions.

Our model of language development introduces a critical further dimension to this argument by clarifying the critical role of emotions in this interactional process of language development. Furthermore, our model shows how affective gesturing is the glue that binds biological and communicative processes together in the development of language skills.

We explained at the outset of this chapter that for a child to develop language, she must possess a biologically endowed nervous system with a few rudimentary capacities and lots of potential. At each step of development, this potential requires highly specific kinds of emotional interactions, which in turn lead to new abilities and new potentials. Indeed, recent evidence shows that when the human nervous system is engaged in this type of emotionally mediated growth, it's also likely laying down new pathways; this happens as a result not of a genetic blueprint but of some fundamental capacities that are now being defined and redefined and organized and reorganized at new levels through interactions with the emotional world.

This argument lies at the heart of a debate that has galvanized the psycholinguistic community. In 1996, Saffran, Aslin, and Newport published a groundbreaking article in *Science* reporting that eight-month-old infants can extract word-like strings of phonemes from the statistical properties of the input; that is, they can distinguish between syllables that regularly hang together from those that are randomly juxtaposed.[41] Here, as Bates and Elman argued in their commentary following the article, was clear evidence that young infants can identify the structure of language simply by being exposed to it.[42]

Not surprisingly, generativists responded en masse that Bates and Elman had confused the mechanisms involved in word learning with those involved in grammar acquisition; that is, that language as such cannot be learned simply on the basis of its statistical properties. The problem raised by generativist critics here is essentially this: How does the infant bridge the gap between pattern recognition per se and as complex a phenomenon as language? As a pattern recognizer, the baby is able to recognize patterns in all sensory areas. These patterns can be visual, spatial, auditory, tactile, motor. Obviously the auditory-vocal experiences for a child hearing and making sounds are critical for language development. But how does something as basic as recognizing patterns in the sounds that one hears and makes become the basis for meaningful language? How does a child transform her basic recognition of and expression of patterns, some of which are instrumentally learned, into a meaningful system of rapid presymbolic and symbolic communication at various levels of meaning and reflection?

It's this latter question that lies at the heart of the debate over the Saffran et al. findings. This question highlights meaningful language as far more than the simple generalization of instrumental behavior. After all, various levels in shades of meaning and reflection would appear to transcend any known mechanisms of learning. We have demonstrated, however, that there are several intervening steps that occur in the early months of life that lead a baby from simple capacities for pattern recognition to a complex symbolic system. Specifically, we identified (see Chapter 2) four presymbolic levels and two early symbolic levels that are critical in this transformation from pattern recognition to full language. It is therefore experience that is the critical factor, but it is a specific type of experience: It is affective interactive experience. But it is more than simply such experience; it is six levels of interactive emotional experience that, as we saw, have taken millions of years to evolve and are now part of each baby's experience in the first two to three years of life.

The full development of language also requires the integration of auditory and vocal patterns with visual-spatial motor and other sensory processing patterns (olfactory, tactile). This is because language, and even presymbolic gestural systems, are bound up with the full range of one's sensory experiences. For example, when one talks about mother or, for that matter, an apple, under normal development one can feel that mother or, if one so chooses, visualize and almost taste, smell, touch that apple. Hence the symbol is bound up with a full and rich multisensory affective experience. This is what we mean by "knowing something," whether it's knowing mother or the apple.

These six basic levels of affective experiences that characterize the pathway to language development not only transform sound perception and vocalization patterns into meaningful language but also, as we described earlier (see Chapters 1 and 2), enable the child to integrate all her experiences—auditory, visual—into integrated patterns at the presymbolic and symbolic level. Looking at the totality of these experiences affords a very different perspective, therefore, of the multiple factors involved in language development.

—10—

Emotions and the Development of Intelligence

OUR FOCUS ON AFFECTIVE or emotional transformation has led not only to a new way of looking at the growth of intelligence but also to a new way of thinking about intelligence and the pathways to it. From our developmental point of view outlined in Chapter 2, intelligence is the progressive transformation of our emotions from global reaction to sensations to high-level reflective thinking. The early stages that we described, dealing with co-regulated emotional interactions leading to symbols, are the cornerstones of this process. This focus on emotional transformation redefines what we mean by intelligence.

A DEVELOPMENTAL MODEL OF INTELLIGENCE

In theory and practice, we have tended to underfocus on the emotional generative aspect of intelligence, the creation of the intent and ideas. In addition, we focused more on putting intentions and new ideas into an analytical frame of reference without realizing that our very capacity to construct a logical frame of reference was itself a product of emotional interactions.

Most modern cognitive theorists following a long tradition in education and, to some degree, Piaget, have focused primarily on the impersonal and analytic rather than the emotional generative or truly reflective thinking aspects of intelligence. The emphasis has been on putting intentions and new ideas into an analytical frame of reference. For example, schools place enormous emphasis on teaching children to organize and sequence their ideas. Children are expected somehow intuitively to come up with the ideas they are analyzing and framing or to judge the relative degrees of importance of the ideas they are sequencing.

As we have mentioned before, we were first alerted to the importance of the generative aspects of thinking and intelligence in our observations of children with and without developmental challenges. When we observed and talked to children with strong self-awareness and reflective thinking skills, we realized that most of them also showed positive self-esteem, demonstrated a capacity for moral judgment, were analytical in their reasoning, and did well in school and with their peers. We sought to understand what helped them become this way, and, therefore, spoke with them and also with children who exhibited opposite personal characteristics. We learned that the traits we commonly label as intelligence, social skills, and morality were based on the child's ability to use his emotions to think out problems in various areas.

For example, when we asked a group of eight-year-olds abstract questions, such as what they thought about justice or fairness, their comments were revealing. Some of the children responded with a rote listing of people who behaved "fairly," such as a particular parent or teacher or television character. However, others gave far more reflective answers, for example, "Well, when I hit my brother after he hit me. It was unfair for me to be punished, but when I hit him first it was fair for me to get a punishment. If I bump into him by accident, it's not fair to be punished, but if I do it on purpose, it is fair."

Not surprisingly, when we looked at the two groups of children more closely, those who gave us the rote list tended to be the ones who were experiencing more problems in their relationships and in their schoolwork. The children who gave us more creative and reflective responses tended to do better in these social and intellectual areas.

We then took a second look at the more reflective responses and discovered that they had two components. This was true whether our test question focused on fairness or any other abstract quality, such as honesty, friendship, or freedom. The first component was that the children's responses always started off with a personal anecdote, an account of *lived emotional experience.* The second component was that the children put these experiences with abstract concepts into some sort of analytic framework and context.

When we later asked this same question of adolescents, they were able to list more categories (five different types of fairness, for example) and supplied an even more worldly-wise analytical framework. But in every instance, and at every age, this lived emotional experience was evident in the more sophisticated replies. The children who didn't have a lot of lived emotional experiences—due either to caregiver patterns or to biological challenges that interfered with interaction—tended to be the ones who responded with concrete lists. We also observed that even children with severe developmental problems, including autistic patterns, could become more creative and reflective when they were exposed to more one-on-one, affectively rich, and progressively more challenging affective interactions with their caregivers, which systematically led to these needed experiences.

To help children with limitations, we discovered that we have to mobilize their emotions and generative abilities to help them learn to create ideas and to become emotionally intentional and interactive. When we created or used natural situations of strong affect to generate intent and ideas, these children often became creative, logical, and reflective.[1] Among very competent adults, those who combine both generative and analytic thinking often make the most original contributions to their fields.

Because generative ideas emerge from emotions and intentions, the historic dichotomy separating reason and emotion may be partly responsible for the lack of emphasis on the generative part of thinking. In the developmental model presented in this work, we attempt to redress this oversight and give proper weight to the generative component of intelligence. The logical, analytical aspect of thinking, however, also stems from emotional interactions, as we showed in Chapter 2. After all, logic and a sense of reality come from reciprocal emotional interactions. These interactions establish the boundary between what's inside me and subjective and what's outside me and more objective; that is, the ability to separate fantasy from reality.

The ability to create ideas from one's experience and to reflect upon or understand those ideas in a broader context or under logical scrutiny operates best in areas where one's actual emotionally meaningful experiences are extensive. Intelligence, considered to be a high level of cognitive ability, is so often contrasted with talent, usually defined as outstanding facility in an expressive field. However, highly accomplished musicians, writers, and visual artists can be every bit as intelligent in their fields of expertise—that is to say, as capable of understanding and reflecting on music or poetry or painting—as a brilliant mathematician is in mathematics. Regardless of the field, fine differentiation and a grasp of relationships are the essence of intelligence. A person who is both intelligent and talented is skillful in some—and sometimes in many—areas of endeavor. Intelligence, however, goes beyond talent in that it involves a systematic understanding of why and how things work—of how certain colors react with others, or why a particular equation describes a phenomenon, or why a given note produces the desired emotional tone.

Nearly every field of human endeavor is thus susceptible to intelligence, although standard IQ tests do not reflect skill in many of these areas. Some, like higher mathematics, law, or philosophy, involve ability for symbolic abstraction. Engineering and science permit the exploration of extremely complex relationships. Literature, music, and the visual and performing arts allow exquisite subtlety of emotional expression. Many other fields not generally considered "intellectual" or "creative"—everyday endeavors such as carpentry or child care or gardening, for example—can nonetheless draw considerable levels of intelligence from expert practitioners.

To tap into high levels of intelligence, intelligence testers generally try to concentrate on cognitive skills in certain highly symbolic fields. High intelligence is thus equated with the ability to do well at manipulating words, numbers, or shapes. Over the years, testers have built up a huge body of data about a grab-bag of skills; indeed, the traditional tests stay in favor because of this useful history and its accumulated database. The tests are used not because they reflect the latest thinking on true intelligence and its relationship to the skills measured but because more sophisticated yet quantitative measures of intelligence are very difficult to design.

Individuals who possess great skills in narrow and often highly symbolic fields often show great ineptitude in areas involving judgment, personal relationships, and aesthetics. Because these individuals have only a narrowly based intelligence, they embody neither the full range nor the highest levels of intelligence and are therefore limited in the array of abilities that a humane and civilized society should encourage and extol.

DEFINING INTELLIGENCE

Prevailing definitions of intelligence continue to evolve. The theorists Howard Gardner and Robert Sternberg have shown that people can possess multiple forms of intelligence—musical, kinesthetic, and social. Their concepts are much more nuanced than definitions based on traditional IQ.[2] There is a need, however, to develop our understanding of intelligence further so that we can elucidate the nature of intelligence in whatever field it is exercised. What are the critical mechanisms and processes that lead to and constitute intelligence in its most basic form? The literature on the emotional and social roots of cognitive capacities, although supporting a relationship between emotions and cognition, has not identified the critical developmental pathways or processes that unite them.[3] In this book and other works, we have explored these relationships as a basis for a new framework and definition of intelligence.[4]

The ability to generate or create a full range of ideas in the areas of one's human emotional experiences and then reflect on them and organize them into a logical framework is, we believe, an appropriate definition of intelligence. Both this creative ability and reflective and sequential reasoning are necessary ingredients of intelligence and both are developed through emotional or affective experience. Rather than measuring intelligence with one cognitive yardstick, we must find ways to evaluate it according to its depth and breadth.

For instance, an extensive and accurate vocabulary is a *sine qua non* of any writer or lawyer, as is outstanding manual coordination for a surgeon or sculptor, or excellent skills of spatial analysis for an architect or fighter pilot. But these abilities are only preconditions for developing high intelligence in those fields, not intelligence itself. They will become part of deep intelligence only when a person uses them to gain knowledge and experience in a particular field and then operates creatively and reflectively in that field. According to our definition, therefore, a "gifted" dilettante will never, no matter how high his IQ scores, achieve a deep level of intelligence. Only deep and extensive knowledge based on extensive personal emotional experience and learning can generate high levels of creative and reflective thinking for that field.

Depending on the degree to which the emotional and intellectual transformations we have been describing occur, we can observe varying degrees of breadth, creativity, and reflectiveness as well as varying degrees of limitations or constrictions. When there are compromises, we may see significant deviations in the levels of emotional health and intelligence growth attained. These can include

- magical and irrational thinking;
- impulsivity rather than reflectiveness;
- polarized (all-or-nothing thinking) rather than integrated reasoning;
- rigid concrete thinking (where only a few possibilities are held onto) rather than broad-based reflection;
- thinking restricted to considering only the "here and now" rather than thinking that takes the future into account;
- various limitations in the breadth and scope of possibilities considered or analytic principles used;
- thinking constricted to any limited or narrow domain.

THE FUNCTIONAL/EMOTIONAL APPROACH TO INTELLIGENCE

We have been discussing the central role of emotions in helping to create the two capacities necessary for intelligence—(1) the capacity to create ideas and (2) the capacity to reflect on those ideas. We showed how emotions are essential for creating ideas because it's through our wishes or desires that we generate new ideas. We also showed how emotions are essential for reflective thinking because it's through the stages of emotional transformation described earlier that we develop the capacity to reflect on new ideas (to think).

Emotions also enable us to use each of our cognitive or "processing" capacities in a functional, meaningful (intelligent) manner. For example, consider something as simple as an eight-month-old who is reaching for his father's nose to get him to make a "toot-toot" sound. This requires emotional intent, the wish to reach for that nose. Without such an emotionally fueled desire, a child would have no reason to employ his motor skills in purposeful action. In other words, the early sense of intentionality or causality stems from an infant's emotional intents. Similarly, the desire to vocalize back when his mother speaks or sings stems from a relationship that embodies the desire to have fun, make sounds, and receive another sound back. Similarly, searching for a hidden object to learn about physical space is also based on the emotional interest and intent to discover where that rattle went when it fell to the floor. Such activities are often the sign of a new emotion, such as curiosity.

All the different aspects of human development that contribute to intelligence are integrated by the emotions. As we will see in Chapter 12, our emotions unite and integrate language, motor, perceptual motor, and visual-spatial processing capacities, as well as influence our ability to modulate sensations. Interactions that are emotionally meaningful usually involve using many developmental capacities at the same time. Not infrequently, a child will be reaching for Daddy's nose (motor patterns), vocalizing in delight (vocal language patterns), and responding to Daddy's vocalizations and encouragement with more vocalizations and reaching. If, at the same time, Daddy is moving his nose from one side to the other and the child keeps finding it and reaching for it, he is also adding on a visual-spatial and perceptual motor component. Under the guidance of his emotional intent within one interactive sequence, the child is practicing language, motor, and visual-spatial capacities all at the same time. He is getting them all to work together as an integrated "mental team." Such multicapacity interactions are the genesis of real intelligence.

Just as we earlier defined emotions as "functional/emotional interactions," we have defined intelligence as "functional/emotional intelligence." It is functional because it employs the abilities (such as language and motor skills) that children need to function in a useful manner to master one's world. It is emotional because the emotions provide the initial intent and, later, through their various transformations, the complex interactions that lead to higher and higher levels of creative and logical reflective thinking.

In Chapter 2, we demonstrated how emotional interactions progress through sixteen stages. Each stage transforms emotions and intelligence. We traced this process through different levels, from the catastrophic expression of intense emotions to higher levels of logical and reflective thinking.

At each level of emotional transformation, emotional affective interactions create a higher level of integration, or a coming together, of the different components, such as language, motor skills, perceptual motor capacities, and visual-spatial thinking. As indicated above, the components are thus harnessed by our emotions to function as an integrated "team" (like a well-rehearsed ballet company or a first-rate basketball team). Consider another example: A five-year-old engaged in pretend play might very well use lots of words to describe the actions of his characters (language); build pretend forts, castles, and new weapons systems (visual-spatial and motor); and answer questions about why one group of characters represent the good guys and the other group the bad guys (reflective thinking). The whole process may well stem from the excitement and pleasure of communicating and interacting with Daddy and conveying an inner fantasy, a personal emotional view about how good and evil interact in the world. In formulating and discussing a business plan with his boss, a young adult might very well be simultaneously using these same capacities—language, motor, visual-spatial, and reflective thinking.

As functional emotional intelligence is building to higher levels, it is also broadening. Individuals with lots of experience in solving problems of a social and emotional nature may achieve a high level of intelligence in these realms. A mathematician or physicist may achieve high levels of intelligence in symbolic and conceptual realms. Some areas of intellectual activity, such as literature and social science, require a fair degree of breadth of experience. Other endeavors can benefit from breadth of experience but may not require it to the same degree.

Origins of a Shared Sense of Reality

A critical component of functional intelligence is a strong sense of reality. In Chapter 2 we saw that opportunities for continuous emotional signaling are essential for establishing a sense of reality, reality testing (the ability to distinguish fantasy from reality), the capacity for organized thought (rather than fragmented thinking), and an organized sense of "self." Children without this experience have only episodic reality contacts with the world. Continuous emotional contact with the world outside us (with other people) occurs through ongoing emotional interactions. This process anchors us to external reality and to our inner world at the same time and enables us to tell the difference between them and separate the two.

It also leads to a shared sense of reality that makes joint endeavors with others possible. Even though there are differences in our nervous systems, there are many similarities in the way human beings process sensations such as sounds or sights. Human beings also have more experiences that are similar than one might think. They tend to share certain features associated with each of the stages outlined earlier in Chapter 2. For example, most people have relationships and interactions with others that promote purposeful emotional signaling and, therefore, a basic sense of "causality." As they negotiate with others, most people learn to sequence, to perceive patterns, to solve problems, and to use ideas. To some degree, most people interact with words in a way that supports some degree of the logical use of ideas. Although individuals differ in levels mastered, for the most part emotional interactions create *similar structures for thinking.* At the same time, the "content" of each person's logical ideas will differ. One person will think about food; another will think about anger. When individuals have significant dysfunction in the physical makeup of their nervous systems or in their families and the interaction patterns that undermine these basic structural aspects of thinking, we generally see a decreased sense of reality and a decrease in a shared sense of reality.

As we saw in Chapters 3 to 7, during the course of evolution, cultural practices gradually enabled prehumans and humans to master these structural features of thinking, thereby providing us with a shared sense of reality. In addition to our similar biologies, therefore, our similar emotional interaction, mediated by cultural practices dating back to nonhuman primates, maintains our capacity, in a relative sense, to see the world as others do and to solve problems together.

This shared sense of reality, in turn, supports our social and political institutions. As we will see in Part IV of this book, however, if too many individuals grow up with deficits in the way their central nervous systems process information, or with extremely dysfunctional family and interaction patterns, the "consensus" of what constitutes reality could easily slip away. The very constructive, near infinite variations in the "content" of our lives can only survive in the context of shared structures for thinking and problem solving.

Understanding Concepts

As an example of the role emotions play in intellectual development, consider the way children learn or "could" learn new concepts. To give a concrete example, one of us (S.I.G.) worked with a little girl who was verbal and could read, but was having a hard time with more abstract reasoning. Her mother gave many examples of things her daughter could not do. I asked whether or not the little girl would be able to understand a concept such as "taxes," and mother quickly responded, "No. That's exactly the type of thing she's having trouble with at school and at home. She doesn't seem to be able to understand abstract concepts like that." I made a bet with the mother that in five minutes we could enable this little girl to understand what taxes were.

I asked her whether she liked pizza. She said she did. I then said, "Okay, if we're going to order pizza, how many pieces should we have in the whole thing?" She said, "Oh, I like to have eight pieces in the pizza." She could also do some basic adding and subtracting, so I said, "Okay, let's pretend you have a pizza!" and we took a piece of paper and drew the pizza, cut into eight slices. I then said, "Now, your brother [who was present at our session] wants to steal all your pizza! I'm going to be a make-believe policeman and protect your pizza, but you have to give me something if I'm going to protect you and your pizza from your brother, who wants to steal it. How many pieces of the pizza will you give me to protect the whole thing?" She said, "I'll give you two pieces." I said, "Okay, I'll take the two pieces and I'm going to protect the rest of the pizza." Then I said to her, "You know, what you're giving me, the policeman [the two pieces of pizza], is called 'taxes.' That's what we give to policemen in the real world to protect us from somebody coming in and robbing our house." She seemed to grasp it. To see whether she really had, I said, "What other things would you be prepared to pay taxes for besides getting your pizza protected?" She said, "I would pay to have the streets cleaned"; and because she was aware of the prominence of the military on TV, she said, "I would also pay soldiers to protect us, too." When I asked how much of her pizza she would give for clean streets, she said, "One piece"; and for soldiers to protect us she said, "One piece." At the end of this brief exchange, the child's mother agreed that her daughter was now beginning to understand what taxes were—that taxes were what you paid the government for certain services, such as police, soldiers, or clean streets.

Now, of course, a child can't be quickly taught to abstract beyond his or her potential at a given moment, but it's hard to know the child's potential until we invest the particular learning task with heightened affect. The heightened affect provides the child with a personal stake and investment in thinking. The child will then apply the highest level of thinking he has available. When a child has a personal stake in the task, he can reason about that issue at a higher level than other issues where there isn't the personal stake. Protecting her pizza from her brother enabled all this child's available facilities to reason. These emotional stakes enable us all to understand certain concepts more quickly. Take the concept of "plot." If you ask a young child just learning to react whether he knows what a "plot" is in a story he is reading, he will have only a vague notion. However, if you begin asking, "What is your favorite TV show?" Once he names it, you can ask, "What's that one about?" The child will probably get right to the core of it. A natural question such as "Hey, what's that about?" is the beginning of an experience that underlies understanding the concept of "plot." At retelling with relish the action of some favorite shows and comparing them, it becomes easier for the child to understand what the word "plot" means. When he grasps that "plot" is simply another way of talking about what the story is about or what the author intended the story to be about, he has mastered a new concept. This approach we just described is derived from our understanding about how concepts are formed in the first place. Children, who have lots of affectively rich experiences in their families, can re-create their personal experiences to understand new concepts. If one asks a child how to solve a new problem in math or how he figured out what the author was saying in the novel, he will often reason it out through personal examples and general principles and then apply these to impersonal experiences. He will use two steps—a generative step, which is emotionally rich and creative, and an analytical step, which then makes sense out of this emotionally rich experience. Many children carry out these steps on their own. We are all amazed by the gifted teacher who can bring physics, math, or literature down to a visceral personal level. This person can be contrasted with a teacher who just works on formulas and has children simply memorize mathematical operations or word definitions without grasping exactly what the formulas or words are revealing about how the world works.

In summary, understanding concepts involves a sequence of steps that begins with emotional interactions. Applying this insight suggests an emphasis for education that may enable children to achieve levels of reflective thinking of which they may not seem capable. Although each will still have his own limitations, we will never know where those limitations are or are not until we explore the most emotionally rich ways of presenting new ideas. Once again, the road to high-level logical thinking is not through containing our emotions, but through their regulation, differentiation, and transformations during the course of development.

Implicit or Procedural Knowledge

Much of what we learn is outside conscious or symbolic awareness. Such presymbolic, or subsymbolic, processes—for example, a toddler learning how to relate to a particular caregiver or learning the difference between acceptable or unacceptable behavior—are sometimes referred to as "implicit" or "procedural" knowledge. It is knowledge of the world that is organized, for the most part, without the use of symbols. Most of us, for example, have a sense of what's dangerous or safe without necessarily thinking about it symbolically.

Implicit or procedural knowledge, however, has not been sufficiently conceptualized or systematized. What are the different types of procedural or implicit knowledge?[5] Are there different levels of procedural or implicit knowledge?

We would suggest that the four presymbolic levels of emotional interchange we have been describing in earlier chapters constitute a useful framework for conceptualizing and systematizing procedural or implicit knowledge. These include shared attention and regulation; engagement; two-way purposeful communication; complex, co-regulated, affective problem-solving interactions; and the formation of a sense of self.

Procedural knowledge can, therefore, become quite complex. For example, it involves the patterns of co-regulated emotional interaction that lead to a sense of self and enable the toddler to integrate emotional polarities as well as form expectations and conceptualize patterns of interaction. Much of the toddler's and preschooler's understanding of the basic themes of life—the nature of closeness and dependency, the scope of acceptable assertiveness and aggression, the behavioral patterns that lead to approval versus disapproval, and the boundaries of safety and danger—are all learned significantly before symbols dominate the horizon. Also, these implicit or procedural patterns make symbol formation possible because they involve co-regulated, reciprocal affective interchanges that enable the child to separate perception from action, tame and regulate catastrophic affect (that push for action), and form symbols.

We propose, therefore, that procedural or implicit knowledge be thought of as involving the increasingly complex emotional interchanges between infants, toddlers, and caregivers that lead to the negotiation and successful mastery of the four stages of presymbolic development we described in Chapter 2. It constitutes a much more important component of human development than is often realized.

Theory of Mind

There has been growing interest in how children form a "theory of mind," that is, the ability to understand another person's perspective.[6] A growing body of research has elevated this construct so that it is now considered an important cognitive ability. Understanding another person's perspective can involve a fairly straightforward perceptual challenge, such as being able to use one's own perceptions and experience to imagine another person's perception and experiences and predict where another person would search for a hidden object. At a more complex level, having a theory of mind may involve understanding how another person feels in a situation and how that person might feel differently from oneself because of his or her unique emotional experiences. We all have different degrees of empathy (different degrees to which we can understand other people's feelings).

How children develop a theory of mind and how they develop it at its highest level is a most interesting question. It is generally recognized that advanced cognitive capacities, such as taking someone else's perspective, have a developmental sequence and timetable (three-year-olds often only use their own perspective, whereas four-year-olds begin taking the perspective of others, at least with regard to simple perceptual tasks). Yet all the steps in this developmental sequence have not been sufficiently worked out.

We would propose that the origins of a theory of mind are identical to the origins of symbol formation. As toddlers experience the exchange of emotional signals between themselves and another person, they are getting a sense of someone else's emotions in relationship to their own emotions. For example, the toddler feels and expresses annoyance with a loud grumbling vocalization. Dad responds with a slightly more annoyed tone, "What do you want?" Through this type of interaction with Dad, the child experiences a pattern of mild annoyance leading to slightly greater annoyance in the other person. Assuming this pattern continues and the child is asked, "How do you think Daddy feels when you are annoyed with him and grumble at him?" the child might well say, "I think he feels annoyed, too, or even a little more annoyed." The child's ability to understand his father's feelings will be based on his experience of his father's emotions in relationship to his emotions. Without such primary exchanges of emotional signaling, the child would have no idea—other than perhaps through some form of deductive logic—how another person would feel in relationship to some behavior or emotional expression of his own. In other words, the ability to put oneself in another person's shoes emotionally is based on having some sense of how one's own emotional expression interacts with the emotions of another person. During the second six months of the first year of life and the second year of life and thereafter, emotional interchanges are providing the substrate for these types of insights.

Catastrophic emotions tend to disrupt the development of empathy rather than facilitate it. Because the catastrophic reactions lead to direct action, including immobilizing panic or withdrawal, they undermine co-regulated emotional exchanges and empathy. The intensity of the catastrophic emotions is often so severe that even the experience and recognition of a simple pattern, such as "I get angry and he gets angrier," may fail to materialize. To be able to conceptualize and reflect on the perspective of another person takes even more than just forming symbols of emotional patterns and being able to describe them verbally. Reflecting on the perspective of another person requires the child to form logical bridges between his emotions and ideas and the emotions and ideas of another person. It involves establishing reality testing and reflective thinking abilities, as described earlier. Reality testing, as we described, emerges from a boundary forming between the child's emotions and the emotions of another person. The child constructs this boundary by experiencing his own emotional intentionality in relationship to the emotional intentionality of another person. Little disagreements, power struggles, and debates, as well as negotiations around feelings involving pleasure, dependency, or assertiveness, help the child define this emotional boundary. The more parents and children negotiate their different emotional intents in the context of a supportive, calm, intimate relationship, the stronger this boundary becomes.

In general, constructing a theory of mind involves each of the levels of emotional exchange we have described. At the presymbolic levels, it involves the exchange of emotional signals. At the early symbolic levels, it involves the symbolization of interactive, emotional patterns and the forming of logical bridges between one's own emotional symbols and those of another person. At the more advanced levels, it involves reflections on one's own and other people's perspective and feelings.

The highest levels of empathy and emotional perspective—that is, the highest levels of theory of mind—involve our highest levels of emotional interaction. For example, to compare the feelings of oneself and another person involves multicausal and comparative thinking (Level 7). Comprehending subtle differences in degree, texture, or tone of feeling between oneself and someone else involves affectively differentiated gray-area thinking (Level 8). Being able to form a judgment about the appropriateness of one's own feelings and someone else's feelings in a given situation requires an even higher level. For example, deciding whether or not another person was angrier than expected in a certain situation and whether you were less angry than expected in that situation requires thinking from an internal standard and a sense of self, Level 9. To achieve even higher levels of empathy—for example, understanding what it feels like to have children or to make a lifelong commitment to a spouse—requires new life experiences that will emerge only during various later stages in the course of life. In short, the psychological and cognitive abilities involved in forming a theory of mind have their origins in the same processes that lead to symbol formation. Theory of mind reinforces an important principle that we have been stressing throughout this work. This is the principle that cognitive, or intellectual, skills cannot be thought of as simply impersonal problem-solving abilities. From their earliest manifestations in infancy to their advanced level in adulthood, they stem from complex emotional interchanges at each of the stages of development we have been describing.

Theory of mind has been presented as a mental ability in its own right, as opposed to simply being a part of a more general ability for reflective thinking or empathy. It has been presented as if it were a purely cognitive construct. What is not realized in both of these assumptions is that theory of mind is actually a derivative of the developmental capacities described earlier. There are different levels of the "theory of mind" that are related to or derivative from the different levels of f/e development. But just as intelligence can be either broad or constricted and may involve some processing capacities but not others, similarly, theory of mind capacities have different levels and are exercised in various ways.

For example, a child with very good visual-spatial capacity may be readily able to identify how another person would perceive a physical object from his vantage-point on the other side of the room. This same child, however, may be completely unable to anticipate how another person would feel in a given situation. The opposite also holds true. In other words, theory of mind tends to be a series of discrete cognitive and emotional skills geared to developmental levels and the different processing capacities organized at these levels, as well as the breadth of emotional experience (e.g., the child who can empathize with love but not with anger, or anger but not with love).

THE ROOTS AND BRANCHES OF INTELLIGENCE

Our developmental perspective on intelligence suggests that many traditional skills which have been called "intelligence," such as having a good vocabulary or memory or being able to match designs, may actually be only roots or branches of intelligence. For example, an excellent memory certainly helps one think and, therefore, it might be a useful root. It is not the only root, however, and in its own right does not constitute intelligence. Similarly, a memorized definition of a word or concept may be a useful branch on the intelligence tree.

In recent years, emotional intelligence, popularized by Daniel Goleman,[7] has been added to the spectrum of multiple intelligences, mentioned earlier. Emotional intelligence used in this way refers to an individual's social and interpersonal skills. Such skills, however, are still only one branch on our intelligence tree.

Concepts based on recent neuroscience research, for example the work of LeDoux and Damasio,[8] which we will discuss in the next chapter, also tend to describe different roots and branches of the intelligence tree. For example, they divide the subcortical emotional discharge of the passions or emotions from cortical reflective thinking processes. Our definition of intelligence, however, involves an integration of subcortical and cortical processes. The continuing separation of these processes most likely represents pathologic development. In contrast, the integration of these processes most likely represents adaptive, healthy development. A failure to make this distinction and take into account the transformations emotions go through has contributed to these erroneous notions of what constitutes intelligence and emotions.

True intelligence, what we are calling "functional/emotional intelligence," integrates all the different roots and branches of the "intelligence tree," including social and emotional skills, language skills, and visual-spatial thinking skills, as well as a range of cognitive and academic abilities. It integrates these skills into creative and reflective thinking. The roots of the intelligence tree include basic capacities such as vision, hearing, movement, memory, and selective features of perception. The branches of the intelligence tree include many of the different types of abilities described as intelligent behavior, such as social or relationship skills and mathematical or musical abilities. The trunk of the intelligence tree and what integrates the roots and branches together are the functional/emotional developmental capacities. As described in Chapter 2, these developmental capacities progress to higher and higher levels during the course of adaptive growth and, as they do, they define the individual's intelligence.

MODEL OF INTELLIGENCE THAT CAN BE APPLIED ACROSS SPECIES DURING THE COURSE OF EVOLUTION

Understanding the emotional roots of intelligence suggests a more universal model of intelligence that can be applied across species. This model would look at the levels of emotional transformation in terms of the stages outlined earlier. The foundation for intelligence would lie in the basic ability for emotional signaling involving increasingly differentiated interactions. There is thus a universal process in the development of intelligent behavior in different members of the animal kingdom. We suggest the following hypotheses:

- Intelligent behavior involves the degree to which an animal (or group of animals), including humans, can exchange differentiated, emotional interactions to communicate and solve problems in increasingly complex ways.
- The more areas in which an animal can apply these problem-solving interactions (e.g., visual-spatial problems, perceptual motor problems, vocal-verbal problems, social problems), the greater the breadth of intelligence.
- The greater the ability for a continuous flow of differentiated emotional interactions, the greater the animal's ability to form complex social networks.
- The greater the social networks, the greater the ability to deal with basic needs and, therefore, the greater the ability to employ mental resources for the development of higher levels of intelligent behavior, including the creation of new tools, technologies, and possibly even symbolic capacities.

We see a graduated ability to enter into co-regulated, affective, problem-solving interactions among higher-level mammals, such as dogs. Bonobo chimps have pretty elaborate systems for nonverbal cueing, but not nearly as elaborate or continuous as those used by human beings. Bonobo chimps tend to have short bursts of affect signaling, but not a continuous flow, as many humans have. In addition, humans can make many more facial expressions, affect-based vocalizations, and fine motor gestures. Human thinking evolved from far more complex affect signaling systems. But the basics are there in bonobo chimps. In our work with bonobo chimps with colleagues in Atlanta, we are exploring whether we can help chimps to be more symbolic through more emotional interactions. We've got them doing imaginative play and being more involved in long chains of signaling. Obviously, there will be limitations. Human beings, however, are not different only because they symbolize to higher levels, they're different because they are more emotionally nuanced and continuous in their interactions.

— 11 —

How Emotional Signaling Links Emotion and Cognition and the Brain's Subsymbolic and Symbolic Cortical Systems: Implications for Neuroscience and Piaget's Cognitive Psychology

GROWING SUPPORT FOR THE notion that emotions play an important role in aspects of cognition, such as decisionmaking, is emerging from many sources, including brain imaging studies and observations of individuals with central nervous system lesions. Based on these studies, emotions are, in part, viewed as providing goals for actions, expectations and related imagery for perceptions, and content for representations. They are also viewed as conflicting with rational thought at certain times. Emotional and cognitive capacities are thought of as separate but related processes.[1]

Missing from this emerging emotional and cognitive framework has been an understanding of the developmental pathways and mechanisms through which an infant, child, and adult connects emotional and cognitive processes at each stage of development—the how and why of this process. As we saw in Chapters 1 and 2, when we observed early development in infants and young children, we found that emotional and cognitive processes were not part of separate but related systems, but part of the same basic process. In fact, we observed that at each stage of development, emotional interactions led to cognitive capacities and orchestrated their operation.

In this chapter, we will see how emotional signaling provides the missing link between what traditionally has been viewed as the world of emotion and the world of cognition and what many modern neuroscientists are describing as the subsymbolic and symbolic systems. We will also observe how emotional interactions serve as a critical foundation for language, visual-spatial thinking, motor planning and sequencing and executive functioning, and other aspects of sensory processing. Furthermore, we will explore how understanding the early stages of emotional development leads to a vital revision in Piaget's pioneering theory of cognitive development and a model for consciousness that looks at its multiple levels (i.e., multiple prerepresentational and representational levels). As discussed earlier, we will further show how emotions, which are often viewed as belonging to a part of the brain that evolved relatively early in evolution, evolved to new levels during the course of evolution. Similarly, we will underscore an important theme of this book—that in the life of each individual, emotions develop to higher and higher levels of organization as an infant progresses through the stages of life. Understanding the transformations emotions go through during each stage of development can help guide and interpret a rapidly growing body of neuroscience research.

Many researchers believe we process a number of emotional patterns subsymbolically outside of conscious awareness. We are also proposing, however, that emotions serve to integrate the different processing areas of the mind, such as attention and sequencing, visual-spatial thinking, sensory processing, and language.

To understand this process, we must first examine a dilemma faced by current neuroscientists. How do subsymbolic systems that process emotions such as fear, as well as others, become part of an integrated system with cortical symbolic processes? Joseph LeDoux, who has mapped out the subsymbolic neuronal pathways involving emotions such as fear, asserts that cortical symbolic processes are somewhat separate from the subsymbolic systems that mediate emotions. He believes that the subsymbolic systems have to be approached through direct biological or physiologic methods or, as he sometimes puts it, some types of "psychological tricks."[2]

We have evidence to suggest that this view of the mind and brain is not correct, however. Instead, we propose that the emotional signaling system we have described in Part I integrates subsymbolic and symbolic mental processes and, therefore, may be the "link" between the subsymbolic systems of the amygdala and the symbolic systems of the cortex.

EMOTIONS TAMED INTO SIGNALS

In the earlier discussion of the origins of symbols, we hypothesized that if a perception is freed from its fixed action pattern it can become a free-standing image, acquire emotional experience, and become a symbol. We also hypothesized that catastrophic affects, such as fear or rage, push for direct action and are often part of these fixed perceptual motor or action patterns. As emotions become used for interacting with others, however, vital changes occur. First, perceptions no longer need to lead to direct action; they can become part of interpersonal negotiations. The other change is that catastrophic affects, such as fear and rage, which push for immediate action, can become "tamed," modulated, or regulated through reciprocal emotional signaling. Therefore, as we saw, emotions such as fear or rage needn't build up to higher and higher intensities; at their earliest manifestation, a soothing emotional signal from a caregiver can modulate such an emotion. Further interaction can make the original emotion of fear or rage part of a modulated exchange. The fear or rage is now a regulated emotion, even though it may still retain some of its distinguishing physical characteristics.

Through such exchanges between a caregiver and a child, emotions such as fear or rage or anxiety can be used as "signals." Signals are very different from catastrophic emotions requiring direct action (such as "fight or flight," shut-down, and/or various associated physiologic states). These signals can generate emotional and social interactions. They can further the continuous flow of emotional interchanges. They can, as we saw, modulate and regulate emotions, behavior, and mood, as well as help form a sense of self. Signals can also be transformed into a mental representation or symbol (an idea). Constructing a symbol or an idea eventually enables infants to know consciously that they are scared or anxious; they can now reason and figure out the "why" of a feeling rather than let it overwhelm them. The ability to know an emotion—*to study one's own perception of it*—is proportional to the degree to which it is tamed and transformed into an interactive and representational form.

If an emotion such as intense fear remains catastrophic, it pushes for direct action; there is awareness of the physiologic states or expectable actions, but not a deep awareness of the associated emotions or feeling. For example, someone might say, "My heart is beating fast. I need to get out of here because I'm scared" (and then the person runs). But people who can fully represent the emotion and reflect on the feeling can describe in depth how it feels to be scared and associate it with similar fears and other relevant feelings and experiences. Put another way, tamed, modulated emotions enable the mind to use its working memory to process the feeling at a deep level, access related or associated long-term memories, and plan effective coping strategies.

The key to the conscious awareness of emotions is, therefore, the human ability to tame the emotion through emotional signaling and make it into a signal. The degree to which this can occur also determines the degree to which the conscious perception of a "feeling" will be similar to the primary experience of the mong them LeDoux, make a distinction between an emotion that is a subsymbolic physical state and the conscious perception of that emotion, which is subjective and is called a "feeling."[3] They believe that the feeling may not be a very accurate description of the original emotion, which can't be fully known consciously and may not be directly accessible. They conclude that talking therapies attempting to explore basic emotions verbally may therefore be of limited value and that the modification of basic emotions such as fear may need to occur through more direct biological interventions.

As we have discussed, however, if the original basic emotion can be transformed into an interactive signal and, thereby, be "tamed," it *can be known.* The key to psychotherapy is to understand the developmental pathways involved in processing emotions. In a previous work,[4] one of the authors showed how, once an emotion is transformed into a signal, therapists can help patients use emotions for interactive signaling and, eventually, regulate, represent, and work with them.

In addition, once emotions are not simply experienced or expressed, but are transformed into signals, they can also become part of a pattern, that is, an interactive, emotional pattern. In this way, complex emotional patterns involving various modulated emotions can also be represented or symbolized and, thereby, become consciously known. "Fear" in this context is no longer a single emotion but part of an interactive pattern. This interactive pattern may involve fear and then soothing, and then less fear and more soothing, and then even less fear and more soothing. What was initially a catastrophic emotion—fear—is now an interactive emotional pattern that involves both the child's emotions and the caregiver's emotions. Just as a baby exists only in the social context of a baby/caregiver relationship, as Winnicott pointed out,[5] similarly an emotion that has become an interactive signal exists in the context of its interactive pattern.

Transforming catastrophic emotions into signals enables a child to integrate subsymbolic and symbolic systems. This is more difficult in situations of stress. Stress hormones tend to weaken the functions of the hippocampus, the part of the brain that deals with conscious memories and communicates with the symbolic system of the cerebral cortex; at the same time, stress hormones tend to strengthen the functions of the subsymbolic systems dealing with emotions (such as fear) through the amygdala. Stress is, therefore, thought to lead to greater fear or anxiety and less awareness or conscious control. The ability to transform catastrophic emotions—for example, fear or rage—into signals, however, also tends to reduce stress and facilitate problem solving.

EMOTIONS IN CONTEXT

Sometimes, emotions are studied outside their social settings. When we study a phenomenon, research reduces its complexity. But sometimes, reducing the complexity of a phenomenon can alter it to the point that it bears little resemblance to the phenomenon originally requiring study. Emotions need to be understood as they ordinarily develop. Within a few months of life, if not earlier, emotions are part of relationships and interactive signaling systems. This is true for many types of emotions ranging from fear and rage to love and respect. For example, higher-level emotional patterns, such as those involving "love" or "respect," almost by definition involve physiological states, social exchanges, and symbolic meanings. Complex emotional patterns or complex feelings can't be defined without all their critical features. Bonding, pleasure, and sexual excitement are very different from mature human love. Mature love and empathy can't be reduced to separate elements.

This fact is more than just of theoretical importance. It provides another clue to how emotions lead to symbol formation. As emotions become part of interactive patterns, the child learns to perceive the pattern. She doesn't simply experience her heart beating or a desire to run away; she experiences fear, then soothing, and eventually mastery. These emotional patterns include expectations and the anticipation of feelings. For example, a toddler learns that her joy leads to happiness in her parents. She also may learn that her anger leads to parental disapproval or disquiet. These patterns now define the emotion. The emotion no longer exists as an isolated entity and, since affective interactions begin very early in life, perhaps it never really did. The notion of an isolated emotion may be an artifact of our methods of investigation.

Emotional patterns become more and more complex. As they do, they help the toddler further separate perception from action and form freestanding images, which become symbols. Over time, freestanding images and emerging symbols become associated with increasingly complex affective interactions. In fact, it is because of this gradual process that a child of twenty-four or thirty months has a sense of what love is. Love is not an isolated emotion. It is not a feeling based on the interpretation of a physically induced emotional state. It is a representation—a symbolization—of a complex emotional interaction that spans a fairly long time interval. As such, it captures the many aspects of the experience of love, the feeling tone, the memory of the relationship's physical sensations associated with interactions, the associated sensory, motor, and language experiences, and the family and social context. As our twenty-four- or thirty-month-old further develops, new emotional experiences will contribute to a growing and changing definition of "love." "Love" will come to represent devotion, caring, anger, ambivalence, and many other complex characteristics based on the unique experiences of the person. Other complex feelings, such as empathy, respect, and compassion, will also be based on growing experiences.

The representation of complex emotional patterns can occur at various levels of consciousness, depending on the levels of emotional interaction going on between caregivers and children. In fact, our naturalistic observations of developing infants and young children suggest a continuous process of gradual increases in consciousness and awareness.[6] The levels described in Chapter 2 are an attempt to capture various points in this continuous process.

The picture painted by modern neuroscientists such as LeDoux and philosophers such as Dennett of a symbolic system that perceives and interprets subsymbolic systems that are outside of consciousness is not, therefore, consistent with what we observe during children's development. Interestingly, their view is very similar to Freud's early topographical model of the mind. This model posited three systems—the unconscious, the preconscious, and the conscious. He later abandoned this model for a more dynamic one.

In summary, during the course of human development, emotions are not unitary, simple entities. They are part of dynamic interactions or systems. They are transformed from catastrophic affects into signals and eventually become represented or symbolized. There are many different levels of symbolization. Critical to the process of symbol formation, however, is the transformation of all types of basic emotion into interactive signals.

THE DUALITY BETWEEN EMOTIONS AND REASON

The duality between emotions and reasoning, as described earlier, has a long tradition in philosophy and psychology and is preserved by many neuroscientists working on emotions today. For example, Antonio Damasio, M.D., Ph.D., states that "the creation of a body state (of emotions) is automatic, largely preset by our genes." He elaborates that, in addition to conscious, rational thought, "another system, probably evolutionarily far older, acts even before the first one. It activates biases related to previous emotional experience."[7]

With regard to his study of patients suffering damage to certain regions of the frontal lobe and who lost the ability to appreciate negative outcomes, Damasio states that "despite maintaining normal intelligence . . . and though they can reason logically, their decision-making ability is flawed. . . . They can no longer sense, for instance, embarrassment, guilt, pride, or shame."[8]

As we have been discussing, however, our observations suggest that the foundation for intelligence, including creative and logical thinking, stems from adaptive emotional functioning (also see Chapters 2 and 10). Individuals who are unable to experience basic feelings would, therefore, not be able to think logically or function intelligently. For example, logical thinking, which includes separating fantasy from reality, depends on the ability to use emotional signals to probe reality. Through emotional and symbolic signaling between a child and others, the child constructs the boundary between fantasy and reality and develops the ability to think logically.[9] Deficits in emotional signaling tend to undermine our sense of reality and logical thinking.

We would, therefore, propose for consideration an alternative and more clinically-consistent interpretation of Damasio's research findings on patients with frontal lobe damage. They have deficits in emotional regulating as well as deficits in sequencing and planning actions, both functions of the prefrontal cortex. As a consequence, they have difficulties with fine-tuned, subtle, co-regulated emotional signaling, logical thinking, higher levels of reflective thinking, and overall functional intelligence.

In clinical work, we observe that individuals vary in their ability to symbolize their emotions. Some can signal across the full range of human emotions, including love and fear and anger, and others can signal only with certain emotions, such as love but not anger or vice versa. Some can maintain emotional signaling and regulation even when under stress. Others can't. Some individuals can label feelings but not reflect on them. Others can label and reflect on feelings. Still others can reflect at high levels and look at subtle shades of "gray" in their feelings; compare their feelings; and explore the relationship between their own feelings and those of others in the past, present, and future and in the context of a stable sense of self and internal standard (see Chapter 2).

As an individual learns to represent the full range of emotions, he constructs a dynamic system that operates as an integrated whole. Therefore, the old view of emotions pushing for discharge and conscious logic or reason trying to deal with them is not what we see clinically in individuals who have progressed to the highest levels of emotional organization.

At the highest levels of emotional organization, for example, an individual will experience a great deal of closeness with his spouse, empathize with his children, be able to experience annoyance and even anger without yelling or hitting, and be able to handle a variety of stresses with a sense of awareness of the feelings generated, but without being overwhelmed. Such an individual will also be able to evaluate his feelings, wonder to himself, "Gee, why am I feeling more loving to my wife today than usual?" or, alternatively, "Why am I more annoyed at my family today?" He'll be able to reflect on possible reasons, ranging from being annoyed at his own parents, or annoyed at people at work, and so forth. When such a person feels a strong emotion, like anger and rage, it doesn't overwhelm him and lead to hitting, but it operates at the different levels we've been describing. It operates as part of a pattern, where anger and concern for others coexist as well as levels of representation and problem solving.

The dualistic view of a separate symbolic-cortical system sitting on top of a subsymbolic system is based in part on a pathological model. When the connection between these two systems has not been made by the development of emotional signaling, the gap between them results in psychological problems. In Chapter 2, we showed that certain individuals who can't fully signal with their emotions tend to put intense emotions into action (impulsive behavior), somaticize them (headache), or engage in polarized (all or nothing) thinking patterns or in fragmented thinking. For example, if they are impulsive, there is a greater likelihood they will diagnosed with a conduct disorder. If withdrawn, they have a greater likelihood of being diagnosed with a pervasive developmental disorder. If very fearful, a diagnosis of anxiety disorder may be likely.

In contrast, individuals tend to evidence a high degree of emotional health if they have fully mastered the skills of emotional signaling and have progressed to age-expected level of symbolization and reflective thinking across the full range of emotional patterns. They tend to have relationships characterized by warmth, intimacy, empathy, and flexible coping strategies. They can effectively pursue their goals, plan, and solve problems. These individuals tend to be more consciously aware of their emotions and there is less distortion or difference between emotional patterns and their associated feelings (the conscious awareness of their emotions). In other words, they are more aware of the physical aspects of their emotional patterns and can consciously interpret them.

As indicated, healthier individuals tend to experience their emotional worlds as an integrated, dynamic system. Physical aspects of emotions such as a tightening of the stomach muscles or a quickening of the heart are not experienced as separate from the feeling that represents them. The physical sensation and the feelings are integrated characteristics of the same basic experience. These individuals are able to exchange emotional signals with others while engaging in symbolic discussion that quickly regulates and modulates emotions (e.g., patterns of down- and up-regulating, as described earlier). In addition, in well-integrated individuals, the emotions tend to remain modulated and interactive and usually do not build to catastrophic levels. In rare circumstances when they do, the individual tends to regroup and, through emotional interactions with others, get back into modulated, regulated patterns. In other words, even under intense emotions, their emotional systems are integrated and dynamic and show lots of homeostatic balance.

It's understandable that many researchers would retain the duality between emotions and reason. Even though they believe, as Damasio does, that emotions are necessary for evaluating events and making decisions (for example, by weighing negative or positive consequences), primary emotions are seen as part of a different and evolutionarily earlier system than conscious, logical thought. This view fits the introspective experience of many individuals as they feel anger or lust and exert conscious control and try to figure out how best to behave. What's not always appreciated is how much individuals vary in their ability to be aware of emotions and reason and to integrate them. Individuals who have fully integrated emotion and reason use emotions and feelings as the instrument of their reason; for example, when they use empathy or compassion or generate a creative solution to a problem. Of course, such individuals, like all human beings, can become angry or fearful, but these emotions do not conflict with their conscious reasoning. Rather, these emotions and their associated feelings become part of larger emotional patterns. Such an emotional pattern might involve a variety of emotions, anger, concern for others, empathy, and altruism. For example, an emotionally healthy, mature, reflective individual confronted by a competitive business colleague who is making a joke at his expense may respond by wondering why his colleague is feeling threatened that day, and might further think about how to put his feisty colleague in his place with a gentle joke and, at the same time, show some compassion for the as-yet-unidentified stress. This would be a complex emotional response involving competition, empathy, compassion, and strategic thinking. Alternatively, a less reflective, mature, or emotionally put-together individual might respond with rage and yell, or attempt to contain his anger by clenching his fists and mouth. This person would likely experience a conflict between the rage and the conscious need to be in control.

These two brief descriptions illustrate an important point. Regulated emotions that are part of nuanced patterns of interaction are generally associated with more integrated patterns of emotional response and higher levels of reflective thinking. Less regulated emotions tend to be part of conflict situations and less reflective patterns. Often, the more fully the individual masters emotional signaling across the full range of human emotions, the more adaptive his emotional patterns and the more these patterns serve as a highly adaptive instrument of "reason."

If these transformations that emotions go through in the pathways leading to symbol formation are not appreciated, it is understandable why the traditional dualistic view of emotion and reason would persist. Our model shows them to be integrated and working together, though at varied levels from one individual to another. In other words, both during the course of evolution and in the life of each individual, emotions were and are transformed to new levels of organization. At their highest levels, they operate in complex, adaptive patterns that orchestrate and define our intelligence.

HOW EMOTIONAL SIGNALING INTEGRATES THE DIFFERENT PROCESSING AREAS OF THE MIND

In addition to integrating subsymbolic and symbolic systems, emotions also foster and integrate such capacities as attention and sequencing, visual-spatial thinking, language, and sensory modulation. As we said, in various types of developmental and learning disorders, including autism, antisocial behavior patterns, and various learning disabilities, this process may be compromised. Emotions begin their pivotal role by helping an infant connect early sensory and motor patterns. For example, a baby turns toward the sound of his mother's voice in part because her voice is interesting, pleasurable, and regulating. This sensory, affect, motor connection leads to patterns of engagement, purposeful communication, and shared social problem solving—all of which further integrate the infant and growing child's various "processing" capacities. Consider a few examples.

Attention and Sequencing (Executive Functions)

Our higher-level language and thinking, including theory of mind, are founded upon our basic abilities for attention and sequencing. Early emotional interactions help a child learn to pay attention and plan and sequence actions, for example, executive functions. Many have wondered why we are seeing more children with attention and motor planning problems. One reason might be fewer opportunities for long chains of problem-solving interactions in child/caregiver relationships. The ability to carry out several action steps in a row, that is, to plan and sequence actions and, in that way, attend to and complete the task at hand, begins with simple reciprocal interactions. The emotional recognition that one's actions can have an impact on someone else is the foundation for sequencing, that is, the ability to carry out many steps in a row where each one is related to the previous one. This is exactly what interactions are. Smiles lead to smiles, frowns to frowns, sounds to sounds, movement to movement. The harder it is for the child to continue a long chain of interaction, the more practice that child requires with long, regulated sequences of interactions.

As a child graduates from simple interactions to more problem-solving interactions, the ability to plan and sequence also improves because the child is now sequencing longer chains of emotional signaling. As a child learns to use ideas to regulate behavior, she can map out sequences with visual images and verbal ideas. As logical thinking comes in, the child can plan strategies for executing actions and consider the alternatives. The entire system of planning and executing problem-solving strategies, though, depends on the child's ability to link many pieces of behavior together.

The degree to which these patterns can be elaborated in different areas of life increases the degree to which the child can plan and execute actions and focus, attend, and solve problems in these different areas. For example, some children are very good at social negotiations, but not very good at fixing a toy, figuring out how to find a hidden object, dressing, or washing up.

It would be useful to study caregiving patterns associated with different abilities to attend and sequence. Our hypothesis would be that less reciprocal caregiving is associated with greater attentional and sequencing problems in children. Optimal attention and sequencing capacity would be associated with caregivers capable of long exchanges in many functional areas—for example, motor, language, visual-spatial, social—and who tend to support the child's gradually increasing assertiveness. Such a caregiver would begin with a continuous response pattern, but gradually shift to a more intermittent pattern as the child became more self-sufficient. If the child is talking, the caregiver would gradually shift to letting the child say more and more before commenting or asking a question, even though the caregiver would be nodding rhythmically in a close emotional rapport all during the child's comments.

Visual-Spatial Thinking

Visual-spatial processing is obviously necessary for one's sense of direction and for feeling secure in relation to space. It is also vital for math and science reasoning, which involve spatial relations and seeing the "forest for the trees," or big-picture thinking, as well as for abstract thinking of any kind that requires "seeing" the whole pattern and how the parts fit into the whole.

Hide-and-go-seek games, treasure hunts, and other search games strengthen this basic capacity. Take a child who goes on little nature hunts with her mom and dad. They explore the flowers, search for hidden toys, and find books around the house. All this is done with chatter or wordless signals back and forth. In this way, the child invests the spatial world with meaning. A child with biologically very weak visual-spatial processing skills will shy away from searching for a toy and will grow more frustrated because it won't be easy for her to search for things. She might go to a new house and get lost, but this problem improves with more experience in visual-spatial tasks. A parent who says enthusiastically, "Let's find that toy! I know we can do it," or who picks a toy that makes a noise so that when he hides it, the child can hear the noise, can further the child's progress. The child receives two sensory inputs—seeing where the toy might be and hearing the noise. The pleasure the child feels with each success fuels the progress.

The worse the processing ability, the more problem-solving interactions are needed. Take the example of a child who can't see. She needs a picture of her world, but it will have to be a "picture" through other senses. Even though she doesn't have sight, the child can still be a good visual-spatial thinker. With toys that make noises or emit smells or provide interesting touches, and through exploratory trips around the room, the child will respond to a parent who says, "We're on the right track! We're heading there!" There are many members of the animal kingdom who work by sound or smell. They develop very good spatial road maps.

For the hearing-impaired child, more visual, motor, and tactile support is needed. This same principle applies to children who have trouble understanding the spatial dimensions of their own bodies, for example, integrating their left and right sides. Here, too, back-and-forth emotional interactions involving the human body, games such as "touch my nose and I'll touch your nose," facilitate body-oriented visual-spatial problem solving.

For the active, inattentive child, attention will be mastered with longer and longer chains of interaction, coupled with modulation patterns in which a caregiver changes the rhythm of interaction or the intensity of response. Clearly the more salient the content of these responses, the more these skills become integrated into the child's cognitive abilities.

Sensory Modulation

The ability to modulate or regulate sensations is also critical. Some children (and adults) hold their ears and get overwhelmed with sound; others are sensitive to touch and get overloaded. Others still are underreactive. If you watch these who are oversensitive, you see that they panic easily and are overwhelmed. Those who can regulate their sensations remain in control, turn down the stereo, and say, "Shhh, it's too noisy in here." The more they move from catastrophic patterns to regulated interactive patterns, the more adaptive they become. The more dysregulated the sensory system, the more regulation and soothing is needed through emotional interactions. Some caregivers just naturally fall into the pattern of reading the child's signals and helping the child. Others overwhelm the child's system. Some educators tailor their responses to the child's individual differences, toning their voices down for the child who's covering his ears. Others confront the child and engage in power struggles. They make the noise louder and overwhelm the child, and they often believe the child is just "being manipulative." The more the caregiver creates opportunities for sensitive emotional interaction, the better the child learns to regulate and to engage in self-calming.

When we put children who are very sensory overreactive into such interactions, they do not remain at the mercy of catastrophic affects. When people are anxious and in a panic state, everything bothers them. They can't sleep at night. They hear that little noise outside. They hear the trucks loading and unloading. They're in a heightened state of vigilance.

Terrorism can put an entire country in the vigilant state, where many people are more overreactive. At any age, soothing interactions help a person go from catastrophic emotions to a more regulated state.

Symbolization with All the Senses

We have been describing the process of forming symbols predominantly from the point of view of a visual symbol, that is, an image, that acquires meaning as perception is separated from action through emotional interchanges and problem solving. This process, however, occurs, or we should say, can occur via each sensory pathway. As it occurs through a particular sensory pathway, the individual is able to symbolize that realm of experience. As it occurs through all the sensory pathways, the symbols that are formed become integrated, multisensory, emotional symbols. If, however, only certain pathways become involved, the development of symbols is uneven; for example, a person can symbolize what he sees, but not what he hears or vice versa.

If development is proceeding well, each realm of experience related to specific sensory, emotional, and motor skills can become involved in intimate exchanges. For example, toddlers can exchange gestures, looks, and glances to express different emotions using primarily what they see. In the auditory-verbal realm, a toddler also, in all likelihood, is gesturing with vocalizations. Just as her facial expressions, body posture changes, and motor gestures convey intent, her vocalizations can be used in this way. Eventually, as she uses words, her words can be used to signal intent and respond to another person's overtures. As the vocal-verbal system becomes part of continuous interaction, the tendency for perceptions processed through this system to be tied to immediate action (hearing a sound and reacting immediately with actions, such as attacking, fleeing, or avoiding) can be replaced with registering the sound, holding it in mind, and using it as a signal or basis for additional mental operations. Just like the visual image, the sound and/or word not tied to immediate action can then acquire more meanings through experience and become a full-fledged symbol.

Olfactory experiences involving various types and degrees of smell can undergo a similar process whereby, rather than serving as a stimulus for immediate action, they come to serve a signal function and have meanings associated with them.

Motor experience itself, the experience of one's body moving in space and carrying out sequences of actions, can undergo this same process. As it becomes used more and more in interpersonal exchanges, the experience of movement does not necessarily lead to further immediate movement (just as, for example, an athletic thrust towards the goal line in a football game simply leads to immediate acceleration). Instead, as movement becomes more regulated, and modulated, we see the marvelous feats of great athletes who seem to choreograph movements midstream, stopping short, moving to the left or right, and circumventing the opposition. Human beings have an enormous capacity for elevating movement out of the realm of automatic action and into the realm of planned sequences. In fact, we would argue that the ability to plan and sequence complex actions is related to the basic ability to use actions in a co-regulated reciprocal problem-solving manner and then elevate it to the level of symbolizing actions.

Visual-spatial experiences that go beyond the registration of affective and motor gestures and involve motor planning and complex problem solving also undergo a similar process. Typical toddler tasks, such as searching for the hidden toy, finding her mommy in a different room, copying shapes, and the like, can become part of varying degrees of co-regulated problem-solving interactions. Consider the difference between a child who lines up her toys or does the same puzzles over and over again and, in that way, carries out very stereotyped visual-spatial problem-solving tasks; compare that child to one who engages her father in a back-and-forth problem-solving adventure in building a city out of blocks. In building the city with blocks, father and child interact in a back-and-forth way, figuring out new visual-spatial constructions. Here, too, rather than engaging in fixed visual-spatial problem-solving sequences, they operate in a flexible and creative manner, mediated initially through co-regulated, reciprocal visual-spatial interchanges. Ultimately, such a child has the ability to create on her own, perhaps becoming a gifted architect.

The pathway to symbolization and ultimately creative and reflective endeavors in each realm of the senses depends on this early experience of interchanges with others. It's difficult, if not impossible, to achieve these skills just by practicing them oneself with the inanimate world. Although some degree of mastery might be achieved this way, it won't have the same creative richness as when a reciprocal partner is involved.

Interestingly, when we look at visual-spatial experiences and motor coordination and planning experiences, we observe that some children, neurologically, have the tendency more towards fixed patterns. Some children, for example, retain some of their primitive reflexes and when they look right or left, they have to turn their bodies. They are unable to move their eyes without moving their heads, or move their heads without moving their bodies. With practice, however, they can learn to separate these actions and develop more adaptive motor strategies. Because of their neurological challenges, these children require even more opportunities for emotional interactions in which they can learn to use their motor and visual-spatial capacities in highly responsive and subtle ways. Adaptation demands a flexible system that can respond in a nuanced manner to its environment (rather than in a fixed, more global manner). One might argue, as we have in the chapter on evolution and primates, that the ability to engage the environment in a flexible, nuanced, reciprocal manner is what has separated human beings from nonhuman primates and other mammals. Most can engage in some interactions involving gross motor activities and even some visual-spatial problem solving, but most do not have the capacity to exchange subtle emotional signals, maintain long chains of problem-solving interaction, and include a range of vocal signals as part of the process. In fact, as we argued earlier, human beings have far greater capacities for subtle textures of emotional signaling than even their nearest cousin.

Typically, during early development, and fortunately so, human toddlers are using all the skills described above—motor, visual-spatial, visual, auditory-verbal—in their daily relationships with adults. For example, in trying to ask for juice, or in flirting with a parent, the toddler is likely to flash a big smile, look a little needy, tug at the parent's hand, pull him towards the refrigerator, bang on the door, vocalize "j-j-j," and point to the juice. For their part, parents are likely responding with motor, visual, and vocal signals. Together, they're engaged in a complex, multisensory, emotional, motor-processing pattern of exchanges. In healthy families, these exchanges are continuously taking place so smoothly and effortlessly that they go largely unnoticed. Perceptual-action patterns give way to exchanges of symbols, and the participation of all the senses renders the symbols rich and many-faceted. We tend to take this integration of experience involved in symbol formation for granted; it seems simply a part of our nature to combine words, visual images, and motor actions.

For many, however, the process doesn't occur in this optimal fashion. For example, children with delays or challenges in one sensory realm or another may be able to elevate one sense to the symbolic level but not another and, therefore, experience uneven development. Many adults, for example, can use visual images symbolically, but not auditory-verbal ones; some adults, although highly verbal, easily get lost on their way to the bathroom. In working with children with autistic spectrum disorders and other special-needs conditions, as well as with children experiencing severe learning disabilities, we have found that the key to enabling them to make significant progress is to help them elevate the processing skill that is delayed or challenged into the realm of emotional interaction, and through this interaction into the realm of symbols.

The process we are describing doesn't occur largely through a prescribed and genetically mediated set of biological events. The biological events, to be sure, create a substrate. Highly complex experiences involving all the senses, such as those we have been describing, are necessary. These experiences involve all the senses in a critical step, that is, co-regulated, reciprocal, affective interchanges.

From Attention and Engagement to Symbol Formation: How Emotions Organize the Different Parts of the Mind at Each Level of Development

As we have been describing, symbols are formed for each type of sensory experience and motor capacity. We have also been describing how emotions develop through stages or steps. They move from simple expressions of annoyance, discomfort, and pleasure to complex emotional interactions involving empathy and mixtures of feelings, such as love and respect. As emotions are transformed from one stage to the next, they also organize a growing infant's and child's basic capacities to comprehend their sensory and motor experiences, that is, to take in and comprehend (process) sights, sounds, and movements, and to modulate sensation. As each new level of emotional (affective) transformation occurs, the child's basic capacity to take in and comprehend sensations becomes organized at a higher level.

For example, consider how the child organizes what he or she sees. Initially, the affective interest in the outer world enables the child to look at the caregiver's face and begin recognizing distinct features, such as a nose or mouth. Synchronous movements with caregivers, and the growing interest in the outer world this facilitates, enable the infant to begin using her senses—hearing, seeing, and movement (turning)—in coordinated ways. This occurs over time with greater and greater flexibility, for example, turning the head without moving the whole body and, eventually, looking only with the eyes. By the second stage, the child invests particular images with specific emotions—for example, she invests her mommy's face with pleasure and joy. The pleasure and joy with the caregiver facilitates an emotional investment in constructing a picture of what will become a whole face and, eventually, a whole person, imbued not just with visual images but with a whole range of feelings and experiences. It is this pleasure and joy that motivates the construction of a more complete set of images because it indicates a live "other" outside oneself that's bringing such nice feelings to oneself. By the third and fourth stages, the child is emotionally able to explore valued objects in space by using purposeful back-and-forth emotionally mediated movements to judge distance and even to discover that what's temporarily out of sight still exists. Such cognitive feats depend on reciprocal affective signals and motor gestures that create the foundation for the cause-and-effect exploration of the physical world. By the fourth stage, the infant can crawl and then walk to Mom across the room, take Daddy by the hand to find a favorite toy, recognize and feel a sense of warmth and security when seeing Mommy smiling and nodding in approval from across the room. The initial construction of space, a room in a house, or a playground in the neighborhood, is through the creation of a spatial road map based on multiple co-regulated, reciprocal emotional interactions—rich emotional experiences—in different spaces. In other words, the child creates a spatial map based on his or her ability for engaging in emotional multiple problem-solving interactions with others and with objects in the different spaces of his world. The first picture of his spatial world is an affective one because the child invests his experience of what he sees with his desires and feelings.

By the fifth and sixth stages, the child, while continuing these complex emotional interactions, learns to create symbols or ideas made up of visual-spatial images based on wishes, desires, or affects. For example, she can make a corral for play horses or a special tea room for a tea party. Higher-level reflective visual-spatial thinking, such as coming up with a new design for a magical doll house that can change into a castle, will often follow; later on, it may include creating new concepts in math or physics. All these capacities are based on origins involving affective investment of space at higher and higher levels of organization.

The ability to plan and sequence actions also becomes transformed. Initially, the rhythmic, synchronous coordination of movement between the infant and caregiver, as well as the rhythmic, synchronous coordination between the caregiver's voice and the infant's movements (and later vocalizations), creates a relationship between the infant and the outer world. Over time, these motor-mediated, synchronous actions convey shared affects, such as pleasure. The infant's ability to use his motor system to relate to the outside world enables him to go from the first steps involving turning and looking to find the caregiver's voice to a second step involving searching out the person who gives pleasure and organizing the facial muscles into a big smile to a third step involving two and three actions in a row, such as reaching for the toy on Mommy's head and offering it back to her. In a fourth step, the toddler goes to even more complex feats, such as pointing at Daddy as he walks in the door, running up to him, taking his hat and putting it on, and then gleefully putting it on Mommy (a four-to-five-sequence action plan involving considerable motor planning under the guidance of emotions, such as curiosity and assertiveness). As the child learns to create symbols to express desires and feelings, he or she can now plan novel action sequences to solve new problems, for example, figuring out how to build a new Lego house that will serve as a good hiding place for the princess fleeing from the evil dragon. At subsequent stages, children can use their fertile imaginations to create even more complex action plans to solve a variety of challenges, such as creating a new invention to clean up their toys.

Language and the comprehension of sounds also goes through this type of sequence. For example, sound and language are initially expressed through synchronous vocalizations as part of the first stage, self-regulation. They then become part of an expression of pleasure as the baby responds with joyful sounds to the caregiver's pleasurable vocalizations (now part of the expression of a specific set of emotions). At the third level, vocalizations become more purposeful and part of two-way, back-and-forth, reciprocal interchanges. A variety of reciprocal vocalizations reflect not only the baby's growing ability to make more sounds (and process more sounds) but also the *desire,* mediated through affects, to communicate different needs, feelings, and interests. Why else would the baby use the different sounds, or later, the words she can make? In fact, as discussed, children with autism only begin using sounds and words meaningfully and with grammatically correct sequences after we've helped them learn to experience and express their affects and desires in a purposeful manner—for example, they will push us from the door they're persevering to open or close with a "Go away, Daddy." By the fourth level, they're involved in long chains of back-and-forth problem-solving interactions (co-regulated, reciprocal, affective exchanges). These exchanges can be perceived as patterns, some of which constitute the baby's own beginning idiosyncratic language that he can use, for example, as part of back-and-forth affective interchanges to communicate specific needs and desires. Eventually, these patterns can be used to construct words, which also can be used as part of back-and-forth interchanges ("Juice!"). At the fifth and sixth stages, the child can express his wishes and feelings symbolically in words or pretend play, and eventually create logical and abstract verbal concepts.

In a similar fashion, at each level of affect transformation, the child learns to modulate sensations more effectively. For example, as she becomes emotionally invested in the outer world and engaged in her first pleasurable relationship, the baby is no longer at the mercy of her own catastrophic affects when being overloaded. There is now a soothing partner who can help calm her. Over time, her soothing partner can help her engage in affective exchanges, which enable the "up-" and "down-regulating" of affective states and the maintenance of a more modulated pattern of reaction and mood. For example, as the child becomes excited, Dad's voice becomes more soothing and he mellows and softens his responsive gestures rather than intensifies or speeds them up (he helps the toddler to "down-regulate"). In other words, the toddler and his caregiver are now dance partners where the caregiver can help maintain a more regulated rhythm of interaction. As the toddler creates symbols and expresses feelings, she can label and describe the feelings associated with different states of sensory arousal and, eventually, not just regulate but plan ways to create desired states. For example, now the child can say, "Let's play some soothing music" or "I need a back rub."

As can be seen, the child is transforming each of these processing capacities into higher and higher levels of organization. In addition, the child tends to engage in these developmental levels, up to the highest one of which he is capable, at the same time. By looking, listening, relating, gesturing, solving problems, using ideas, and thinking all at the same time, the child is connecting the developmental levels of experience together. The child is also connecting or integrating all his capacities for information processing. In daily life, the child tends to use sight, sounds, movements, and sensory and emotional modulating capacities together. For example, when she is involved in lots of co-regulated, reciprocal exchanges, the child is often gesturing with motor actions, vocalizing, sizing up what she sees, and co-regulating her state of arousal with her caregivers. As these occur together, the child is connecting or integrating her different ways of taking in and comprehending sensations and planning actions (information processing domains). Therefore, appropriate interactive experiences enable the child to connect all her processing domains together at the same time she's organized them at higher and higher levels of emotional transformation and organization. In other words, rich interactive experiences help the child connect the different parts of her mind *horizontally* (across all the processing domains) and *vertically* (at the different levels of developmental organization).

In contrast, children with special needs may, for biological reasons, have uneven ways of processing sensation and, therefore, have special difficulties in connecting or integrating all the ways they take in and understand information about the world. This is a problem that we see as especially prominent in autistic spectrum disorders, where it's hard for the children to integrate their emotions with their motor planning and sequencing capacities; therefore, it's hard for them to connect all their processing capacities together. That's why, in our treatment program for children with autism, we have created special experiences that support the connection between affect and the different processing capacities, as we mentioned earlier.

Children with special needs, however, often are unable to engage spontaneously and fully at all the developmental levels and with all the "processing" domains. In general, we have found that children with special needs do best when their interactive partners figure out what types of interactions and what levels of emotional interaction and types of information processing are particularly difficult and make a special effort to engage these areas as well as the ones that are more readily a part of the interaction. We try to inspire the child to look and listen to us, relate with pleasure, exchange emotional gestures, solve problems, and use ideas and words—at the same time. We also challenge him to look, listen, comprehend, and plan actions at the same time we are helping him connect his processing domains.

In summary, in human interaction and all types of human learning, it's most useful to be engaging and interactive at all the developmental levels we have been describing. These include the following:

1. Basic regulation and modulation, which often include taking in sensations and connecting sensations, affect, and motor patterns (e.g., turning toward a pleasurable voice), as well as rhythmic and timed synchronous motor gestures at a covert level (infants can readily be observed moving their arms and legs in rhythm with their caregivers' movements and voices as part of this basic level of interaction).
2. Pleasurable affective engagement and a growing, shared sense of humanity with another person.
3. Back-and-forth (reciprocal) affective and motor gesturing that convey intentions, needs, or desires.
4. Long chains of co-regulated, reciprocal, problem-solving interactions that define a sense of "self" and "other" and enable the child to construct and understand patterns and negotiate and further regulate behavior, mood, and attention.
5. The creation of symbols that represent the experiences and "meanings" of all the previous levels and can now be used for symbolic interaction and the construction of new and creative levels of experience.
6. The building of bridges between symbols to create the capacity for logical thinking and higher levels of reflective and abstract reasoning; this includes the capacity for testing reality, separating reality from fantasy, and organizing a sense of "self" and "other" at higher and higher levels involving a broad range of symbolic and presymbolic experiences.

It's useful to harness all the sensory processing domains at the same time, as well. These include:

1. Auditory processing and language
2. Visual-spatial processing
3. Motor planning and sequencing
4. Sensory modulation

HOW EMOTIONAL SIGNALING MAY INCREASE NEURONAL CONNECTIONS IN THE BRAIN

As the developmental processes we have been describing are occurring, it's interesting to note the likely parallel going on in the way the brain is growing. Parts of the brain register sensations and experience (looking or listening), and other parts of the brain organize experiences into patterns. These include the parts of the brain that connect perceptions with the parts that sequence actions with the parts that generate and regulate emotions or affects. Development proceeds to include the higher cortical centers that symbolize and interpret experience and enable further planning and anticipation as well as the creation of new experiences and imagery. From our clinical observations, it appears that emotional or affective transformations serve as organizers of the central nervous system. They create connections between the lower and higher brain centers and coordinate the parts dealing with the registration of experience with the parts that plan and sequence actions and the parts that symbolize and interpret experience. (The chart below shows some of the parallels between the six basic stages of emotional or affective transformation and the growth of the brain.)

TABLE 11.1 The Six Essential Developmental Stages and the Growth of the Brain

Developmental Level	*Required Experiences*	*How the Brain Supports and Grows in Response to Each Developmental Stage*
Stage One: Being Calm and Interested in All the Sensations of the World	Help the baby look, listen, begin to move, and calm down.	Neuronal connections are occurring in the areas of the brain that process sensory information and help the baby initiate movements (i.e., primary sensory-motor cortex, thalamus, brainstem, and cerebellar vermis) and in the areas that support emotional interest in the world (i.e., amygdala, hippocampus, and cingulate cortex).
Stage Two: Engagement (Falling in Love)	Woo the baby into engaging with you with pleasure and delight.	Further activity in the areas supporting emotion, integration of visual, sensory, and motor areas, and right-sided neuronal connections supports the recognition of patterns (sights, movements) and promotes

(continued on next page)

TABLE 11.1 *(continued from previous page)*

Developmental Level	*Required Experiences*	*How the Brain Supports and Grows in Response to Each Developmental Stage*
Stage Two: Engagement (Falling in Love) ***(continued)***		emotional relating, expressiveness, and signaling (i.e., parietal, temporal, primary visual cortical regions, frontal eye fields, basal ganglia, cerebellar hemispheres, beginning of cerebral cortex as well as continuation of limbic system).
Stage Three: Affective Intentional Two-Way Communication	Follow the baby's lead and challenge him to exchange gestures and emotional signals with you about his interests.	As the baby processes patterns and initiates more selective responses to environmental clues, growth in areas that support sequencing and reading and expressing gestures and emotion (two-way communication) are more active (i.e., increases in frontal cortex, including dorsal prefrontal areas).
Stage Four: Multiple Reciprocal Affective Interactions to Solve Problems and Discover a Sense of Self	Become an interactive partner with the toddler as he learns to use a continuous flow of gestures with you to pursue his interests and meet his needs.	Cerebral cortex is more active and continues so. Left-sided neuronal branching surges as toddler sequences sounds and occasional word(s) to problem-solve. Right-sided growth continues together with the ability to figure out larger patterns in the world and interact with a wider range of emotions.
Stage Five: Creating Symbols (Ideas)	Enter the child's make-believe world as a character in his dramas. Engage him in long conversations about his interests, desires, and even his complaints.	Left-sided neuronal branching becomes denser as child comprehends, uses, and sequences more words and masters some of the basics of grammar. The visual-imaging parts of the brain grow as the child begins to engage more and more in pretend play. Both sides of the brain are becoming more specialized as language is rapidly being acquired.
Stage Six: Building Bridges Between Symbols (Ideas)	Challenge the child to connect his ideas together by seeking his opinion, enjoying his debates, and negotiating for things he wants.	Brain undergoes growth spurt, metabolizing glucose (sugar) at twice the adult rate. Increased activity occurs in areas of the brain that deal with the creation and comprehension of words, and connections among words. Increased activity continues throughout childhood and then gradually shifts to adult rate.

In our view, emotional signaling, which separates perception from action and leads to the formation of symbols, also fosters the growth of new or better functioning pathways in the brain. There is considerable evidence for this hypothesis. A body of research in the neurosciences shows that experiences, especially early in life, but continuing throughout life, influence the structure of the brain, especially neuronal connections and the chemical processes they involve. For example, in 2000, Eric Kandel of Columbia University won the Nobel Prize for his body of work revealing the changes that occur in neuronal connections as experience creates long-term memory. He showed how "experience" affects the regulator genes which, in turn, control gene expression and neuronal structure.[10] PET scans on adult musicians show that they have more neuronal connections in the area of the brain that controls the hand movements they use in playing their instruments. PET scans also show that emotionally meaningful experiences are associated with activity in many areas of the brain at the same time, in comparison to impersonal cognitive tasks, which are associated with fewer areas of the brain. Greenough's well-known research with animals[11] shows a direct connection between specific types of "enriched" experience and greater connections between neurons in areas of the brain related to that experience. Remarkable research shows that enriched experiences can help an area of the brain that typically is used for one function to be used for another.[12] For example, the visual center can be used for language if the language centers have been removed. These are only a few examples of the growing body of evidence pointing to a relationship between "experience" and the physical structure of the brain.

These emerging findings in neuroscience research prompt one to consider what would happen to neuronal connections if each new generation of children had more opportunities than the previous one for exchanging emotionally meaningful sounds and sights and movement patterns. This would not involve rote sensory stimulation, but only meaningful, pleasurable, well-regulated interactions. Would each generation develop more connections in critical areas of the brain within the range of their existing genetic potential? Of course, we have no way of knowing the upper limits of an individual's genetic range unless we create an infinite number of environments to tease out the top of this range. Since we can't do this impossible experiment, it's reasonable to assume from current evidence that during prehistoric times early men and women had more range than they were using. Today, we likely have more range than we are using. It's also reasonable to assume that during the course of evolution (and, currently, over long time intervals), our cultural practices—including child-rearing practices—involving the exchange of ever more complex emotional and meaningful signals, sights and sounds, and movements have been and are improving (with expectable, short-term vacillations). Have these practices, which improve the quality and variety of experiences, lead to expectable changes in the physical structure of the brain? Will such improvement continue?

There is no evidence that such physical changes, if they have been occurring, become encoded in our genes. Rather, to the degree they occur, they depend on each generation's having access to critical learning experiences. If these learning experiences keep improving in the adaptive, balanced, regulated way described earlier—that is, not simply stimulating memory—we may gradually be able to improve our critical intellectual, emotional, and social skills. As long as favorable cultural and child-rearing practices continue, we may see the development of greater neuronal connections in the critical areas of the brain that support these skills. If, because of a change in family patterns or exposure to toxic chemicals that affect the nervous system and learning, vital interactive learning experiences stop or become compromised, we may very well see movement in the reverse direction.

What are the parts of the brain most likely influenced by emotional signaling? Our hypothesis is the following: Emotional signaling initially supports increased neuronal connections in a number of areas (e.g., the cerebellum because it deals with rhythmicity and timing), but especially the prefrontal cortex because this area is directly involved in planning and sequencing actions, problem solving, and emotional regulation. The prefrontal cortex can, in turn, support longer and more complex patterns of sequencing and thus become better adapted to solving problems. Over time, meaningful emotional "experience" of this type and improved neuronal connections probably facilitate each other in an adaptive cyclical process.

The development of the prefrontal cortex and its functions, including emotional regulation and signaling, also supports symbol formation. Symbol formation, in turn, may lead to neuronal connections in the area of the cortex involved with different types of symbols. For example, the increasing use of visual images and symbols to solve visual-spatial problems likely leads to increasing connections between neurons in these cortical areas. Constructing and using auditory-verbal symbols likely leads to increasing neuronal connections in the language areas of the cortex. Similar processes likely occur for touch, smell, taste, movement, and emotions.

Symbols are often used together. As this occurs, we would expect to see increasing neuronal connections between different symbolic cortical areas. In fact, as we explained earlier, we believe our emotions mediate the connections between different types of experiences and, therefore, between many different areas of the central nervous system.

Even before symbols are formed to a significant degree, however, the higher cortical centers for such skills as language, visual-spatial processing, and motor planning are probably already being supported (through increased connections between neurons) because of the nature of emotional signaling. These often involve clusters of sounds and motor gestures, touch, smell, and the negotiation of space (through vision). As these multisensory emotional signals are used for more elaborate problem solving, the child begins to decipher "patterns" and expectations. As these patterns of sounds, sights, and other sensory experiences are constructed, they would likely be supporting neuronal connections in the cortical areas concerned with pattern recognition, and vice versa. These developments likely precede what will occur as a result of symbol formation.

We would, therefore, hypothesize that emotional signaling, and the pattern recognition and problem solving it facilitates, is involved in integrating the pathways of the central nervous system including subsymbolic systems and cortical symbolic systems.

As we mentioned earlier, emotional signaling also likely involves connections between the limbic system and the prefrontal cortex as well as other areas. Signaling with regard to positive emotions may involve connections with the left prefrontal cortex. Signaling with regard to negative emotions may involve connections with the right prefrontal cortex. The degree of these connections may contribute to a person's ability to use positive and negative emotions to signal and, in turn, to modulate, make discriminations, and to represent as well as to reason about negative and positive emotions.

Furthermore, we would hypothesize that those species that evidence greater ability to use emotional signals in problem solving have a more developed prefrontal cortex and more connections between the prefrontal cortex and limbic system and other areas.

RESEARCH SUPPORT FROM COGNITIVE NEUROSCIENCE

Recent neuroscience research supports elements of the theory we have been presenting. An important part of this theory is that the infant uses emotions to explore the world and that she double codes experience. She codes it according to both its visceral affective or emotional properties and its cognitive properties (which also involve affective aspects).

Although Joseph LeDoux suggests the more traditional dualistic picture of emotions and conscious reasoning ability,[13] some of his specific research findings support the theory we are proposing. He has shown that when the emotion of fear is processed by the amygdala, which is part of the limbic system, the perception travels from the thalamus (where the perception is first registered) to the amygdala in 0.15 milliseconds. For the perception to travel from the thalamus to the cortex, however, it takes 0.25 milliseconds. According to LeDoux, this explains why we may react emotionally with fear to seeing a snake more quickly than we can become consciously aware of what we are reacting to. In other words, we feel scared before we know why or can find ideas or words to express it to ourselves and others. There are also more pathways from the limbic system to the cortex than from the cortex to the limbic system, suggesting that emotions such as fear influence thinking more than thinking influences emotions.

This research on the primacy of the emotional pathway to "knowing" at a subsymbolic or subcortical level supports a major hypothesis in the theory we are presenting. It also supports our theory of the "double code."[14] Because emotional pathways are faster and more direct, they become, as we said, the director or orchestra leader or architect of our cognitive capacities. Therefore, in healthy development these emotional pathways have a direct and rapid influence, but do not influence cognition in the way LeDoux and others suggest. As we have discussed, emotions do not have to conflict with reason or logic. As they become transformed into emotional signals through co-regulated interactions, emotions can guide cognition. They can integrate different aspects of the mind and serve as its orchestra leader.

Further support comes from the work of Jocelyne H. Bachevalier, Ph.D., who has shown that procedural or implicit memory, which involves subsymbolic emotional systems, functions very shortly after birth, but declarative or explicit memory, which involves symbolic cortical processes and conscious awareness, gradually starts operating by four years of age.[15] Therefore, as we have been proposing, subsymbolic emotional systems are operative in the early months and years of life and symbolic cortical processes become operative relatively later. What has not been realized before, however, is how these subsymbolic emotional pathways can be used to "know and learn" from the world. We have been emphasizing how, as the infant learns to use her emotions as signals, she transforms her emotions into modulated tools of knowing and learning way before symbols are formed. They enable her to "double code" information, construct patterns, solve problems, form a presymbolic self, and over time develop higher and higher levels of intelligence.

Using brain imaging studies, Richard J. Davidson, Ph.D., has shown that the left prefrontal cortex "may be important in inhibiting activity in the amygdala and dampening response to negative emotional events, and particularly in shutting off the negative response quickly once it has been activated."[16] This would support our hypothesis that the prefrontal cortex is important for the type of emotional signaling that can regulate catastrophic or extreme emotions, as well as our hypothesis that such signaling involves pathways that integrate the prefrontal cortex and the limbic system.

Mary Carlson, Ph.D., has conducted research that reveals important findings about the importance of early emotional interactions for cognitive and social development.[17] She shows that touch and emotional nurturing support healthy development and that their absence undermines it. She also shows that the deprivation of needed emotional relationships and interactions undermines healthy development and that children cared for by parents who used state-run daycare showed similar patterns as children in depriving orphanage settings. This suggests that interferences in emotional nurturing interactions and regulation do not have to be as extreme as has been seen in some orphanages.

For example, Carlson found that human touch and interaction were vital to keeping infant monkeys from developing autistic behaviors. They were also vital for the development of sensory, motor, and memory pathways and the functioning of hormonal systems that regulate stress responses. In addition she found that:

When we compared the levels of cortisol in the two institutional groups to that of family-reared children, we saw that both institutional groups had abnormally suppressed cortisol levels in the morning. When children's cortisol levels at each time of day were correlated with their behavioral development, we found that those with the highest morning levels showed the best performance on motor and mental tests.

At later times of day, when cortisol should be low or almost undetectable, the most deprived children—the control group—showed significantly more elevated cortisol levels at noon compared to the intervention group. High cortisol levels at noon and in the evening in both of the institutional groups were associated with poor mental and motor performance. Cortisol levels measured by parents at home on weekends were similar to those of American children.

The most surprising and disturbing finding came from measuring the cortisol levels of family-reared children while they were in the bleak and highly structured context of their state-run day care. We found abnormalities similar to those seen in orphanages, and again those with abnormally elevated levels in day care were the ones who had the lowest behavioral scores.

These findings, plus studies of children who experienced the Armenian earthquake and adults suffering from post-traumatic stress disorder, offer evidence of the relationships between abnormal cortisol secretion, hippocampal shrinkage, and related memory deficits.

Clearly, the social networks in which children grow up bear on the development of the neural networks that mediate memory, emotion, and self.

In a study of American children in daycare, Megan Gunnar, Ph.D., of the University of Minnesota Institute for Child Development, found that they had higher cortisol levels (and stress-related behaviors) in the late afternoon than home-reared children.[18] Catastrophic emotions are associated with high levels of stress, which alters cortisol patterns. Regulated, interactive patterns of emotional signaling, in contrast, enable individuals to cope with and reduce stress and to negotiate and explore solutions. When a child does not learn how to regulate and signal with her emotions she may, therefore, be vulnerable to problems at the behavioral, cognitive, social, and hormonal levels. This type of dysregulation often has a negative impact on long-term development.

Bachevalier found that neuronal pathways involved in conscious or explicit memory are also affected by abnormal cortisol patterns, and if these patterns are compromised they can have a delayed long-term negative effect on development.[19]

Three additional studies on early development also provide a high degree of support for our emotion-based model of symbolic thinking. Lois Bloom and Joanne Bitetti Capatides studied causal meanings in children between twenty-six and thirty-eight months of age. Children's initial symbolic expressions of causality were related to social and emotional psychological phenomena more than to impersonal physical phenomena or events.[20] Another study of mothers' responses to the emotional expressions of children between nine and twenty-one months of age showed that mothers who responded to their children's emotional expressions in a reciprocal or interactive manner and tended to respond to the child's underlying intentions, and the situation or context, as well as to what the toddler might do to change the situation provided a rich experience in establishing symbolic meanings.[21]

A study of the parents of children with autism, by Michael Siller and Marian Sigman, found that caregivers of children with autism who were more responsive in their synchronization and continued interaction during play with their children enabled them to develop better language and "joint attention" over a period of one, ten, and sixteen years than did children of caregivers who were less emotionally responsive and interactive.[22] Even though autism has clear biological basis, there is evidence that children with autism are sensitive to their emotional environments. Interestingly, we have found that children with a range of biologically based "processing difficulties" and special needs are especially sensitive to their environments because their biology makes it hard for them to process and regulate sensations and to plan actions; they depend on their environments, that is, their caregivers, to help them learn these skills. As we have discussed, we have found that an "affect"-based developmental approach can help these children gradually learn how to relate, interact, signal, think, and communicate.[23]

As can be seen, cultural and family practices lead to high levels of emotional, intellectual, language, and problem-solving capacities. They can also likely lead to changes in the physical structure of the brain. Experience-based learning and physical changes in the brain tend to support each other. Both depend on vital cultural and family-mediated interactions that have enabled us since prehuman times to develop and sustain our vital capacity for emotional signaling.

THE MISSING LINK IN PIAGET'S THEORY OF COGNITIVE DEVELOPMENT

As we examine the role of emotional signaling, it becomes more evident that Piaget, the pioneer of modern cognitive psychology, didn't sufficiently explore the most important element in thinking—the role of affect. While Piaget was interested in children's imaginations and pretend play and the parallels between emotional and cognitive development, he did not fully observe the critical role of affect in the pathways leading to intelligence. We have found that affect is vital for understanding all levels of thinking as well as brain development.

Affect is also crucial in trying to understand the processes that enable a child to construct progressively more complex mental operations that enable her to understand her world. As we have been demonstrating, progressively more complex modulated emotional signaling is a vital factor in these developmental processes.[24]

Let's examine Jean Piaget's comprehensive theory of cognitive development and explore its missing components. Piaget thought that the first steps in logic a baby mastered were means-ends relationships; that is, causal thinking at the level of sensory-motor interactions with the physical world. For example, a baby learned to pull a string to ring a bell. If one cut the string, the baby would stop pulling it because it didn't ring the bell. Piaget said that this showed that an eight- or nine-month-old baby understood causality with what he called "means-ends relationships." What Piaget didn't understand was that causal thinking starts much earlier, when each time the baby smiles the mother smiles back. The first lessons in logic are not learned through motor actions, with one's arms interacting with the physical world, but through affective, emotional signals such as one's smiles and coos of delight and joy, and then further smiles or coos that lead to the caregiver's emotional responses. At every stage of development, these emotional interactions lead the way.

Therefore, the first problem solving is not sensorimotor problem solving, as Piaget thought, but co-regulated, emotional interaction through the emotional system. In addition, meaningful symbol formation (what Piaget called the "semiotic function") does not come about as an heir to sensorimotor processing, as Piaget postulated. As we described earlier, it comes about because emotional interactions enable the child to separate perception from action. Freestanding perceptions become invested with more and more emotional meaning.

An emotion-based theory of intelligence, however, can incorporate important insights from Piaget's theory.[25] Piaget brought the field of cognition to a more advanced plane by introducing a constructivist model. In this model, he emphasized that children learn through their own actions on the world (rather than passively). He described the parallel processes involved in cognitive and affective development but did not sufficiently explain the relationship between them. He was not aware of the observations or insights, as far as we can tell, that have led us to conclude that emotions are the architect of intelligence. Piaget's work has influenced child care and education (not sufficiently) as well as research and theory development.

Nonetheless, by focusing on external perceptual-motor behavior and later on perception, and by not fully appreciating the role of emotions—for example, to consider aspects of their development but not as the instrument and architect of learning—Piaget missed a critical factor in human cognition.

In *Intelligence and Adaptation: An Integration of Psychoanalytic and Piagetian Developmental Psychology,*[26] Stanley Greenspan presented a model to conceptualize the emotional basis for cognition, along with its impersonal components, and suggested that such a model offers a much more complete theory of cognitive development. Since then, we have accumulated additional evidence for this position. We have observed that typically developing infants, infants with autism and various types of cognitive and learning deficits, and infants in high-risk environments (e.g., multiproblem families) all need to use emotions to learn about the world, and interactions that help them master early affective interactions are especially helpful.[27]

There is good reason for the infant's first using emotions to probe the world. The infant has control of her facial muscles before she can control her arm and hand muscles to make purposeful actions with them. The emotional expressions, therefore, become the first probes into the world and continue throughout life. The interactions that lead to symbol formation occur at the same time as the advanced stages of sensorimotor intelligence, as Piaget described. For each new cognitive structure, however, there are emotional interactions that lead the way. Emotionally learned "lessons" then become applied to the impersonal world. In other words, emotional knowledge of the world leads to principles that are applied to the impersonal world, rather than vice versa, as Piaget suggested.

The emotional probes that the child uses to explore the world lead to cognitive capacities in many realms of experience, including verbal, visual-spatial, and motor. Interestingly, this model of cognition helps account for the unevenness of cognitive development that's been pointed out relative to different levels of experience. This fact is often pointed to in some critiques of Piaget's general model (e.g., why children with extra experience in certain areas, such as visual-spatial problem solving or verbal problem solving, often do not fit with Piaget's hypotheses about a sequence of cognitive structures).[28] We have observed that children differ in the degree of emotional interaction they have with caregivers in different processing realms (auditory, visual-spatial, perceptual-motor, tactile) as well as in their basic strengths in each processing realm and, therefore, in the relative development they attain in each processing realm.

Therefore, there are differences not only between children but within each child in his different processing realms.

HOW EMOTIONAL SIGNALING EXPLAINS PIAGET'S CONCEPTS OF SYMBOLIC THINKING

According to Piaget, the semiotic function involves the ability to represent something (a signified something: object, event, conceptual scheme) by means of a "signifier" that is differentiated and serves only a representative purpose: language, mental-image symbolic gestures, and so on.[29] Whereas in the sensorimotor period a thing could be represented by a part of itself (the mother's voice indicates the mother), *symbols and signs* are signifiers that are differentiated from their significants (e.g., the sound of a word or a visual design) and become available to the child with the appearance of the semiotic function.[30]

Piaget's theory of how the semiotic function comes about, however, does not explain why or how symbols develop. According to Piaget, the semiotic function, which makes representation possible, is heralded by five earlier developmental patterns in the second year of life. These include deferred imitation; symbolic play, or pretending; the drawing, or graphic image; the mental image, or internalized imitation; and the verbal evocation of events not occurring at the time of these processes. Imitation is especially important. According to Piaget, symbols are created by the same process that gives rise to deferred imitation. Piaget's descriptions of these related developmental processes are useful, but of themselves do not explain symbol formation.

Similarly, Piaget asserts that although symbols are generated by a process in which *accommodation* outweighs *assimilation,* they are then placed at the service of a process in which a liberating assimilation outweighs accommodation. For Piaget, a symbol or thought is constructed from the coordination of actions and not from a figurative or linguistic base. The semiotic function, which finds its roots in imitation, liberates the child from the bonds of immediate space and time and enables her to begin manipulating signals to think. Therefore, for Piaget, the critical process involved in the formation of the semiotic function and the capacity to use symbols to think is the progressive differentiation of symbols from their external points of reference and their use for interiorized operations. This is an eloquent statement of what thinking involves, that is, (1) the capacity to use symbols for "interiorized operations" (thinking) and (2) for creating meanings that are not fixed by an external point of reference. Describing the processes of thinking, however, does not explain how or why it develops.

Piaget's theory of assimilation and accommodation also describes elements in the process, such as the relative freedom from external points of reference, but doesn't fully explain the process. Similarly, Piaget's focus on imitation and deferred imitation, while identifying developmental steps in the process leading to thinking with symbols, doesn't explain the how and why of the process.

As we have been describing, the construction of true symbols and their use in thinking depends on several stages of emotional transformation. These enable infants, toddlers, and children to use emotions to signal intent rather than take full action; to tame and modulate catastrophic affects pushing for fixed actions; and to move from fixed actions to problem-solving emotional interactions with meaningful persons and objects. These steps enable the child to separate a perception from its fixed actions and, thereby, construct a freestanding perception or image. The formation of freestanding images as they become invested with emotional meaning, in turn, becomes a foundation for symbolic thinking. As emotional signaling becomes more complex and regulated, it leads to further and further separation of perceptions or images and their originally, relatively-fixed emotions and actions. *In this way, complex regulated emotional signaling leads to more and more differentiation between the symbol and its external points of reference, allowing for greater and greater interiorized operations.* In our "affective" explanation of symbol formation and thinking, we offer an explanation for the how and why of this process. As affective signaling separates perceptions from actions, freestanding images become possible; that is, the structure for symbol formation and interiorized operations is now available. As further interaction associates meaningful experiences with these internal images, symbols are formed.

There is another feature of reflective thinking that needs to be accounted for. It is responsible for maintaining the needed balance or equilibrium (Piaget's model) between assimilation and accommodation. This involves being able to maintain a solid grasp of the external points of reference of "reality" while constructing a fuller and fuller thinking-based (interiorized) view of reality. In other words, it's easy to become so abstract that one loses grasp of the external points of reference for reality—that is, become lost in one's own logical system, however coherent it is vis-à-vis it's own assumptions. Similarly, it's easy to be so tied to the external points of reference for reality that one becomes too "concrete and literal."

How is this vital balance maintained between the external points of reference for reality and the interiorized differentiated processes involved in thinking? We propose that continuous emotional signaling between people maintains this balance. The reciprocal emotional symbolic responses of a caregiver to a child's symbolic play establish a boundary between what's inside the child and what's outside, leading to an initial symbolic capacity for reality testing. Reality testing is formed from the reciprocal affective interactions between the affective "intent" of "me," or the self, and the affective intent of another person, that is, the "not me."[31] Similarly, these exchanges of emotional signals during symbolic play lead to the ability to connect symbols together in a causal manner, for example, "My angry dolls caused her sad doll to feel even sadder." Emotional signaling, as we have been discussing, therefore, not only leads to symbol formation but maintains the vital balance needed to sustain true symbolic thinking.

Piaget's model identifies important steps and processes associated with impersonal thinking and cognition. The insufficient understanding of the role of affect, however, is a "missing link." This missing link is most obvious in trying to understand the processes that enable a child to form symbols and progressively differentiate and yet maintain a vital balance between her internal world and the processes that enable her to think from her external points of reference.

CONSCIOUSNESS: HOW AFFECTS CONNECT THE MIND AND BRAIN

Exploring the phenomenon of consciousness provides further understanding of the role of affects in the functioning of the mind and the brain. Suppose that a sleep lab equipped with all the latest technology in MRIs and ECGs were to map every event that occurred during someone's dream. For the sake of argument, let us suppose that they could accurately determine when the dream began and ended, whether the dream was pleasurable or produced anxiety, even whether it was similar to a dream that the subject had had before. No matter how refined our technology might become, however, the one thing it will never be able to tell us is what the dream was about; for that we must rely on the subject's narratives. And therein lies the crux of the mind/body problem.

The biggest obstacle to understanding the relationship between mind and brain has been to figure out how the two interact. It is now very clear that the brain serves as a vital substrate for the mind. We say this on the grounds that brain injuries or lesions can affect selected mental capacities; or that stimulating parts of the brain can cause a subject to experience particular memories or influence mental phenomena in other ways; or that severing the corpus callosum results in the two hemispheres of the brain operating independently of each other, so that if an image is projected on the right visual field the patient can describe what they see but not if it is projected on the left visual field, even though they can point to a similar object. But the missing piece has been to understand how consciousness arises: how the biology of the nervous system relates to a human being's capacity for self-awareness, or for creative and reflective thinking.

The problem here is that a person's conscious experiences can only be described in their own terms: They cannot be reduced to a description of neurophysiological patterns. We can only understand, say, a creative production as a product of a range of personal experiences, which, to be sure, are organized by contributions by the underlying biologically based nervous system. The key question, however, is: How does this process get started? How does consciousness itself come about?

The problem here is to avoid getting stuck in age-old polemics that revolve around assumptions that simply stay as assumptions: for example, that ultimately we will be able to explain the mind in terms of physiological processes once our understanding of the brain is sufficiently advanced; or that understanding how the brain works will *never* enable us to understand the mind. The way out of this polemic lies, not in perseverating on one side or the other, but in understanding the mechanism through which consciousness and creative and reflective thinking come about.

To understand this phenomenon, we need to understand the early transformations of affect. Affects bridge the two worlds. Affects are partially physical, mediated, for example, in part through the sympathetic and parasympathetic nervous systems (e.g., a tightening of the muscles). Yet they are also mental or psychological, experienced as conscious feelings, such as anger or sadness. Affects begin early in life in part as a physiologic process. We see newborn infants at various states of arousal and showing various types of physiologic reactions involved in distress, pleasure, etc. How do these global physiologic states begin to take on mental or psychological meaning?

The answer, we believe, lies in what we earlier described as the *dual-coding of experience.* For example, a baby touches its caregiver and registers both a physical sensation and pleasure; or she touches something hot and registers the physical sensation and a feeling of discomfort. Experience is, therefore, coded according to its physical and affective properties.

The human nervous system has the capacity, even at birth, to perceive differences in these double-coded sensations. This capacity to discriminate or detect differences improves as the caregiver begins interacting with her infant, exchanging sights, sounds, touches, and other physical sensations and their associated affects because such interactions enable the infant to experience a range of subtle sensations. In addition to detecting differences early, the infant's nervous system can organize sensations, and more broadly, experiences, into patterns. The formation of patterns, however, also depends on and improves with interactive experiences, such as an interactive game with a caregiver. Such a game might involve a pattern incorporating pleasurable touch, sound, and certain comforting rhythms of rocking. As the central nervous system of the infant grows, it is better able to detect differences in physical and affective sensations and organize patterns, but only if interactions with caregivers are sensitively responsive to the different infant cues and are soothing and regulating at the same time. For example, stress can interfere with an infant learning to optimally discriminate among sensations and the formation of central nervous system pathways having to do with emotional discrimination and regulation.

The baby's motor responses are also double-coded because in order to respond, such as turn towards the sound of a mother's voice or to look for her smiling face, the baby has to experience some type of calming, a pleasurable affective component. Thus, a baby looks away from an aversive sound and looks towards a comforting and pleasurable one. With growing motor control she reaches towards a pleasurable touch and away from an unpleasant one. Motor responses quickly move beyond reflexes and become part of an affective-motor pattern. Affect serves as a mediator between sensation and motor response, connecting the two together. This basic unit of sensory-affect-motor response becomes more and more established through infant-caregiver interactions.

But how does the double coding of experience, coupled with sensitive, regulating infant-caregiver interactions and the nervous system's capacity to detect differences and organize patterns, work together to create emotional psychological experience? As we indicated, the infant experiences sensations such as a caregiver's sound as a physical event and, at the same time, as an affect, such as pleasure or discomfort. As the infant detects more and more differences, one sound from another, she is detecting both different physical sensations, such as a louder vs. a quieter sound, and different degrees of pleasure or discomfort. At the same time, however, as indicated, she is detecting patterns. The quieter, more pleasurable sounds may come when she is being rocked, which is also pleasurable. A "pleasurable" pattern made up of experiences which are experienced with similar types of pleasure begin to emerge. As this occurs, she is able to gradually experience differences between this pleasurable pattern and another pattern which is less pleasurable or aversive. This is the beginning of a long process of experiencing more and more specific patterns of affects or emotions.

As the infant separates these physical experiences with affective qualities into more and more separate or subtle patterns, she is, in essence, creating "mental" or psychological phenomenon. The organization and detection of different patterns of physical and affective sensation is a mental or psychological phenomenon. The degree to which the affective qualities of the interaction have a depth, range, and subtlety to them, a very loving, warm, intimate touch, look, and "you are my sweetheart" set of sounds, as compared to a shallow, mechanical, hurried interaction, is the degree to which the mental patterns and consciousness capture sensations that have what we commonly think of as a range of emotional qualities. Initially, the organization and detection of emotional patterns is experienced without symbols or words (i.e., presymbolically). Later on, these patterns can be experienced and described symbolically. These constitute different levels of consciousness.

However, an infant can only experience and, therefore, detect differences in and organize patterns of sensations they actually experience. Therefore, to perceive and construct a category of "intimacy" depends on the degree of nurturing pleasure an infant experiences with a primary caregiver. Superficial affection will be a very different type of experience and lead to a very different type of pattern. To construct a pattern of empathetic responses, later on a child will need to experience a caregiver's empathy. In other words, the depth, quality, and subtlety of one's inner life depends on the depth, quality, and subtlety of one's relationships and their emotional interactions. Infants who are deprived of emotional interactions, as often seen in some orphanages, often do not fully develop the capacity to detect different emotional patterns.

Not just the quality of emotions, such as the depth of intimacy, but the subtlety of the pattern one constructs depends on the degree to which caregivers read and respond to subtle differences in an infant's behavior and emotional expressions. This is because the infant's capacity to detect differences and construct patterns depends on the responses and patterns she experiences to her own expressions. How sensitive are caregivers to her different overtures? If subtleties of pleasure or anger are missed or ignored, she will not be able to experience fine-tuned differences in these emotional states.

Similarly, as indicated earlier, the very capacity to perceive a pattern itself beyond its rudimentary form also depends on emotional interactions with caregivers. There are two caregiver-infant interactional qualities that propel this process forward. One is the sensitivity with which the infant's cues are read and responded to and, in turn, the sensitivity with which the infant responds to the caregiver. The more "sensitive" these interactions are, the more subtle the sensations the infant learns to detect and organize. The other is the regulating nature of the interactions. Unregulated interaction tends to maintain global physiologic states (e.g., fight or flight reactions or an overwhelmed versus a calm, alert infant). Only regulated interactions enable an infant to detect fine-tuned, subtle gradations of sensations because such sensations only occur or are experienced as part of sensitive, regulated interactions. In other words, what we call emotions and the texture of consciousness is influenced by the qualities of distinctly human interactions characterized by regulation, sensitivity, and a variety of feeling states. Children deprived of these types of interactions do not experience a range of subtle emotions, especially those dealing with caring, love, and empathy.

Therefore, through sensitive, regulating interactions with caregivers, the infant's global physiologic states become regulated and experienced more and more as *both discrete physical and affective sensations* (e.g., different types of pleasure or comforting). *In this way, global physical or physiologic states take on the qualities we call emotions. As the infant experiences and organizes these affective states (emotions) into patterns, a mental or psychological level of experience (consciousness) unfolds.*

Thus we see basic central nervous system processes that have a strong physiologic component become transformed into affective processes through the dual-coding of experience, the detection of differences, and the construction of patterns. As this process occurs, experience is forever changed as affects become more and more part of an internal world of patterns and ultimately subjective experience. We described in Chapters 1 and 2 how children typically go through a number of levels of affective transformations leading to creative and logical thinking and then different degrees of self reflection.

We can think of consciousness as existing on these different levels. Beginning with the basic level of physiologic, affective arousal (a sense of aliveness) and developing into subjective awareness of global internal affective states, such as pleasure and discomfort, it progresses into the internal experience of more differentiated affects, such as different degrees of pleasure, annoyance, and so forth. As symbols are formed as an outgrowth of separating perceptions from actions, through the capacity for emotional signaling, consciousness comes to embody self-awareness, including the capacity to label and reflect on affective states or feelings. Consciousness is, therefore, a growing dynamic process throughout the course of life.

The following point is often missed, however. What is the necessary foundation for consciousness? Can computers be programmed to have it or any types of truly reflective intelligence? The answer is NO! Consciousness depends on affective experience (i.e., the experience of one's own emotional patterns). True affects and their near infinite variations can only arise from living biological systems and the developmental processes that we have been discussing.[32]

SUMMARY OF HOW EMOTIONAL SIGNALING LINKS EMOTION TO COGNITION AND SUBSYMBOLIC AND SYMBOLIC CAPACITIES

We have shown how emotional signaling may explain the vital components of the relationship between emotion and cognition in terms of the brain and cognitive theory. In healthy development, primary emotions such as fear and anger (more or less, depending on the degree of emotional health) tend to be transformed from catastrophic emotions and fixed perceptual-motor or perceptual-action patterns into emotional signals and interactive emotional patterns increasingly in the second half of the first year of life and the second year of life and thereafter.[33]

As this happens, more or less, emotions are no longer locked into global action patterns, such as fight or flight or withdrawal. As emotions are used to signal, they can be modulated through emotional interactions with others. The perception of the emotion, such as fear or anger or joy, can be experienced as a freestanding perception or image. This image can then acquire meaning through further emotional interactions. For example, as indicated earlier, the image of a mother becomes associated with a whole pattern of feelings—love, devotion, frustration, anger. As it acquires emotional meaning, the image and its associated emotional patterns become a symbol. As a growing symbol, it can be united with other symbols into concepts and ultimately become part of an integrated system of reflective thinking.

What becomes conscious, therefore, is the pattern of emotional signaling into which the primary emotions have been transformed. For healthy development, which progresses fully into the stage of a continuous flow of emotional signaling, the primary emotions no longer exist in their original form. They are now part of an interactive emotional pattern. In healthy development, even when emotions are intense, they are experienced as part of complex patterns. As indicated earlier, during the second half of the first year and through the second year of life, babies increasingly shift from experiencing the world as discrete emotions and behaviors to perceiving and constructing patterns out of experience. Interestingly, this observation has been confirmed by many researchers looking at babies and caregivers from different perspectives, including cognitive, affective, social, and cultural.[34]

To the degree primary emotions are transformed into interactive signals and patterns that are symbolized, the emotional system operates as an integrated unit. Each emotional experience has many components, all working together as a whole. These include visceral (heart throbbing or stomach churning) sensations, signaling ("I'm scared"), interactive patterns with expectation ("Mom will soothe me and I will feel better"), and symbolic meaning ("What could be scaring me so much?").

To the degree to which primary emotions are not sufficiently transformed into interactive signals—that is, maladaptive functioning—primary emotions may continue to operate in their original catastrophic mode in which perception is tied to global action patterns. This pattern would be associated with different types of disorders.

FUTURE RESEARCH

To build on LeDoux's and other neuroscientists' extraordinary work in mapping the pathways involved in primary emotions such as fear, we would encourage colleagues in cognitive neuroscience to look for the pathways involved in the system of emotional signaling we have been describing. We would suggest looking at the relationship between the emotional signaling system and integrating pathways between the different parts of the central nervous system. We would suggest studying emotionally healthy, well-regulated, reflective individuals in comparison to less emotionally healthy individuals to see whether they have better functioning integrative pathways (perhaps more neuronal connections and better regulatory dynamics) between the subsymbolic and symbolic systems. In such studies, we should look at the person's level of functioning and individual profile of motor, sensory, and affective processing in addition to their diagnosis.

As can be seen, the transformation of emotions into higher levels (emotions into signals that become symbolized) enables emotions to organize a vast array of cognitive operations, including reflective thinking and operational intelligence. This hypothesis can also enrich Piaget's model of intelligence. It is the transformation of emotions into interactive signals that is the vehicle for the progressive differentiation of the self and the object world necessary for reflective thinking as well as the instrument for progressing from one level of thinking to the next.

— 12 —

Emotional Development Derailed: Pathways to and from Autism

THE THEORY WE HAVE been presenting in this work is put to its most rigorous test in our work with children with Autistic Spectrum Disorders (ASD) and children who evidence similar social, language, and cognitive problems due to extreme emotional deprivation in institutional or home settings. Our theory of symbol formation, language, and intelligence holds that a series of critical emotional interactions early in life are responsible for these abilities. When these abilities do not develop, our theory suggests that these critical early emotional interactions have not been mastered. Biological factors can make it difficult for a child to participate in these emotional interactions (as in autism) or, as happens in institutional care, the caregivers themselves may not provide them.

When we looked at children with ASD and children with early emotional deprivation, we found that they had not fully participated in and mastered a number of these critical early emotional interactions.[1]

We then asked a more difficult question. Would it be possible to help children who lack critical symbolic, language, and intellectual abilities to develop them with an intervention program that creates opportunities for these formative early emotional experiences? Would a subgroup of children with ASD whose biological challenges were not too great (but who nonetheless had a biological basis for not mastering these functional emotional developmental capacities) be able to engage in these critical emotional interactions and progress to high levels of relating, symbolic thinking, and empathy? We have been pleased to find that both these questions could be answered affirmatively.[2] With an intervention program that focuses on basic emotional interactions in the context of the child's individual processing differences, we have found that a subgroup of children with ASD are able to learn how to engage and communicate and, therefore, advance to symbol formation, expressive language, and reflective thinking, as well as to develop empathy and enjoy relationships with peers and adults.[3] The fact that providing critical emotional interactions can enable a subgroup of children who typically do not develop high levels of symbolic thinking to develop them in spite of their biological challenges provides important support for the theory of human intelligence presented in this work.

AUTISM AND AUTISTIC SPECTRUM DISORDERS (ASD)

As is well known, autism is a complex developmental disorder involving delays and dysfunctions in social interaction, language, and a range of emotional, cognitive, motor, and sensory capacities. Empathy and creative and abstract reflective thinking (including making inferences), reading emotional signals, and engaging in emotional interactions are generally believed to be particularly difficult for individuals with this disorder. Many professionals believe these abilities are beyond the reach even of a subgroup of individuals with autism, even if they make the maximum progress possible in an intensive therapeutic program. As indicated, we believe and have demonstrated, however, that this belief is incorrect. Understanding the role of emotional signaling in the developmental pathways associated with ASD has led to strategies leading to levels of creative and reflective thinking and empathy formerly thought unattainable.

Although there are no definitive national data, a variety of studies estimate that the rate for autism is from 2 to 4 per 1,000, and for ASD, higher. In some locations, however, the rates are at the higher end of this range. For example, in one study conducted by the CDC, the rate was 1 per 250 for autism and approximately 1 per 150 for ASD.[4] In comparison to these current estimates, the rates ten or fifteen years ago were considerably lower. Although they tended to vary a great deal by the study conducted, the most widely cited rate then was 1 per 2,000 to

2,500 for autism. Estimates from ten to fifteen years ago are not as readily available for ASD. Although some believe these increasing rates are due to better identification and diagnosis, many investigators believe there is an alarming increase in autism and ASD.

It has been assumed that autism is a unitary disorder related to an as-yet unidentified genetic pattern. It is further assumed that there may be multiple forms of expression (i.e., phenotypes) related to this genetic pattern. Current research, however, may suggest consideration of an alternative hypothesis—that of multiple pathways, each with a different genetic pattern and set of clinical features. These different pathways may also, however, share some common features, much like many different medical problems involve "fever" or "inflammation." Therefore, it may be useful to consider a multipath, cumulative risk model. In support of such a model is research showing a range of biological, developmental, and clinical patterns (e.g., different genetic neuroanatomical, neurophysiological, neurochemical, and clinical patterns).[5] Within such a model, perhaps different genetic patterns and prenatal and postnatal developmental processes create vulnerabilities to cumulative prenatal and postnatal risks or challenges.

Current research suggests there may be multiple factors involved in the pathways leading to ASD. Therefore, it may be useful to consider a cumulative risk model. There is a great deal of research supporting genetic influences. Genetic factors and prenatal developmental processes are likely to predispose a child to autism or create vulnerabilities to prenatal and postnatal challenges that have the same effect. These challenges may include infectious illnesses; toxic substances such as lead, methylmercury, PCBs, organophosphates, nicotine, and endocrine disrupters such as Dioxin; and factors that precipitate active autoimmunity in genetically predisposed children, for example, viral infections, vaccines, and allergies. Clinical observations suggest that postnatal factors, although not causative, may also include experiential or physical stress as a contributor to some of the observed behavioral patterns. For example, children with severe sensory hyper- or hyposensitivities, motor planning, and auditory processing problems that are biologically based will tend to be more likely to withdraw from relationships and become perseverative and self-stimulatory in a noisy and chaotic setting. Research should focus on multipath cumulative risk models and explore the mechanisms of action among different etiological and precipitating, intensifying, and/or ameliorating factors.

In our view, the clinical features of ASD are best understood in terms of primary and secondary features. The primary features involve deficits in the ability to engage in interactions involving emotional signals, motor gestures, and vocalizations; difficulty in maintaining these exchanges to solve problems; and difficulty in processing auditory, visual-spatial, and other sensory input and planning and sequencing actions. The secondary features that derive from the more primary impairments involve the well-known language, cognitive, and social problems, as well as tendencies toward self-absorption, perseveration, and self-stimulation.

These challenges can be intensified by inappropriate or inadequate interventions and improved by appropriate interventions and individually tailored family and educational environments. An appropriate intervention must be based on an understanding of a child's unique developmental profile and unique pattern strengths and weaknesses. Research to better understand causes and improve interventions should focus on improved methods to identify individual differences and describe and classify clinical subtypes. Without adequate understanding of individual differences and clinical subtypes, important findings are often missed because there is no way to know to which children they apply.

THE CORE PSYCHOLOGICAL DEFICIT IN AUTISM

When children with autism are compared to children without autism, and their level of intelligence, as measured with IQ tests, is controlled for, there are various autism-specific developmental deficits. These include deficits in the ability for empathy and seeing the world from another person's perspective in both physical and emotional contexts (theory of mind);[6] higher-level abstract thinking, including making inferences;[7] and shared attention, including social referencing and problem solving.[8] In addition, deficits in the capacities for emotional reciprocity[9] and functional (pragmatic) language[10] also appear specific to autism.

As we looked at these various deficits, we began asking whether they might stem from a common pathway. We also looked at neuropsychological models that have been suggested to account for the clinical features of autism to see if the deficits described in these models might also stem from a common pathway. These include models that focus on planning, sequencing, and problem-solving (i.e., executive functioning), information processing, shifting frames of reference (i.e., set theory), and, as described above, the capacity to understand the thoughts and feelings of other people (i.e., theory of mind).[11] Our clinical work with infants and children has shown that all these abilities (i.e., to empathize, form a theory of mind, think abstractly, and solve problems with others, as well as use language functionally and engage in emotional reciprocity) stem from the infant's ability to connect emotions or intent to motor planning and emerging symbols.[12] Relative deficits in this core ability, we found, led to problems in higher-level emotional and intellectual processes. The core psychological deficit in autism, we reasoned, may, therefore, involve an inability to connect emotions and intent to motor planning and sequencing and to emerging symbols. We further hypothesized that biological differences (genetic and developmental) associated with ASD may express themselves through the derailing of this critical psychological process, which, in turn, leads to the symptoms and cognitive, language, social, and motor deficits seen in autism. In other words, this vital intermediary process may stand between the biological factors and the manifest symptoms associated with ASD.

As discussed earlier, a child's capacity to connect emotions to motor planning and emerging symbols begins to become more apparent between nine and eighteen months of age as the infant shifts from simple patterns of engagement and reciprocity to complex chains of emotional reciprocity. Consider a fourteen-month-old child who takes his father by the hand and pulls him to the toy area, points to the shelf, and motions for a toy. As Dad picks him up, and he reaches for and gets the toy, he nods, smiles, and bubbles with pleasure. For this to occur, the child needs to have a desire or wish—emotional intent—for what he wants. He then needs to connect this desire or intent to an action plan. The desire and the action plan together enable the child to create a pattern of meaningful, social, problem-solving interactions. Without this connection between desire or emotions and action plans, complex interactive problem-solving patterns are not possible. Action without desire or meaning tends to become repetitive (perseverative), aimless, or self-stimulatory, which is what is observed when there is a deficit in this core capacity.

It is hard for a child to get beyond simple motor patterns if he cannot connect them with his desires. When we create high states of affect or motivation to help children with autism learn to use their motor actions purposefully, exchange emotional signals, and connect with others, we see their motor planning improve. Under states of high motivation, a child who is capable only of repetition can embark on a two-step sequence. For example, consider a child who obviously wanted to go out the door but was vacillating between touching it repetitively and aimless spinning around near it. When the child was helped to open the door a little to heighten his affect in terms of his desire to go out the door, the child responded immediately. He began gesturing toward the doorknob as if to say, "Hey, open the door." He did this by touching the door and then moving his caregiver's hand toward it. He went from repetitive actions to a purposeful, two-step touch. If the desire or affect becomes too intense, however, it can lead to a tantrum.

If this connection between emotional intent and purposeful action can be established, it can enable the child to begin participating in back-and-forth emotional signaling. If emotional signaling, in turn, becomes sufficiently complex, it can enable the child to separate perception from action, tame overwhelming emotions, and use freestanding images to create symbols. The meaningful use of symbols usually emerges from the meaningful emotional problem-solving interactions that enable a toddler to understand the patterns in his world. Without this connection between emotions and motor planning, symbols are often used in a repetitive manner (e.g., scripting, echolalia).

When a child is able to interact with emotional signals, he can modulate emotions and action; this in turn permits flexible scanning of the environment. The child gets feedback from what he sees and, based on that feedback, explores further. In this way, his visual-spatial world and motor functioning becomes integrated.

The ability to interact flexibly with others and the rest of the environment also makes for associative learning. Associative learning (building up a reservoir of related experiences, thoughts, feelings, and behaviors which give range and depth to one's personality, inner life, and adaptive responses) is necessary for healthy mental growth. Its absence leads to rigid and mechanical feelings, thinking, and behavior patterns, as are often seen in ASD.

Reciprocal, affective interactions and affectively guided problem-solving interactions and symbols are necessary for the unique capacities research has shown to distinguish individuals with autism from individuals without autism (as outlined earlier). The unique capacity for social reciprocity depends on affect guiding interactive social behavior. The ability for shared attention, which includes social referencing and shared problem solving, depends on affect guiding shared social problem solving and explorations. Empathy and theory of mind capacities (see Chapter 10) depend on the ability to understand both one's own affects or feelings and another person's affects or feelings and to project oneself into the other person's mindset. This complex emotional and cognitive task begins with the ability to exchange affect signals with another person and, through these exchanges, emotionally sense one's own intent and the other person's intent (a sense of "self" in interaction with "another"). Similarly, higher-level abstract thinking skills, such as making inferences, depend on the ability to generate new ideas from one's own affective experiences and then reflect on and categorize them.[13]

In observing infants and toddlers heading into autistic patterns and in taking careful histories of older children with autism, we noted that they did not fully make the transition from simple patterns of engagement and interaction into complex emotional and problem-solving interactions. Even affectionate children who were able to repeat a few words or memorize numbers and letters, but who went on to evidence autistic patterns, did not master, for the most part, this early continuous exchange of signals. They were unable to develop empathy and creative and abstract thinking without an intervention program that helped them learn to engage in this flow of emotional exchanges.

In a review of the developmental profiles of two hundred children with ASD, we observed that most of the children shared the unique processing deficit we have been describing. Approximately two-thirds of the children who developed ASD had this unique type of biologically based processing deficit that involved the connection of emotions and intent to motor planning and sequencing capacities as well as to emerging symbolic capacities.[14] At the same time, however, the children differed with regard to other processing deficits involving their auditory, motor planning, visual-spatial, and sensory modulation abilities. These differences accounted for the variety of social, language, motor, and cognitive problems and impairments that accompany this fundamental deficit.

We have labeled the hypothesis that explores the connection between affect and motor planning and sequencing, as well as other processing capacities, the Affect Diathesis Hypothesis. This hypothesis asserts that a child uses his affect to provide intent (or direction) for his actions and meaning for his symbols or words. Typically, during the second year of life, a child begins to use his affect to guide intentional problem-solving behavior and, later on, meaningful use of symbols and language. Through many affective problem-solving interactions, the child develops complex social skills and higher-level emotional and intellectual capacities.

THE EARLY ROOTS OF AUTISM

Because the unique processing deficit we are proposing as part of the pathway leading to ASD occurs early in life, it can undermine the toddler's opportunity for expectable learning experiences essential for many critical emotional and cognitive skills. For example, he may have more difficulty eliciting ordinary interactions from his parents and the people in his immediate environment. He may perplex, confuse, frustrate, and undermine purposeful, interactive communication even if he has competent parents. Without such experience, he may not be able to comprehend the rules of complex social interactions or to develop a sense of self. The child may not learn social "rules" or develop friendships and a sense of humor, which are learned at an especially rapid rate between twelve and twenty-four months of age.[15] By the time this child receives professional attention, his stunted interaction patterns with caregivers have, therefore, excluded him from important learning experiences and may be intensifying his difficulties. The loss of engagement and intentional relatedness to key caregivers may cause him to withdraw more idiosyncratically into his own world and become even more aimless and/or repetitive. What later looks like a primary biological deficit may, therefore, be part of a dynamic process through which the child's lack of emotional interactions has intensified specific early biological processing problems and derailed the learning of critical social and intellectual skills.

The early roots of these problems may make early identification possible. For example, most babies will move rhythmically to the sound of their mothers' voices.[16] We are currently conducting studies to see whether children at risk for ASD evidence a difference in these rhythmic emotional-motor patterns. At each developmental level, there may be opportunities for early identification and preventive intervention.

To foster such efforts we have further developed our theory on the relationship between affect and motor planning. This relationship begins very early in infancy as sensations, affects, and motor responses are used together.

THE SENSORY-AFFECT-MOTOR CONNECTIONS

In healthy development, as the infant progresses through the first year of life, she's continually connecting the sensory system to the motor system. As this occurs, the sensory system provides direction and purpose to the motor system. For example, an infant sees his caregiver's face and turns toward it.

Even early in infancy, however, these sensory-motor patterns have a special sensory component: affect. As indicated earlier, affect operates like an additional extra sensory capacity. A person touches a surface and it feels cold, but also aversive or scary or pleasurable. As we have shown, all sensation or perception is double-coded according to its physical and affective qualities.[17] There are infinite variations to the affective coloring of sensation. For example, there are many subtle variations in the experience of pleasure, excitement, joy, rage, and fear. This degree of variation enables us to use affect to code and store a near infinite amount of information as well as retrieve it later on.

As infants progress during the first year, the capacity to create links between the physical and affect qualities of sensation and motor behavior enables the infant to go beyond basic sensory-motor reflexes and operate more and more in terms of patterns. He can perceive patterns, as well as organize his own behavior into patterns. For example, the baby perceives the pacifier sitting on his mommy's head. She experiences interest and pleasure in this sight and the novelty of its location. She reaches for it. We now have a purposeful motor pattern. As the infant develops, he is able to bring together one or two purposeful units into larger patterns. He sees a toy, experiences delight, points to it, and takes his father by the hand and, through multiple back-and-forth interactions involving perception, affect, and action, persuades his dad to pick him up so that he can reach for the toy. By the second year of life, these patterns lead to a sense of self (the affective, intentional agent of the pattern) and a sense of others (the perception, expectation, and associated affects of the "others"). As we have discussed, this developmental progression enables the child ultimately to form and give meaning to symbols as well as develop higher levels of thinking.

What happens, however, when biological factors (or severe deprivation or abuse) interfere with the formation of a primary connection between the sensory system and the motor system? In such a case, motor behavior is not tightly linked either to the physical or the affective qualities of sensation. When this happens, we observe that infants evidence more aimless motor behavior. Instead of hearing or seeing the caregiver (i.e., a perception) and turning towards her (a motor behavior), the infant may move his head aimlessly. There is more or less no perception or sensation guiding motor behavior. Similarly, instead of hearing his mother's voice and kicking his legs in a rhythmic movement (in synchrony), the child may move his legs more randomly.

The relative lack of a connection between sensation and motor behavior interferes with connections forming between affect and motor behavior. Therefore, as the infant perceives the pacifier on his mommy's head and experiences pleasure in the sight, he is unable to connect it to a motor action, such as reaching for the pacifier.

If the infant could reach for the pacifier, this behavior would let the caregiver know of the baby's perception and pleasurable interest and create the first step in a back-and-forth pattern of communication. The caregiver would likely hold the pacifier closer and, after the infant took it, gesture for it back with an outstretched hand, a big smile, and a welcoming sound.

When a baby is unable to create a back-and-forth pattern of communication, she is unable to fully participate in the primary units of purposeful interaction that give rise to complex interactive patterns. Complex interactive patterns (i.e., a continuous flow of back-and-forth communication) are necessary if a child is to understand the world's social, language, and physical patterns. Such complex interactions enable problem solving, the construction of a sense of self, symbol-formation, and higher levels of thinking.

When biological factors interfere with the formation of the critical sensory-affect motor connection, the infant and child has difficulty in successfully negotiating the developmental pathways leading to these emotional and intellectual capacities. Our observations suggest that this is what happens in many children with autistic spectrum disorders. In children with autistic spectrum disorders, there may not be a total lack of connection between the sensory-affect and motor system but rather a relative one. This relative lack of sensory-affect-motor connections makes it difficult for affect to guide motor actions and, therefore, we observe a compromised ability for complex, purposeful, meaningful behavior. For example, an infant with this challenge may evidence some purposeful and even reciprocal interactions, but often will not progress to initiating and sustaining complex, continuing, reciprocal interactions and shared purposeful problem-solving behavior.

As indicated, we call the model, which formulates the role of affect and sensation in creating purposeful, meaningful behavior (and then symbols), the Affect Diathesis Hypothesis. In this model we formulate a number of stages through which the relationship between sensation, affect, and motor behavior progresses.[18] Understanding these stages has enabled us to identify infants at risk for developmental problems and autistic spectrum disorders early when interventions are likely to be more successful.

At each stage in the sensory-affect-motor connection, we are able to look for the presence, absence, or compromise in this critical capacity. For example, in the first six months, we look for different types and degrees of aimless rather than purposeful, coordinated or synchronous motor activity. We also look for aimless rather than synchronous or purposeful interactions between infants and caregivers. In the second half of the first year, we look for a lack of initiative and multiple reciprocal affective interactions in a row (i.e., back-and-forth patterns of perception, affect, and action). In the first half of the second year of life, we look for a toddler demonstrating a lack of initiating and sustaining a continuous flow of back-and-forth complex social problem-solving. Then we look for the lack of the meaningful use of emerging symbols and a lack of progression to the meaningful and creative use of symbols (i.e., affect-directed conversations) as well as a continuation of the earlier presymbolic difficulties.

This developmental model also informs an approach for early intervention and preventive work. Because the sensory-affect-motor connection in children at risk of autism and related developmental problems is not an all-or-nothing phenomenon (i.e., involves relative degrees of interference in the connection), there is potential for interventions to strengthen the sensory-affect-motor connections. The main "highways" may be relatively blocked, but the side roads are often available to be developed. The way to develop the side roads is to strengthen the connection between sensation, affect, and motor action.

As we have discussed, to strengthen this connection, we have found that creating heightened affect states is especially helpful. As heightened affects are connected to simple motor actions, infants and young children can often become more purposeful. For example, a toddler is moving a little train back and forth repetitively. If the caregiver puts the train on his head, the child may reach for it and vocalize with delight. As affect connects more and more to motor actions, it also connects motor patterns to sensations (affect is part of the double coding of sensations). As the sensation-affect-motor connection is strengthened, we observe more and more purposeful affective behavior. Purposeful affective interactions and motor patterns, in turn, lead to reciprocal affect signaling, a sense of self, symbolic functioning, and higher-level thinking skills.

To function in the ways described above, affect must be relatively pleasurable and regulated (a little annoyance can be helpful, but a disorganized tantrum is not). With pleasurable, regulated affect, there's an opportunity for many purposeful interactions in a row, such as when a baby takes a pacifier, hands it back, takes it again, and so forth. An overwhelming and fearful response, on the other hand, will tend to lead to a one step action, an avoidance response, and, in all likelihood, if the fear were great enough, to fragmented or aimless behavior. Overwhelming negative affects, while leading initially to a purposeful behavior, often overwhelm the infant and child and lead to more disorganized patterns.

Therefore, heightened affect is required to create the connection between sensation and action and help the child move into complex interactive patterns. As indicated, when we have been able to create emotional interactions involving heightened pleasurable affect with infants and children evidencing the early symptoms of ASD, we've observed enormous progress in the most essential capacities—relating and responding to emotional signals and creating meaningful ideas.[19]

The goal of creating the sensation-affect-motor connection is usually approached through a comprehensive, intensive intervention program that tailors pleasurable, regulated interactions to the child's individual processing differences (including sensory over- or underreactivity, auditory processing and language challenges, and visual-spatial and motor planning and sequencing difficulties).

In summary, the difficulty in connecting affect to motor planning and sequencing begins with a challenge in connecting sensory and motor patterns and, as the infant develops, includes creating the sensory-affect-motor linkages that make complex interactive patterns possible. If affect is unable to provide direction and meaning to motor patterns, the child's capacity to engage in a continuous flow of reciprocal affect cueing as part of social problem solving is severely compromised. Because complex affective cueing is a critical step in symbol formation, the creation of meaningful symbols (i.e., symbols invested with affect) also becomes compromised. The steps in this process are summarized below:

- Compromises in simple sensory-motor connections; seen in a baby who is having difficulty moving to the rhythm of his caregiver's voice and with other synchronous rhythmic activities;
- Compromises in purposeful emotional interactions and complex sensory-affect-motor linkages; seen in a baby who is having difficulties in vocalizing to a caregiver's vocalization or in his reaching for a caregiver's nose;
- Compromises in reciprocal affect cueing (circles of communication) and complex sensory-affect-motor linkages that enable complex interactive and perceptual patterns;
- Compromises in the continuous flow of affect signaling, shared social problem solving, and the complex sensory-affective-motor planning linkages that make an early, organized sense of self and the early meaningful use of symbols possible;
- Compromises in the formation and use of meaningful complex symbols because of compromises in linkages between affect, action, and emerging images.

There are many children who do not evidence autism but have developmental problems, for example, severe motor problems, in which intentionality or purposeful action is difficult in its own right. Such problems result in less practice at using intentional behavior and in participating in intentional interactions. In these circumstances, creating purposeful interactions around any motor skill (even head or tongue movements) may strengthen the intent-motor connection and reduce aimless, repetitive behavior. Recent MRI (magnetic resonance imaging) studies suggest that practicing and improving motor skills may enhance the developmental plasticity of neuronal connections.[20]

SECONDARY DEFICITS

In our review of two hundred cases of ASD, although many children shared the primary deficit of connecting affect to processing capacities, their differences in other processing skills tended to create a variety of symptoms and splinter skills, such as whether a child lined up toys (which requires some motor planning) or just banged them, or scripted TV shows (which requires some auditory memory) or was silent. Children with relatively stronger processing skills in one or more areas tended to make rapid progress once they were helped to connect emotions or intent to these stronger skills. Children with numerous processing problems tended to make more gradual progress and required specific therapies, such as speech and occupational therapy.

These observations are consistent with recent neuroscience studies suggesting that different processing skills may compete for cortical access, depending on functional use.[21] They are also consistent with neuropsychological studies of individuals with autism—but without mental retardation—that show that "within affected domains, impairments consistently involved the most complex tasks dependent on higher-order abilities" (concept formation, complex memory, complex language, and complex motor abilities).[22] Higher-level abilities tend to depend more on "meanings" that, in turn, depend on emotional interaction with the world. These observations are also consistent with work on the shifts to a more complex central nervous system organization, including hemispheric connections that occur at the end of the first year of life and the early part of the second year.[23] They may also help us understand why many children with ASD use peripheral vision (they don't look directly at caregivers but seem to look from the side) rather than central vision to scan their environments. The neuroanatomy of the visual tracks is such that peripheral vision requires only one hemisphere, the left or right one, to function. Central vision, however, requires that both hemispheres function together (because some of the pathways cross over and others do not). It would be reasonable to explore the hypotheses that problems in integrating the two hemispheres may contribute to ASD. Integration of the two hemispheres would likely facilitate emotional interaction and integrated central vision, and it's possible that complex emotional interaction with a continuous flow of emotional signaling may facilitate using both hemispheres together. Work showing that the limbic system and hippocampus are developing and forming cortical connections at around eighteen months is also consistent with these clinical observations on the increasingly purposeful and meaningful use of actions and ideas in the second year of life.[24]

We have seen that, during the first year of life, many infants who later evidence autistic patterns could focus on objects, experience some affection and warmth, and even enter into simple reciprocal interactions.

Perhaps they are able to perform these tasks because the basic patterns can be carried out by either side of the brain alone.[25] Complex, goal-directed, reciprocal, emotional patterns, however, may require that both sides of the brain work together for optimal functioning, especially during the early years. Although children appear able to learn to engage in complex social interactions with only one functioning hemisphere, they may require special learning to do so. As we discussed in Chapter 9, emotional interactions enable the child to use all his senses, as well as motor and language abilities, together. These may, therefore, be necessary for integrating central nervous system pathways to form.

This model may also explain why some children make rapid progress and even evidence precocious capacities once they can connect emotional intent to other skills. Perhaps the component parts of their nervous system are developing quite well but lack only the direction and coordination of affect or intent. Other children may have deficits in several component parts. For them, connecting emotional intent to the component parts is only a first step that begins a slower pattern of progress. In all likelihood, the pathways that connect emotion and intent to sequencing of motor, verbal, and visual behavior involve many different tracks of the central nervous system.

Our model of the core deficit can be contrasted with hypotheses proposing that the primary deficit in autism is the child's inability to construct a theory of mind, to understand or imagine another person's state of mind, and to empathize with the feelings of others.[26] In our view, this inability may very well be a product of the more primary difficulty in connecting emotions and intent to motor complex and interaction with others. In our population, the children who made very good progress learned to empathize gradually as they became more emotionally involved in other people's lives. The children who were unable to become involved with others and communicate emotionally in a continuous and spontaneous manner, and who instead relied on scripts or prompts, did not develop the ability to appreciate the emotions of others.

There is also recent research suggesting that children with autism tend to employ different perceptual strategies, such as looking at another person's mouth rather than the eyes, and have difficulties in reading another person's facial expressions.[27] These findings, however, may be a developmental consequence of the more primary, earlier deficit that we have described. Infants and toddlers who do not engage in reciprocal affective gesturing do not have the experiential foundation for understanding the emotional cues of another person, including facial expressions, tone of voice, body posture, and the like. The ability to comprehend emotional gestures comes from many months of practicing this interactive experience. Furthermore, infants and toddlers without this experience may be unable to regulate their emotions and, therefore, tend to be overloaded by emotional cues. If these children are oversensitive to sensations—as many children at risk for ASD are—they experience such cues in an especially intense manner. The eyes tend to convey more emotion than the mouth. Children with these sensitivities may find looking at someone else's eyes too overloading. Therefore, the preference for the mouth over the eyes may, in part, be related to the broader difficulty in modulating emotions.

Research in progress by Morton Gernsbacher and colleagues at the University of Wisconsin-Madison supports this thesis. She has recently shown that individuals with autistic spectrum disorders who were found to have different brain imaging patterns when looking at pictures of the human face had the expectable brain imaging patterns when they were encouraged to look at the face (i.e., their "abnormal" brain imaging patterns were due to their not looking, rather than processing the experience differently). She further showed that these individuals, when encouraged to look at, for example, the expressive eyes, showed a physiologic stress response indicating that it was difficult for them to experience this type of emotion. Dr. Gernsbacher's work, therefore, shows that individuals with ASD may have heightened sensitivity to emotions and, therefore, may attempt to cope by not looking at the source of those emotions. This work beautifully illustrates the complex interplay of perception, emotion, and interaction in human development and functioning.[28]

CLINICAL OUTCOMES WITH THE DEVELOPMENTAL MODEL OF INTERVENTION

In the chart review mentioned above, the children's patterns and clinical course were based on an experienced clinician's observations; detailed notes were organized according to a scale we devised called the Functional Emotional Assessment Scale.[29] The children described in this study were treated with methods based on our developmental model—methods known informally as the "Floor Time" approach, or formally known as the Developmental Individual Relationship-based model

(DIR™).[30] The DIR™ approach is a comprehensive program that involves harnessing a child's affect, tailoring interactions to his profile of motor, sensory, and language processing strengths and weaknesses, and facilitating all his functional/emotional developmental capacities (engagement, affective signaling) up to his highest levels and helping him move to the next level. It often includes intensive home, educational, and therapeutic components.

To describe outcomes, we divided the children's progress into three broad groups. A "good to outstanding" outcome group included children who, after two or more years of intervention, evidenced joyful relating, including preverbal gestures with a variety of emotional cues (appropriate, reciprocal smiling, frowns, looks of surprise, annoyance, glee, happiness, and the like). They were able to engage in lengthy purposeful and well-organized social problem-solving interactive sequences (over fifty circles of spontaneous communication), and share attention on various social, cognitive, or motor-based tasks. They could use symbols and words creatively and in pretend play and conversation. They could construct logical bridges between their ideas and hold a logical, two-way conversation, separate fantasy from reality, and anticipate consequences. Most important, in this group, the children's symbolic activity was related to their underlying emotional intent and desires, rather than to memorized or rote sequences. They progressed to high levels of abstract thinking, including the capacity to make inferences and experience empathy. In general, these children mastered basic capacities, including reality testing, impulse control, the organization of thoughts and emotions, a differentiated sense of self, and an ability to experience a range of emotions, thoughts, ideas, and concerns. They no longer showed self-absorption, avoidance, self-stimulation, or perseveration. On the Childhood Autism Rating Scale (CARS), a standard way of indicating the degree of autism, all the children in this group shifted into the nonautistic range.

Some children in the "good to outstanding" group became precocious in their academic abilities, reading or doing math two or three grade levels above their ages (some perhaps developed their visual-spatial abilities early when auditory processing lagged). Some, even though they had intact basic personality functions, still evidenced auditory or visual-spatial difficulties, which were improving. Most of the children in the "good to outstanding" group, even those with precocious reading or math skills, had some degree of motor planning challenges in penmanship or drawing or in complex, gross motor activity.

A second group made significant gains in their ability to relate and communicate with gestures. They became related to their parents, often seeking them out in a joyful, zestful, and pleasurable manner. Parents commented, "I've discovered a little person inside my child." They could enter into longer sequences of purposeful emotional cueing and interactions, but did not necessarily become capable of a continuous flow of interaction. They could also share attention and engage in social, cognitive, and motor problem solving. In this group, however, the children were still having significant problems in developing their symbolic capacities. Some had the partial ability to use symbols in pretend play and language, but it was significantly below age levels. For example, in this group many children could engage in concrete pretend play sequences, such as driving a car or feeding a doll, and use words for some simple negotiations of their desires ("I want to go outside" or "I want juice"), but were not yet able to construct long, creative, symbolic exchanges; that is, they couldn't have a conversation or elaborate an experience in a play sequence. This group, like the first group, no longer evidenced self-absorption, avoidance of relating, self-stimulation, or perseveration.

A third group continued to have significant difficulties in both the presymbolic and the symbolic realms. They had difficulty with attention and with simple and complex sequences of gesturing. If they were using some concrete symbols in pretend play when props were available, or could use language when they wanted something, this was coupled with a significant degree of self-absorption, avoidance, self-stimulation, and perseveration. In this group, those who had some symbolic capacity (e.g., to sing songs or work puzzles) were unable to imitate and use these abilities in an interactive, communicative manner. Many in this group were making slow progress in their basic ability to relate with warmth to others, but some alternated between progress and regression.

One hundred sixteen of the two hundred children (58 percent) were in the "good to outstanding" outcome group, fifty (25 percent) were in the "medium" outcome group, and thirty-four (17 percent) continued to have significant difficulties. Some of the group with significant continuing difficulties were making very slow progress. In a subgroup of those with significant difficulties, eight (4 percent) were vacillating or losing capacities. It is important to emphasize, however, that this was not a representative population of children with ASD. Although the families who came to see us represented a broad range of challenges, they were also motivated and may have seen some special strengths in their children.

Therefore, it is reasonable to hypothesize that a "subgroup" of children with ASD can make enormous progress. Only future "clinical trial" studies, however, will be able to confirm this hypothesis and determine what percent of children are in this subgroup.

TABLE 12.1 Floor Time Intervention Outcomes

	n = 200 %
Good to Outstanding	58
Medium	25
Ongoing Difficulties	17

SEQUENCE OF IMPROVEMENT OF CHILDREN IN THE DIR™ PROGRAM

Children who made progress with this treatment model tended to improve in a certain sequence. First to improve was the child's expression of emotion and pleasure in relating. Within the first three to four months we would usually see greater joy and positive emotions, along with more consistent relatedness (e.g., seeking out parents and caregivers). Even children who had been extremely avoidant and self-absorbed would, after parents were playfully obstructive for periods of time, begin going over to their parents and signaling them with a look, smile, or pat on the knee. Some parents worried about being too playfully obstructive. "Won't he get mad at me if I get stuck behind the door?" These parents were pleasantly surprised when their children after a while would push them to get stuck behind the door so that they could play the same game again. By creating problems for their children to solve through playful obstruction, they made it possible for the children to "undo" the parents' actions. This was very important for children who could not initiate and sequence purposeful behavior and interactions because of difficulties in motor planning.

Contrary to stereotypes of autism, the children seemed eager for emotional contact, but initially couldn't figure out how to achieve their goal. They seemed grateful when their parents helped them find ways to engage in greater social interaction.

Eighty-three percent of the children, which included children who progressed very slowly, initially showed improvement in the range and depth of their engagement and their pleasure and emotions. Once engaged, many made further gains and moved from simple to complex emotional and motor gestures.

Long sequences of emotional interaction led to the third area of gain—the emergence of functional symbolic capacities. Creative and imaginative symbolic elaboration and the expressive use of language always followed presymbolic emotional cueing and communication. Many children went through a transitional stage in which they used words from video or book scripts and then became more and more creative with their behaviors and gestures. If we overfocused on the words rather than the gestures and affects, we slowed progress. Children who remained rigid and stereotyped in their gestural interactions were often rigid and stereotyped as they learned words; for example, using scripts and ritualized language. Once the children became more flexible and creative in nonverbal interactions—a big smile, trick Dad by hiding the cookie in her hand—they would begin to use symbols more spontaneously and creatively.

As children became more symbolic, many could not stop talking or flitting from one idea to another. It was as though they were excited about their newfound gifts. There was a mixture of fragmented and illogical ideas, islands of fantasy, and some scripting, such as repeating words heard on TV, as well as repeating words to get needs met. Over time, however, 58 percent of the children were able to use their emerging symbolic skills creatively and logically.

Most of the children could express their own ideas much more quickly than they could comprehend the ideas of others. Even children who initially had some understanding of others' language—for example, of simple commands—still had more trouble with incoming information than in their ability to express ideas. They knew what they wanted to say but understood others inconsistently. Even when they could tell you what they wanted ("Go out play" or "Give me juice") or do pretend play sequences with the dolls hugging and kissing, they would find it difficult to answer the abstract "what," "where," and "why" questions ("What do you want to do next?" or "Where do you want to go?" or "Why do you want to go outside?").

Eventually, with a great deal of interaction and dialogue, the ability to abstract and comprehend the ideas of others emerged. Children did not get to this level unless their parents and the therapists focused on rapid, two-way, symbolic communication. For example, it wasn't sufficient to listen to a verbal child and repeat what he said. Caregivers had to challenge the children to process incoming ideas; for example, they used affective tones, visual clues, multiple choices, and statements that inspired complex verbal responses to help the children deal with more abstract dialogues. It required long back-and-forth exchanges rather than short, thirty-second conversations: "Oh, you want to go outside? To do what? To play or kick the ball? Which one would be more fun?" Pretend play in which the caregiver became a character who enjoyed verbal interchange was very helpful as well: "I'm hungry! I need something to eat!"

The children with good-to-outstanding outcomes who became creative and logical were able to hold spontaneous, affect-driven, two-way, symbolic communication. As a consequence, they were able to learn to differentiate their internal worlds.[31] Logical thinking, impulse control, and an organized sense of self emerged.

For many, there were two steps in this process. First, they learned to hold short creative dialogues that lacked a cohesive, integrated capacity for thinking or an organized sense of self (islands of logical dialogue). Over time, they learned to integrate and expand. The islands became continents and a cohesive, integrated sense of self and capacity for logic emerged. As a consequence, their academic abilities also improved as they became more flexible, were able to learn how to use functional logical exchanges, to engage in two-way thinking, to solve problems, and to work together with others. Their peer relationships also improved, but it required a great deal of practice—four or five play dates a week and access to very communicative peers in preschool and school programs. With dynamic, interactive academic learning in a warm, secure, organized setting, many of the creative and logical children developed academic abilities that ranged from average to superior. When they were in overly structured academic settings for most of the day, however, their academic progress was slower and they tended to remain more rigid, concrete, and rote.

For some children, the augmentative use of pictures, signs, and other symbolic equivalents was very helpful. For a small group of children who were unable to make the transition into the symbolic realm, we added more structured techniques to improve skills, gestures, and then use of words. When we combined these with the dynamic, interactive "Floor Time" approaches, these structured strategies were more effective than when used alone.[32]

The pattern of progress described above occurred in the context of a comprehensive intervention program that involves many elements. All the elements are guided by the DIR™ model of facilitating f/e capacities through interactions that are tailored to a child's individual differences. These elements are outlined in the following table. It should also be mentioned that the National Academy of Sciences in their report "Educating Children with Autism" pointed out that there is growing support for the developmental, relationship-based models, with a number of programs, including the DIR™ model, showing positive results for children with ASD.[33]

TABLE 12.2 Elements of a Comprehensive Program

Home-based, developmentally appropriate interactions and practices (floor time)

- Spontaneous, follow-the-child's-lead floor time (20–30 minute sessions, eight to ten times a day);
- Semi-structured problem-solving (15 or more minutes, five to eight times a day);
- Spatial, motor, and sensory activities (15 minutes or more, four times a day), including:
  a. Running and changing direction, jumping, spinning, swinging, deep tactile pressure;
  b. Perceptual motor challenges, including looking and doing games;
  c. Visual-spatial processing and motor planning games, including treasure hunts and obstacle courses;
  d. The above activities can become integrated with the pretend play.
- Peer play (four times a week).

Speech therapy, typically three or more times a week

Sensory integration-based occupational therapy and/or physical therapy, typically two or more times a week

Educational Program, daily

- For children who can interact and imitate gestures and/or words and engage in preverbal problem-solving, either an integrated program or a regular preschool program with an aide.
- For children not yet able to engage in preverbal problem-solving or imitation, a special education program where the major focus is on engagement, preverbal purposeful gestural interaction, preverbal problem-solving (a continuous flow of back-and-forth communication), and learning to imitate actions, sounds, and words.

Biomedical interventions, including consideration of medication, to enhance motor planning and sequencing, self-regulation, concentration, and/or auditory processing and language.

A consideration of:

- Nutrition and diet;
- Technologies geared to improve processing abilities, including auditory processing, visual-spatial processing, sensory modulation, and motor planning.

CHILDREN DEPRIVED OF FORMATIVE EMOTIONAL INTERACTIONS

Going back to the well-known studies of René Spitz, H. McV. Hunt,[34] and others, the results of severe emotional deprivation (a lack of nurturing interactions) during the early months and years of life have been well documented. Severe deficits in critical social, cognitive, and language capacities can result. Some children fail to thrive and may not survive. Some become very self-absorbed and aimless, with little or no language, cognitive, or social capacities, and others become socially promiscuous, antisocial, and relate to other people as "things." Even though the effects of emotional deprivation have been well known for the past forty to fifty years, in recent years we have seen children with these patterns who had been in orphanages in Eastern Europe. We also see children with these patterns who were reared in multiproblem families where mental illness or a variety of other risk factors interfered with nurturing interactions.[35]

The intervention program we developed for children who have suffered this emotional deprivation was fundamentally similar to the program we developed for children with ASD. Although the reasons for the children's deficits were very different in both instances, the goal was to create opportunities for the children to engage in formative interactive emotional experiences. A critical difference in the intervention approaches, however, was that the children in multiproblem families required substantially more family support and intervention in the dysfunctional family patterns. We have described the details of this component of the intervention model elsewhere.[36] We found a clear statistical relationship between our ability to help the caregivers develop their own ability to engage in formative interactions and the outcomes we saw in the children. For example, as caregivers began to learn how to form intimate relationships, their children were able to do so as well. As caregivers were able to learn to engage in emotional interactions, their children became better problem solvers. As the caregivers engaged in symbolic thinking and building bridges between their ideas and symbols (e.g., describing their feelings and reflecting on their feelings), their children showed higher degrees of language development and intelligence.[37] All such findings strongly support the theory presented in this work that formative emotional interactions are critical for symbol formation, language, and intelligence.

Many of the children who were adopted after having experienced emotional deprivation in orphanages often became part of families that could offer a great deal of nurturing care and could readily engage in these formative types of emotional interactions. With these children and their caregivers, our intervention focused on educating and supporting the caregivers regarding the best ways to tailor these interactions to the children's differences in sensory processing, sensory modulation, motor planning and sequencing, and existing emotional capacities. Some children and families progressed very quickly. Others progressed more slowly because either the children presented severe challenges or the abilities of the new caregivers or families were wanting. Overall, we observed patterns of progress very similar to those we saw in our work with children suffering from ASD. When emotional deprivation occurs very early in life, it appears to affect basic processing capacities in ways that are very similar to biologically based challenges.

Perhaps the most important finding regarding children with ASD and severe emotional deprivation is that in both groups the children's progress was proportional to their ability to engage in the six primary types of formative emotional interactions that our model hypothesizes leads to symbol formation, language, and intelligence.

We have also been able to describe the developmental pathways associated with depression, anxiety, aggressive behavior and antisocial patterns, and attention and learning problems. We have developed the DIR™ model treatment programs for each type of challenge.

The developmental model described in this chapter provides a new way of conceptualizing mental illness as well as mental health. It enables us to look at the pathways (and steps) to healthy or adaptive emotional functioning as well as see problems in terms of the stages of emotional growth outlined in Chapter 2. It then helps us look at challenges in terms of the biological contributions (expressed through motor, sensory, or language processing difficulties and environmental and interactive contributions). These challenges are looked at in the context of the goals of each of the stages in the developmental pathway. Each stage in the pathway, therefore, constitutes *an intervening developmental organization that organizes biology and experience.* The identification of these intervening organizations, in turn, makes possible improved interventions and new models for research.

—— PART FOUR ——

The Development of Social Groups

Introduction

IN THIS SECTION WE propose that our developmental model provides us with a framework for integrating studies of group formation and behavior with studies of a child's cognitive, linguistic, emotional, and social enculturation. This framework is not intended to replace but rather to support, and indeed, unify, existing approaches in anthropology and the social sciences. Moreover, the model that we present here will continue to evolve as we acquire further insights into how humans, individually and in groups, develop many of their most vital human capacities.

In the chapters that follow, we trace the developmental needs and capacities of groups and then consider how this framework provides an important additional lens through which to study large group behavior and gain a deeper understanding of cultural and social processes, including the great debate in anthropology between relativists and universalists. In the following chapter, we show how this model provides us with a new way to conceptualize the developmental origins of early human history from the Magdelanian period to the rise of civilizations, as well as a new understanding of the history of *history.* Finally, we turn our attention to the nature of the challenges that the global community currently faces and the new psychology that is needed to deal in an adaptive and reflective manner with the fact of global interdependency. As we will see, the developmental model of evolution that we have presented in this book helps us understand not only the types of capacities that need to be inculcated for future survival but also the forces that are presently impeding this process.

— 13 —

The Developmental Levels of Groups, Societies, and Cultures

With Elizabeth Greenspan

THE DEVELOPMENTAL FRAMEWORK THAT we have outlined in this book focuses on a few essential questions that standard approaches in anthropology and the social sciences do not consider, or at least not to the same degree. The issues that this framework highlights involve the following fundamental questions:

What are the developmental needs of social groups, societies, and cultures?

Are there certain fundamental needs that have to be met for a group to exist and for an individual to operate in a social or cultural context?

How do groups cohere to form a stable entity?

What are the processes that enable the members of a group to come together around common beliefs, shared institutions, kinship or bloodlines, and prejudices or biases?

Does coming together involve participating in some shared practical endeavor, such as providing the essentials for survival (food, shelter, and protection)?

What about the communicative processes of the group: How do they negotiate basic needs with one another?

How does a group mediate its shared sense of reality such that it can maintain a state of harmony in the face of both internal and external stressors?

There is, of course, a very rich literature in anthropology and the social sciences on the characterization of large groups, cultures, and social organizations. But the question of how individuals acquire the ability to form groups and societies, and how children acquire the unique characteristics of their culture, has largely remained a mystery. Moreover, the needs of groups vary according, for example, to the size of the group, its internal complexity, external conditions (physical and political), and so on. Our functional/emotional framework enables us to explore the basic developmental needs of groups and how they meet these needs in order to cohere and function under varying conditions.

Consider one example: For a group to exist, individuals must be able to relate to and engage with one another. Without this ability, individuals will operate as isolated entities. But how do these basic patterns of relating and engaging express themselves in different groups, cultures, and societies? Similarly, a group cannot exist without having a way of exchanging affective signals that enable rapid communication around issues of survival and relationships, such as safety, danger, approval, disapproval, acceptance, dominance, rejection, cooperation, aggression, caring, and affection. These communicative processes, even more basic than the verbal expression and negotiation of these issues, are a defining characteristic of all groups and societies, yet evidence different characteristics in different groups and societies. Some groups have much more refined preverbal affective signaling systems for certain issues than others. Indeed, this is one of the areas in which ethnographers have recorded remarkable variations.

In the pages that follow we will show how developmental needs must be met for groups to exist, maintain themselves, and survive. We will further show how these developmental needs provide a framework with which to understand how groups function and adapt to different challenges. The developmental framework we are proposing is not intended to be a way of classifying or putting values on group processes. Our intention, rather, is to explore the relationship between these processes and some of the major concepts that have been used in anthropology and the social sciences to characterize groups and societies.

Our developmental framework enables us to move beyond certain polarizing debates that have hitherto dominated anthropology and the social sciences. By enabling us to see the universal processes underpinning how a group meets its developmental needs, we are also able to account for the near infinite diversity or differences among group members and among groups and societies. Further, our developmental analysis of groups leads to a new way of conceptualizing the relationship between the invariant and endlessly diverse aspects of human societies.

Various modern schools of thought have argued that certain uniquely human traits account for social and cultural advances, and that insofar as such traits are uniquely human they must be biologically innate: encoded, perhaps, in the human genome. For many years, the leading candidate for such a trait was language, but in recent years theorists have searched for a more basic capacity that can account for our development of language skills and our complex social behavior. The leading candidates today are "mindreading" and "imitation"; that is, the idea that children are preprogrammed to engage in joint attention and to imitate their caregivers at some predetermined age, from whence follows their ability to communicate symbolically and to perceive social behavior in ways that are different from their perceptions of inanimate beings or objects.[1] But we have already seen how a child's capacities for "mindreading" or imitation are themselves a product of earlier stages of development. That is, only once children have progressed through the earlier levels of self-regulation and attention, engaging and relating with caregivers, becoming two-way communicators and then communicating in more complex affective exchanges, do they start to display these joint attentional and complex social imitative behaviors (as compared to earlier basic motor imitations). Thus, far from being genetic traits, these, too, are learned capacities.

THE DEVELOPMENTAL FOUNDATIONS OF CULTURAL PRACTICES

In an important cross-cultural study in language development, Katherine Nelson identified two kinds of language-learning styles in American and Japanese children; she called these styles "referential" and "expressive."[2] *Referential* children largely use common nouns, whereas *expressive* children use words marking actions or interactions. What was particularly interesting about Nelson's research was her discovery that the difference between these early language styles reflects a difference that occurs, even before words are spoken to any great degree, in the manner caregivers interact with their children, which in turn is a reflection of cultural attitudes and practices involved in child rearing. That is, the origin of the difference between referential and expressive language-learners lies in the different values and related practices that American and Japanese cultures use to inculcate in a child a strong sense of individualism versus a strong emphasis on social harmony.

A vivid example of this sort of enculturation is a fascinating experiment conducted with young Japanese and American children in which the child was asked to select a colored chip from a pile of blue chips containing a single yellow one in it. Young Japanese children consistently selected a blue chip, whereas young American children consistently chose the yellow one. Such behaviors are the result of the countless interactions that these children have had with their caregivers: The Japanese children have learnt, through the cold stares and unresponsive behaviors of their parents in response to behaviors that might draw attention to themselves in social situations, that individualistic behavior should be avoided; American children learn, through the warm smiles and vocalizations of their parents in similar contexts, that such behaviors are highly valued. It is through thousands of these sorts of interchanges that the child learns what is acceptable, what is felt, said, and done, and what is unacceptable, not said, and not done.

In recent years, social anthropologists have added considerably to this picture by showing how such cultural practices shape how a child speaks, and intimately connected with this, how a child conceptualizes reality.[3] For example, Penny Brown and Steve Levinson showed how different groups employ different frames of reference to organize their spatial experience. This is not simply a matter of using different terms to carve up space; it is literally a matter of seeing space in radically different ways. For example, according to our spatial frame of reference, if someone is sitting to my left at a dinner table, that person will continue to be on my left no matter what might happen to the room in which we are sitting. But in the spatial frame of reference of the Tzeltal Indians studied by Brown and Levinson, if two individuals from this culture should happen to be eating in a revolving restaurant, they would see the spatial relationship between them as constantly changing as the restaurant revolved.[4]

Our developmental model adds to this important line of research the crucial missing piece linking the child's cognitive, linguistic, emotional, and social enculturation with the development and functioning of her society. As a child is mastering the successive levels of f/e development, she is at the same time becoming a part of her social group in a progressively more differentiated way. In other words, *the growth of a child's mind and her social development are inextricably tied together. Indeed, the growth of a child's mind and the formation of groups are inextricably tied together.*

Early in life, a child's exposure to her group comes largely through her parents and immediate family members. As she grows older she engages more directly in the larger social group. The processes of emotional differentiation and the movement to higher levels of reflection occur in progressively larger social contexts. But it's the basic intertwined processes involved in the child's emotional and intellectual development and her social enculturation that tie the individual to the group psyche. This enculturation process is not a matter of learning how to use genetically determined abilities according to the specific conventions of the society into which one is born; rather, as we have seen throughout this book, the development of these abilities, and indeed, the very development of the brain, is shaped by these culturally mediated caregiving practices.

As the child negotiates the early developmental levels through countless emotional exchanges with her caregiver, she develops an implicit understanding of her society's attitudes towards beliefs and social practices, norms and values, power hierarchies and the kinship system, and so on. In other words, as we described in Chapter 2, long before symbols are used, affective interactions enable the infant and toddler to construct all sorts of patterns; these include the social expectations that define the child's sense of self and reality. Symbols, in part, acquire their initial meaning from this earlier affective reality. As she slowly expands her realms of social experience, from the relationship with her primary caregiver into multiple relations with the members of her family and her peers, the child forms an ever more complex identity until eventually she acquires an identity as a member of her society.

Throughout this process of enculturation, her early emotional gesturing continues to develop. As the child masters symbols and language, she is still communicating at the preverbal level. Symbols keep forming and reforming, amplifying and enlarging, because the emotional signaling system continues to operate in ever new meaningful social contexts. Even as adults, the beliefs, desires, and intentions that we embrace, the institutions, institutional processes, and political figures that we support, are all informed by implicit emotional processes that in themselves define the group's social identity and become a function of it.

The processes at work here are not *uni-directional*; rather, our development is very much a dynamic process in which levels of reflective thinking are influenced by the complexity of emotional interactions that continue to broaden and to differentiate, becoming more accommodating and adaptable. The better an individual can discern and express subtle nuances of meaning, the more complex and differentiated her social identity. Therefore, individuals are tied to the social group by their emotional gesturing and affect probes; these put group members into a constant relationship with their dyadic partners and their nuclear and extended families, as well as with larger emerging social groups. This vital presymbolic equilibrium is a constant source of the emotions that serve as the architect for each person's mind, and equally important, as the architect for the institutions and practices that govern the larger social groups, which, to exist and sustain themselves, must reflect these same affective interests. But because the emotional range of each individual and each group is highly distinctive and endlessly diverse, within the broad range of universals (such as affect cueing, symbol formation, and relating to others) there are near infinite variations.

The larger the group or society, the more it must develop a group identity that can cohere around these variations. For every time a group expands—that is, from dyad to nuclear family to extended family to small community, village, and so on—it becomes more challenging for the group to maintain an equilibrium, simply because it has to contend with more and more individual differences and variations. Therefore, the larger and more complex the society, the more it requires flexible affect gesturing to achieve security, governance, and cohesion.

Not all groups with the same degree of internal complexity possess the same degree of flexibility in their affect gesturing. The more constricted a group's f/e development, the less flexible is its affect gesturing, and thus, the less it can tolerate a wide range of individual variations and differences. Such societies can maintain cohesion in a limited or polarized manner; for example, by investing in a charismatic leader, or around racial stereotypes, or in superficial physical traits. For a modern multicultural society to achieve cohesion, it needs a flexible range of affect gesturing that creates ways of maintaining security, cohesion, and functioning in the midst of the welter of individual and group variations that it encompasses. The society can meet its basic developmental needs in such a condition, however, only if it attains a high level of reflective symbolic functioning and reflective institutions that, by necessity, are informed by the ongoing affects that invest those symbols and institutions.

A good example of the connections between individual development and cultural variations can be found in E. T. Hall's autobiography, *West of the Thirties.*[5] Hall recounts how, as a young man in the 1930s, he worked on the construction of dams in the Navajo reservations. He describes how, when they greet each other,

> White males (and now most white females) grip the proffered hand firmly . . . looking the other party directly in the eye, all of which is intended to convey interest, honesty, and sincerity. But when we would do this to the Navajos, it conveyed quite a different message; a direct, unwavering gaze meant anger. . . . The Navajo greeting does not center on showing relative strength and dominance (as it does with two Anglo males) but is instead a communication in which there is a mutual assessment of feelings and expression. As two men approach each other, eye contact is broken—at about the point where it is possible to begin to pick up the details of facial expression. Once this boundary is crossed, they look past each other, holding the approaching figure in their peripheral field of vision. To look directly at the other is tantamount to swearing at them.[6]

But Hall did not learn this lesson from a textbook or by taking field notes. Rather, Hall tells the story of how, a stranger in a strange land, he slowly began to learn the ways of the Navajo by being thrust into their midst. He was forced to abandon his cultural mindset if he was to succeed, not just in his job, but more fundamentally, in understanding and sharing in their unique mode of being-in-the-world.

It was second nature for a young white male of the period—especially one who had been given the important job of supervising a road crew of Navajo workers—to try to impress them at their first meeting by pulling up smartly in his truck, slamming the door shut, marching over to the workers and, in a loud and firm voice, making sure to look everyone directly in the eye, introduce himself as the new boss. How else was he to establish his authority? And if this didn't work, he had to slam the door shut even louder the next time, raise his voice and stare more fixedly, and perhaps discipline one or two of the more recalcitrant workers. But far from obtaining the sort of respectful attention that he had expected, Hall's behaviors were met with a stony silence that was filled with tension.

Hall slowly understood that his behavior was construed by the Navajo as highly aggressive and even a little bit unbalanced. The fact that all the white managers behaved in this fashion only confirmed for them that their supervisors all came from a somewhat deranged race. He learnt that, if he were to obtain the attention and cooperation he needed from his Navajo workers, he had to arrive at a work site as unobtrusively as possible and allow everyone an opportunity to adjust to his presence before quietly beginning to discuss the day's work; he also realized that he should keep his voicc down and avoid direct eye contact. And he learned these lessons, but not by consciously employing the kinds of "participant observation" techniques that are currently taught to anthropology students, and certainly not from his superiors, who were oblivious—or perhaps indifferent—to the reaction that their behaviors elicited. Rather, he learned these lessons by being sensitive to the reaction that his behaviors produced: a disinterest in his presence expressed through impassive facial expressions and posture; indeed, it was so profound that, as Hall put it, "I could live or die, and it wouldn't have made a particle of difference."[7]

It was only years later that, equipped with the tools and mindset of an anthropologist, Hall could articulate these fascinating insights into Navajo culture. At the time he was unaware of all that was going on. Thus, for example, he describes how he came to learn that even the way one walked was important: "The agency people clearly thought you should stride around like a white man, someone in charge. But I was spending most of my time with the Indians and, because I was fortunate enough to have what the linguists call a high adaptability factor, found myself walking as the Indians do without my being aware of it."[8] In other words, because of his high adaptability factor, Hall was able to assimilate the messages being signaled to him at the preverbal level. It is for this reason that, as he wrote in his autobiography, "Navajo culture became a part of my very being. Already, I found myself avoiding eye contact when close to others or talking to them. Already, I was synchronizing my body movements with the Navajo rhythm and tempo, which I found to be smoother and more coordinated than my own (most whites move rather jerkily). Already, I was learning how to enter and leave situations and how to comport myself in ways congruent with those of the Navajos."[9]

Hall's high adaptability factor is the sort of construct that every anthropologist can immediately resonate with, even though they are unable to explain precisely what is going on or how one acquires such a skill. Indeed, for anthropologists it is almost as if this were some sort of special gift possessed by only a few ethnographers for reasons unknown;

maybe genetic! But the f/e model of development enables us to see quite clearly why Hall was able to pick up, intuitively, on the subtle rhythms and cadences of Navajo life. The process that Hall was going through in his entry into Navajo society was exactly the process that a young child goes through in the early stages of development that we have outlined. That is, like a young Navajo child, Hall had to go through the early stages of attention and engagement, emotional signaling, symbol formation—using Navajo symbols and words—and connecting symbols together to define reality in a Navajo fashion. Like a young child, he was going through these stages of Navajo enculturation by learning, unconsciously, through affect gesturing, which of his behaviors were met with stony silence and which were met with quiet approval or laughter; what could be talked about and which things were considered sacrosanct.

As another example, Hall quickly learned that "to a Navajo, a person's name is sacred and is endowed with power; they do not abuse that power by calling people by their names to their faces."[10] But he was never told this explicitly; such a spoken prohibition would, in fact, have been antithetical to Navajo norms of appropriate behavior. Rather, he learned from the tense facial expressions and avoidant behaviors that greeted his requests to be told someone's name that such a question was never to be asked in Navajo society. This is precisely what Navajo children of the period had to learn. They had to learn that their ceremonial names were never to be told to others for fear of misuse, or even witchcraft; that one should never mention the name of the recently deceased; and that it was *very* rude to ask someone directly, "What is your name?" And children were taught all these lessons through preverbal signals, long before they could even speak.[11]

It was because Hall had very flexible affect gesturing skills (Level 4) and was so solidly entrenched in his capacity for emotionally differentiated gray-area thinking (eighth stage) and his ability to think from a stable internal sense of self (ninth stage) that he was able not only to adapt to this completely new mode of being but also to become aware of this fact. As he himself remarked: "I did not see these adaptations as giving up any part of my own personality, which is a common fear of many Americans and their excuse for not taking other cultures seriously."[12] That is, because he was solidly entrenched at the lower and the higher levels of f/e development—although still a young man—Hall became aware that he was intuitively perceiving and responding to the preconscious affective signaling the Navajos conveyed through gestures, facial expressions, vocalizations, body posture, and even gait.

The process that Hall went through applies not just to how one behaves, communicates, and perceives reality but also—and, perhaps, especially—to the more abstract social constructs studied by anthropologists, such as kinship. A Navajo child of the period had to learn that what marked a Navajo individual was the complicated matter of one's clan. A Navajo's clan affiliation came from the mother's line, but the child was born for the father's clan and in addition had the clan affiliation of his or her maternal and paternal grandfathers, for a total of four clans. One could not marry into one's clan (which was considered incestuous) and one was under considerable moral obligations to one's fellow clan members to provide food and other subsistence items and to protect one another. But the manner in which the child learned these norms was entirely "intuitive," which in our terms means through presymbolic affect gesturing. That is, a young child was not presented with a manual detailing "How to Behave with the Members of Our Clan." Rather, the child learned implicitly, from the myriad interactions with clan members, the subtle obligations and expectations that extended to fellow clan members.[13] One can easily imagine a child's experiencing the disturbingly impassive facial expressions that Hall experienced when he transgressed social norms; and similarly, the same sorts of subtle smiles, head nods, friendly gestures, and animated movements when, like Hall's, their actions conformed to their culture's sanctioned modes of kinship behavior.

In this manner, our developmental model helps us perceive the different stages or steps through which the enculturation process progresses. And as can be seen in Hall's memoir, or Mead's classic *Coming of Age in Samoa,* or Whiting and Edwards's *Children of Different Worlds,* such an argument applies not just to the implicit manner of communicating, walking, thinking, and behaving, but also to the emotions that they feel or the manner in which children think about their physical and social worlds. Hall remarks how the difference between the thinking of the Navajo and of ourselves is greater than "the difference between Homer's thought processes in *The Iliad* and those embedded in the script for *Star Trek.*" But the problem for cultural relativism from the beginning has been that of explaining why it is that, underlying these extraordinary cultural variations, children everywhere experience interest, happiness, surprise, sadness, anger, disgust, and fear. Indeed, they all seem to display the same basic facial expressions of emotion, such as smiling or showing fear. Moreover, as mentioned in earlier chapters, joint attention and the social smile emerge in nearly all children around the age of two months; at around three months children everywhere start to react when they hear their names; and from six to nine months they start to shake their heads to indicate no, and so on. Without an explanation of how these steps develop, cultural relativists have failed to persuade a large contingent of nativist theorists in psychology as well as anthropology that these very general traits and behaviors that are present in virtually every child in every known society are not biological phenomena.

Our model supplies a critical foundation for cultural relativism by clarifying how the basic stages of development outlined in Chapter 2 lead both to the various universals that have been identified by structuralist anthropologists and to the endless cultural variations recorded by ethnographers. To see how such an argument works, consider one of the most basic of all universals: All groups engage in some form of communication. Indeed, the very notion of a group whose members could not communicate with one another is a contradiction in terms. This was the basic insight that Thomas Hobbes articulated when he warned in *Leviathan* that should society ever be allowed to collapse its members would revert to a solitary state in which no one could communicate with anyone else. Where Hobbes was wrong, however, was in likening this condition to the natural state of animals, for as we saw in Chapters 3 to 5, nonhuman primates do indeed possess quite sophisticated preverbal communicative skills that are essential for the maintenance of their surprisingly complex societies.

Such a point has greater significance for the debate between cultural relativists and structuralists than might at first appear, for it is precisely because preverbal signaling is such a primeval phenomenon—precisely because of the fact that these learned capacities predate the earliest human and even the earliest hominid groups—that this is a universal of human society. That is, the caregiving practices that promote the development of such universals have been passed down and thus learned anew, not just from one generation to the next but even from one species to another, over the course of millions of years. Hence, far from being irreducible, universals are themselves the products of more basic developmental processes.

Furthermore, the very nature of the processes underpinning universal traits and behaviors are such that they allow for infinite cultural variations around different emotions, beliefs, and social practices. Consider, for example, how in American culture one discourages someone from having a conversation without looking one in the eyes by, for example, raising one's voice, adopting stern facial expressions, or making aggressive gestures. But Navajo culture does exactly the opposite. Hence, the basic universal operating here, namely, co-regulated affect gesturing, provides the means for cultural variations. Similarly, consider the sorts of affirmative and affirming preverbal responses that greet a young child in America who seeks to be the center of attention versus the sorts of negative preverbal responses that would have greeted a Navajo child's attempt to engage in similar behaviors. These looks and head shakes and vocalizations don't just convey to the child which behaviors are desirable and which are unwelcome; *they become a core part of the child's personality.* Throughout life, the individual will convey to others, through these same sorts of gestures and head nods and vocalizations, which behaviors or values or attitudes are desirable and which are not.

The same point applies to the various universals identified by structuralist anthropologists, namely, every society has language, or a family structure, or a kinship system, or a folk psychology. These are not somehow innate traits of human society, behaviors programmed into the human genome that compel individuals to band together in groups around which they must then organize some kinship system or folk psychology. Rather, such universals are the product of the core developmental processes that enable individuals to exist in groups.

In other words, *it is the pattern of reciprocal, co-regulated affective interactions in the early stages of development that helps the child differentiate her own individual personality and helps the group determine its collective personality.* Some emotional inclinations are part of long co-regulated chains that are creative in exploring new directions; others are part of shorter chains of interaction that tend to be more rigid and more limited. Thus, the f/e framework looks at the range and stability of those interactional patterns that lead to catastrophic affects in a child versus those that lead to modulated and regulated affects; for example, where there is a cultural sanction to respond to certain behavior by withdrawal versus a cultural norm that encourages responses to other behaviors in a rewarding fashion.

The way societal myths and rituals evolve from these implicit processes is that, as individuals progress to symbolic levels, they need to provide symbolic meaning for what they already know implicitly. That is, implicit processes are elevated to the symbolic level where the culture's implicit beliefs are expressed and elaborated. For example, the culture that is anxious about themes of sexuality may form and invest meanings in certain symbols by creating stories or rituals that make sexuality taboo and that thereby justify and give meaning to what is already *felt* and maintained through affect gesturing to be a basic truth. The culture that creates affect gesturing to inhibit aggression or assertiveness and encourage dutiful behavior may create myths and stories that support that particular theme. A culture that exchanges emotional signals that support adventuresome behavior and curiosity often finds myths and stories to support a more adventurous outlook, such as the Horatio Alger story in American culture.

Thus, our f/e model will enable us to understand the constructs studied by anthropologists and social scientists, such as semiotic functioning, the content of rituals or myths, kinship systems, legalistic structures, and socioeconomic status, that have been formulated to characterize groups. By seeing the intertwined nature of individual and group processes, we can also shed light on one of the major problems that has long stymied social theorists: how to account for group behavior that is often quite foreign to how the individuals involved normally behave. For example, it is now widely documented that individuals can suffer a loss of their personal boundaries—a loss of control over their feelings and thoughts—when they come together in a crowd, particularly in regard to certain emotions that seem to spread like wildfire, such as fear and anxiety, anger, or excitement. Thus we see crowds behave in an irrational manner even though the individuals involved normally behave in a perfectly logical and coherent manner.[14] Understanding the developmental needs and levels of groups provides another lens through which to comprehend why the group may function at a fragmented or polarized level, even when many individuals are capable of functioning at a reflective level.

THE DEVELOPMENTAL NEEDS AND PROCESSES OF GROUPS

As we saw in Part II, at each juncture in human evolution, individuals not only developed new skills in communicating and thinking, they also developed progressively more complex social relationships, social groups, cultures, and societies. The two processes involved here supported each other. The ability to relate, communicate, and think made social organizations possible and workable. Complex social organizations provided economies of scale that fostered physical protection, more effective means of providing the basics (food and shelter, as well as safety and security), and allowed time for thinking to be used for innovations, such as new inventions or technologies that, in turn, led to more economies of scale and more time for thinking, planning, and implementing increasingly complex social, cultural, and economic structures, as well as for exploring higher levels of reflective thinking.

Traditionally, anthropologists and social scientists have thought about groups in terms of the macroprocesses and systems that enable these larger social organizations to function. These (including kinship systems, family structures, myths and rituals, classification systems, power hierarchies, and socioeconomic status), however, emerge from more basic elements in the formation of groups, cultures, and societies. We have seen in the preceding sections that we can begin to describe these more basic developmental foundations that help explain how groups, societies, and cultures evolve and maintain themselves around these systems and structures.

The steps that we have seen in the development of higher-level, symbolic thinking in individuals provide clues regarding the basic developmental foundations of groups. The links between these needed foundations and the levels of individual development, as well as the stages of human evolution laid out in Part II, will be readily apparent.

1. First, to exist at all the individuals in a group have to perceive and relate to each other sufficiently to provide the attention, safety, and security that enables the group's survival.
2. Second, to cohere, the individuals have to be able to form ongoing relationships that support some degree of a shared sense of reality. By definition a group is made up of multiple relationships that can sustain themselves.
3. Third, to deal with the basics of safety, danger, acceptance, and rejection, those in a group need a way of communicating and negotiating with one another that's very rapid. The collective actions and interactions required when there is imminent danger or conflict between group members requires very rapid "implicit" mutual understanding, made possible by rapid emotional gesturing. We saw in Chapter 4 how chimpanzees, gorillas, bonobos, and even monkeys possess this capacity, which accounts for their ability to live in large troops with fairly complex social dynamics.

4. Individuals in a group need to be able to come together to solve problems and work collaboratively. We saw a particularly striking example of this phenomenon in our discussion of the sophisticated collaborative hunting strategies of chimpanzees in the wild (see Chapter 4).
5. To advance to higher levels of social structure, the group benefits enormously by having symbolic abilities that improve the complexity of communication. This enables them to move into higher levels of collaborative or reflective thinking. We saw evidence that bonobos in the wild possess the rudiments of this capacity, which strongly suggests that our earliest human ancestors had also arrived at this level of group development.
6. After groups learn to symbolize, create bridges between symbols (think logically), separate fantasy from reality, negotiate rules, and create organizations, they begin to move up to a higher level where there is gray-area, or differentiated and multicausal, thinking. In Chapter 5, we saw that *Archaic Homo sapiens* had reached this level of group development. Without this ability it's hard to have any sort of advanced social organization. If groups get stuck at this level and don't go beyond, they become mired in rigid categorizations that lead to prejudice, bias, and polarized thinking.
7. To move beyond this level, a group has to create a shared identity and internal standards or ideals as well as institutional processes that support these ideals. We will see in Chapter 14 how the peoples of the Mesolithic and Neolithic periods were already developing these characteristics, and indeed, that this development was critical for their transition from small food-gathering groups to food-producing cultures in which larger groups of people (numbering in the thousands) began to live in one place in settlements that reflected a conscious design.

These basic developmental steps do not compete with existing explanatory models. Rather, as we will see in greater detail below, they provide a developmental road map to the formation of these important explanatory systems and fill in the missing pieces, such as how imitation, language, or group identification develops in the first place.

The selection of these specific group developmental needs is based on what a group requires for its own survival, adaptability, and growth. The ways in which groups meet these needs, however, should not imply a conscious group choice to pursue a particular goal, such as safety or security for its members. Rather, it reveals aspects of how the group functions implicitly. For example, a group that exposes its members to danger reveals the characteristics of the group's functioning with regard to safety and security.

Each of these developmental functions that characterize a group can occur at varying levels of adaptation. For example, group cohesion and a shared sense of reality can occur around a narrow definition of commonality, such as skin color, or a broader one, such as being part of a race. The former leads to polarized beliefs ("us" versus "them") about other groups, whereas the latter is far more inclusive, but, of course, is far harder to organize and sustain. The developmental level or level of adaptation does not, however, imply a universal value judgment. Different patterns may be more or less adaptive. For example, a group that organizes around polarized beliefs may be quicker to mobilize for a war, whereas a group that organizes around reflective, democratic institutions may become involved in long debates and take a longer time to mobilize. What is clear, however, is that some societies struggle with the basics of regulating behavior and forming a cohesive group, but others work on a collective self-reflection that generates values and institutions that can accommodate different beliefs and grow flexibly with emerging challenges.

These group capacities are in part hierarchical in that some mastery of earlier capacities is necessary for more advanced ones, although there is often partial mastery of various levels. For example, security and cohesion are basic to the definition and survival of the group and provide a foundation for reflective institutions. In addition, each capacity is applied to many societal and personal themes differentially. For example, individuals or groups may deal with dependency at an organizational-structural level that is more complex than the one they use for dealing with aggression.

No specific member of the group—who may operate at a relatively adaptive or nonadaptive level—defines these societal structures or organizations, even though the level of individuals making up a group has an important influence. Rather, as individuals come together they form new organizations that have their own character and operate at various developmental levels.

YARDSTICKS OF GROUP DEVELOPMENT

A group, community, society, or several societies working together can be characterized according to the following developmental dimensions.

Group Security

Providing physical protection from attack is perhaps the most basic of all social functions. All the groups that we have looked at in this book, ranging from the marmosets and tamarins to modern nation-states, have developed preverbal affect gesturing to negotiate their safety. As we saw earlier, just as monkeys learn, through the behaviors and vocalizations of their caregivers and other adults, how to make distinct calls to warn the members of the troop of an impending attack,[15] so, too, we warn our infants—and each other—with facial expressions, vocalizations, and behaviors that such-and-such an action or situation is potentially dangerous. This is not a developmental milestone that, once mastered, is rendered redundant; rather, it remains necessary for survival and well being.

Without this provision, other forms of social and cultural development become difficult if not impossible. Societies range along this dimension from the chaos of civil war to the order of highly stable states. Within this large range we find various scantily governed nations and assorted police states in which citizens vanish without a trace for indeterminate reasons; the United States, where rates of murder and violent crime are exceptionally high for the industrial West; Canada, which is like the United States in many ways but substantially safer; and Japan, which sees a minuscule incidence of homicide despite what might seem to be an overwhelmingly dense population.

Cohesion and Shared Allegiance

Humans are remarkably sophisticated in the sorts of preverbal and verbal techniques they employ to form bonds of common humanity. We use items of clothing, jewelry, and body decoration, distinctive ways of walking and talking, distinctive kinds of gestures and looks, and distinctive kinds of belief systems and social structures to build our sense of shared allegiance. Nor are we the only species capable of such cohesion. As we saw in Chapter 4, chimpanzees also build a strong sense of group allegiance.

In some societies, individuals feel a strong sense of mutual obligation. Aware of themselves as a society, they recognize a shared destiny and the duty of each individual to contribute to and, if necessary, sacrifice for the common welfare. In other societies, individuals feel no such sense of unity and shared commitment. Their allegiance may be to some small group, region, or subculture that is at odds with others or with the society at large, or they may function as atomized individuals looking out for themselves and perhaps a small group of close relatives or friends.

Intentions, Expectations, and Shared Assumptions: The Formation of a Collective Character

Covert emotional communications define the shared assumptions a group uses to function successfully, such as the control of aggression, power, sexuality, and the like. These tacit or implicit processes underlie more overt communications and are a significant component of a group's character. Through its caregiving practices, a society communicates these to its young. They deal with basic themes such as safety versus danger, security versus fear, acceptance versus rejection, approval versus disapproval, humiliation versus pride and respect, power and assertiveness versus helplessness, and sexuality and respect for the body versus shame and embarrassment.

We saw a fascinating example in Chapter 3 of how such preverbal communicative practices play a critical role in nonhuman primate societies. Recall what happened when a juvenile male bonobo was caught trying to take a piece of sugar cane before the alpha male had given the signal to commence feeding? What is so interesting in this example is how the alpha punishes the offending juvenile: He stands fully erect, has a stern look on his face, and slowly raises his right arm, but without making a sound. The juvenile responds by submissively approaching the alpha. He receives a token slap on the back, but this symbolic action clearly has its intended effect: not only on the juvenile, who forlornly retreats to the edge of the clearing, but on the other members of the group as well, whose mood becomes noticeably subdued.

In humans, too, preverbal communication conveys to the rising generation a strong and lasting sense of what's "correct," thereby transmitting the most fundamental features of its culture and beliefs to its youngest members. The learning at this stage is so profound that it results in something very close to what we think of as "values." To an important extent, it also structures an individual's sense of self.

We are or are not, for example, people who show emotions openly, or who maintain a "stiff upper lip," or who adopt a decorous demeanor for the sake of propriety. We are or are not people who ask questions about why things are the way they are, or who permit sexual innuendo, or who place filial obligation above personal fulfillment. People who do these things differently are—well—different from us.

Throughout life, the gestures and expressions of others in response to our behavior impart with a power greater than any words could have which areas are acceptable to discuss and which are not. Societies and subcultures lavishly embroider certain themes and pay only scant attention to others. The Yiddish-speaking communities of Eastern Europe developed a vast gestural and verbal vocabulary for complaint and disparagement that allowed an oppressed people to release their anger, anxiety, and disappointment through humor rather than action. Densely populated, socially stratified, and relatively homogenous, the island nations of Britain and Japan instead evolved a finely graduated system of courtesy and protocol to lubricate social life. African Americans may express a gamut of emotions, from bitter sadness to spiritual exaltation, in their rich and exquisitely nuanced tradition of vocal and instrumental music.

These covert processes, which are imbedded in group procedures dealing with basic needs, as well as in the content of emerging symbols (beliefs), are different in different cultures and they can lead to varying degrees of adaptive or maladaptive organization. For example, some groups or cultures employ complex gestures, elaborate procedures, and stable reflective institutions to regulate aggression and integrate this dimension of life with dependency and caring. Other groups do not have elaborate implicit communications involving affective signaling to deal with certain themes and so employ the mechanism of splitting emotional polarities when confronted with intense affects; rather than seeing another group as sometimes good and sometimes bad and adopting an integrated perspective that unites the parts into a whole, these cultures tend to split experience into one or another dimension (believing at that moment that only one dimension exists). Such splitting can justify enormous hostility when another group is characterized as all evil. Groups that tend to maintain nurturing processes that support adaptive dependency in the face of intense affects (such as anger, fear, and humiliation) facilitate more integrative organizations. Other aspects of the collective character that are formed and maintained by covert processes include the degree of individual or group rigidity, the tendency to project intentions onto others, the tendency for rapid unpredictable shifts in attitudes rather than stable perspectives, and the tendency towards impulsive action rather than delay and caution.

Symbolic Expression

The themes that are elaborated by a particular society in nonverbal expression also predominate at the level of symbols. Symbolic expression can occur at different levels. For example, a society's level of symbolic expression is partly reflected in the relationship of its symbols to its deeper values. Are symbolic objects perceived to carry power in and of themselves or as representations of important abstract values? Is the nation's flag, for example, sacred in itself, or is it a symbol of "the republic for which it stands"? The difference is clearly seen in the contrast between the Nazi regime's burning of books to suppress ideas it found objectionable and the decision of the U.S. Supreme Court that burning the American flag constitutes an act of protected symbolic speech. Though desecration of so dear an emblem deeply offends many Americans, the Court nonetheless insisted upon separating the physical object from the ideas it represents.

As we examine the levels of symbolic expression a society uses, we can see some of the following patterns. Symbols can be extremely concrete, barely separated from a direct bodily reaction to a situation or stimulus. Thus an individual might mutter "When he said that, I could have hit him" in place of actually taking the swing, or "My stomach feels tight" instead of taking flight. Shouting curses at an abortion doctor may take the place of shooting him. Such uses of symbols, however, are so close to behavioral discharge and so polarized that they cannot support any subtlety.

Only slightly more conducive to abstract thinking are symbols that are fragmented, inconsistent, idiosyncratic, and discordant with the reality they purportedly represent. Such symbols do not fit any coherent system of significance but exist as islands of meaning separate from people's understanding of the world. Communist regimes, like others, literally rewrote history—including the references made in official speeches and documents and the textbooks used in schools—to match current ideology rather than the common memory of those who had lived through the events in question, clearly violating citizens' shared sense of reality. Because families continue to transmit to their children a more basic nonverbal presymbolic emotional "truth," these symbolic fabrications vanish immediately upon the collapse of the dictatorships that enforce them. After three generations of Soviet rule, traditional elements in Russian culture supposedly extirpated decades ago—the church, for example, and nationalism—reemerged as if by magic. Older and more fundamental emotional verities had survived beneath all the symbolic manipulation and showed themselves again as soon as it had ceased.

This sort of fragmentation and splitting contrasts with a far more integrated system of symbols consistent with a society's pattern of values as well as with one another. In societies whose symbols cohere in this way, a rational approach to even controversial incidents becomes possible. For example, the Israeli government instituted an investigation into its military's handling of the massacres of Palestinians at the Sabra and Shatila refugee camps. Although the killings were perpetrated by Lebanese fighters, the Israeli military was in effective control of the large settlements of refugees. Israeli investigators came to the politically inconvenient conclusion that Israel bore responsibility, if not direct blame, for the bloodshed because the officers in charge did not exercise due caution and discipline to ensure the safety of the civilians under their authority.

The characteristics of a society's symbolic system can also be viewed according to its degree of polarization and rigidity, on the one hand, or flexibility and integration on the other. Southern slave society, for example, viewed people with any degree of African ancestry as inherently inferior to those of purely European ancestry. Such polarized thinking dissociates one's own group from certain undesirable qualities and projects these traits on the other, rendering its members frightening, loathsome, or both. All forms of prejudice against groups of people perceived as undifferentiated masses defined by a single characteristic exemplify this kind of thinking.

Slightly more adaptive than polarized symbol systems are those that allow a few strictly defined categories of thought and behavior. The American electorate's sharp swings in presidential preference reflect such constricted thinking. Candidates are perceived by voters as exemplars of particular ideas, ideologies, and character traits rather than as fully rounded human beings. However, few chief executives—and none in recent history—can fulfill the expectations raised by their campaigns. Voters then tend to overreact to a president's perceived failings and to choose his opposite as a successor, setting themselves up for yet another disappointment. The mass media reinforce such polarized notions by covering politics and government as a contest between rivals rather than as a set of problems that require solutions.

At the symbolic level, therefore, we can observe the degree to which groups or societies

1. use action rather than symbols to convey intentions (e.g., acting out aggression rather than negotiating the aggressive, but symbolized, intent);
2. use concrete symbols to represent intentions even when they are intense, but easily lapse into fragmented symbols;
3. become solidified into polarized, all-or-nothing symbolic organizations;
4. remain rigid in their beliefs in the face of change; or alternatively,
5. symbolize the full range and intensity of collective intentions in a flexible and stable manner.

Institutions that Encourage Symbolic Expression and Reflection

Symbolic expression permits groups to reflect on problems and decisions. Some societies discourage self-critical evaluations, branding them, depending on the context, as unpatriotic, traitorous, heretical, or counterrevolutionary. Others have elaborate procedures for weighing issues of importance to the society at large—consider, for example, the Koerner Commission report that examined the causes of American racial unrest in the 1970s, or the rigorous self-examination by which the German state has worked to rid itself of the remnants of Nazism.

The legislative apparatus in many democratic states exists for exactly the purpose of making decisions by deliberation rather than by fiat. The process prescribed by certain European parliaments or the United States Constitution is explicitly designed to force reflection, to make it essentially impossible for a nation to make an important decision without discussion, and to compromise among the various power centers of society. Although not always employed as intended, these reflective processes involving debate, and often compromise, illustrate the direction of reflective societies.

Functioning consistently at this level is not easy in large groups because it demands considerable social maturity. But stable reflective structures lead to a far more balanced and judicious approach to issues, as is obvious from the fact that democracies that require a majority vote before they can take action almost never go to war with one another.

Economic Stability and Growth in Reflective Individuals and Institutions

Even economic trends may be influenced by these patterns. Reflective innovators create economic opportunities and carefully weigh investment decisions. Followers, some of whom may be at less advanced levels of symbolic reflection (they are more concrete), may make less intelligent investment decisions, swayed by impulse and poorly thought-out expectations. They can't weigh and integrate all the relevant factors as well as they should. Perhaps business cycles reflect the gradual entry of this second group, each member marginally less efficient than the one before. An atmosphere of economic success may further encourage this second group into the market, leading eventually to a downturn in the cycle. If a society becomes overly characterized by individuals and groups that lack high levels of reflection, economic instability may well increase.

An economic principle based on understanding human development can, therefore, be hypothesized. As a market becomes seasoned and the innovators take the best opportunities, the marginal utility of each subsequent investment unit tends to go down because there are fewer opportunities available for profitable investments and fewer reflective individuals to weigh and judge subtle alternatives and make balanced decisions. This leads, ultimately, to a downturn. The rate of decline is related to the reflective abilities of the individuals and institutions in that society. Societies characterized largely by individuals and groups that are vulnerable to polarized, all-or-nothing thinking rather than subtle, balanced, integrated, reflective thinking will be at risk for more sharp downturns, particularly as investment opportunities decrease and the opportunities that are left demand more judgment. In contrast, societies characterized by reflective individuals and institutions will tend towards more gradual downturns and greater potential for corrections. If there are a large number of good investment opportunities, this factor also slows the downturns. Investment opportunities, however, are also a product of the reflective level of the society's individuals and institutions. For example, creative reflective individuals and institutions tend to produce new ideas, technologies, and opportunities. This pattern therefore suggests that the developmental level of the individuals, groups, and institutions characterizing a society significantly influences the relative marginal utility of each new unit of investment, especially as markets become saturated.

Stability Through Change

Sustainable societies are able to change while retaining core values. Many modern nations have gone several times from peace to war and back while preserving democratic institutions and practices—for example, holding elections in wartime and, in the most extreme case, in the midst of civil war. In a country that lacks the reflective processes that enable stability in a crisis, however, leadership changes only with the overthrow of the ruling clique. New policies arise not from electoral landslides but from radical upheavals that can result in a new constitution, flag, and sometimes even name.

Breadth of Emotional Themes

In addition to differences in the maturity of institutions, societies also vary in how they handle emotional themes. How rich and varied are its ideas for dealing with love, anger, competitiveness, obligation? How do its literature, art, music, movies, theater, television shows, and news coverage deal with these themes? If a group has a large number of words or symbolic images for representing and discussing an area of experience, clearly it can deal more precisely, and possibly more reflectively, with that array of feelings than a society able to avail itself of only a few roughly differentiated symbols.

A group's capacity to deal with and symbolize emotional themes is especially evidenced in its child-rearing and educational practices, what attitudes it has toward dependency or aggression, and how it embraces differences embedded in both covert interactive and overt symbolic processes.

UNIVERSALS VERSUS CULTURAL DIFFERENCES: COMMON FOUNDATIONS

The model we have been presenting for the basic needs of groups helps us to fill in certain missing pieces in the vast literature that exists in anthropology and the social sciences concerned with the diversity and the universality of human societies. As the dichotomy between cultural relativism, which emphasizes diversity, and structuralism, which emphasizes universal patterns, attests, this polarized literature has left certain basic questions unanswered. For instance, how do the relatively universal features of human societies allow for variations—different social rules, different kinds of language, rituals, and myths—while retaining their "structural" features over the millennia? How can near infinite variations share stable structural characteristics?

Refusing to accept the possibility that human behavior could be governed by biological imperatives, cultural relativists have focused on the cognitive, social, linguistic, and emotional variations that can be observed in societies and cultures around the globe. Structuralists have focused on the constructs underlying cultural diversities. Because they lack a developmental perspective, however, structuralist schools have fallen back on a preformationist model of how these universal features of human thought and behavior were naturally selected and passed down from one culture to the next in the form of the human genome.

Our developmental perspective addresses both of these questions, namely, how do these universals arise, and how do they lead to the myriad cultural differences that have been observed? Each of the "universals" is itself a product of a developmental process that involves "learning" passed on from one generation to another through cultural patterns influencing family or small group interactions. To be sure, these learned patterns emanate from a complex interaction between biology and experience, each of them mutually influencing the other. But the biological contribution as such is not, as we have shown, a preformed advanced behavior but the capacity for pattern recognition and construction, both of which develop through experience and learning.

Endless variation occurs within each of these dynamic developmental processes leading to language, basic social patterns, or kinship systems. The following diagrams illustrate the contrast between our perspective and the sharp polarity that exists in anthropology and the social sciences.

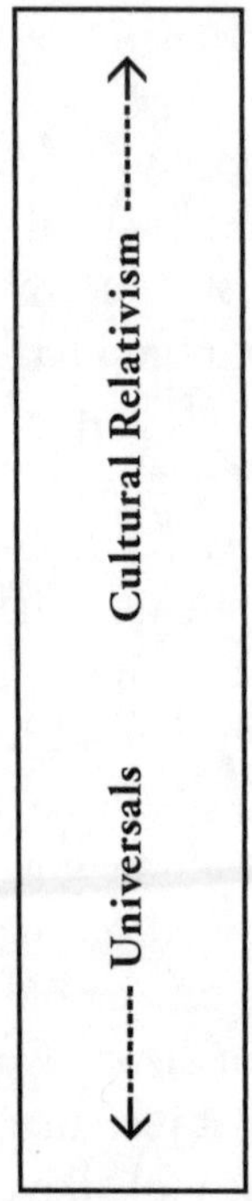

←······ Universals Cultural Relativism ······→

Customs; rituals; art; proxemics; clothing and ornamentation; shelters; political and social organizations; emotions; thoughts; speech acts; greetings; names

Natural languages; social systems; social stratification; cognition; symbolic representation; myths; childrearing practices; gender and racial practices

Kinship systems; mythologies; primitive classification systems; marriage; power hierarchies; socio-economic organization

Social behaviors (e.g., aggression, sexual, play, tool use, coalitions, deception); families; language; culture; folk psychology (theory of mind)

Biology; emotions; facial expressions; gestures; body postures; joint attention; mindreading; intentionality; thinking

GENES

FIGURE 13.1 Traditional Polarity

If looked at in the way suggested by the two arrows to the left, the list seems to underline two irreconcilable extremes. Now compare this with the following illustration of the foundation that our developmental perspective provides. The point to notice here is how, by providing such a foundation, the dynamic of the field has changed. This new model suggests a progression from universal aspects of behavior to the widely diverse practices across cultures; further, the rationale for basing the model predominantly on a genetic substratum has been removed.

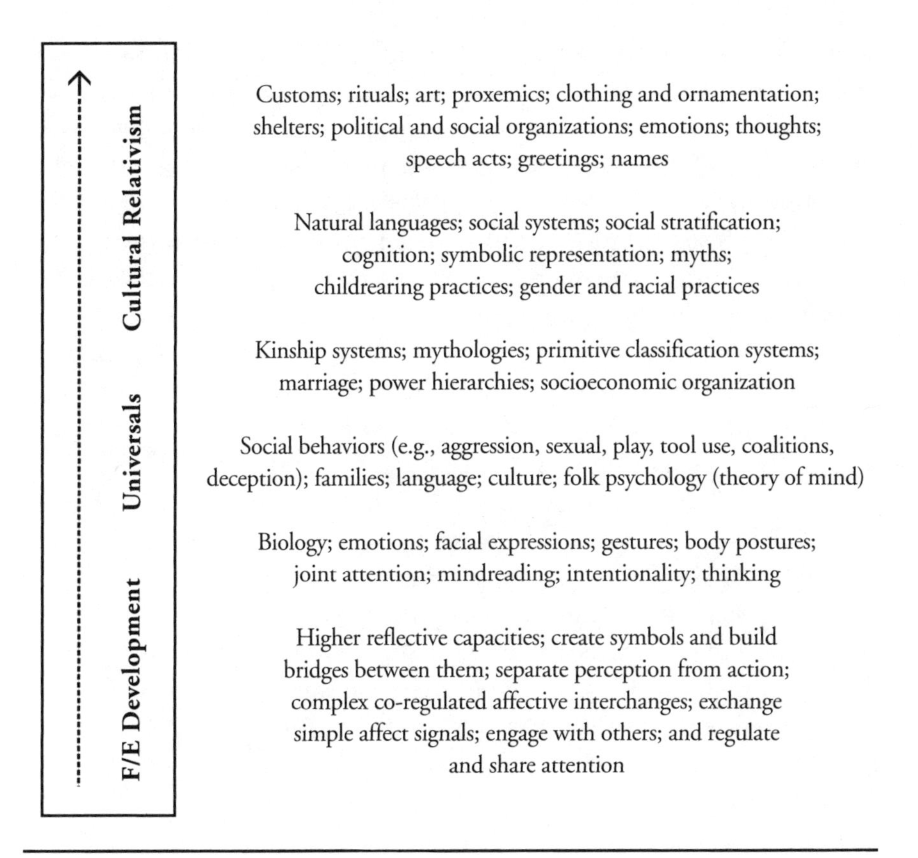

FIGURE 13.2 Dynamic Developmental Perspective

THE ANTHROPOLOGICAL SEARCH FOR THE NATURE OF HUMAN NATURE

The origins of the diversity versus universality conflict in the history of Western thought are remarkably ancient. In the fifth century B.C., sophists argued that the differences in human cultures being reported by Greek travelers were not, as had previously been assumed by pre-Socratic philosophers, the result of nature alone *(physis)*—that is, handed down by the gods—but were in part the result of custom *(nomos),* or the varying manner in which different societies responded to their unique conditions. Western thinking about human nature has been dominated by this issue ever since.

During the Age of Reason, the *physis/nomos* distinction was transformed into the great Rationalist/Empiricist debate. The main source of contention between these views, exemplified in Descartes and Locke, is whether the human mind works according to innate built-in principles or is a blank slate written on by experience. Not surprisingly, the persistent failure to resolve this issue has led many to search for a solution that combines the two; but then, that still leaves us with the same problem that the sophists grappled with: the problem of ascertaining which aspects of human nature are innate and which the result of environmental contingencies. Indeed, some modern thinkers are convinced that, armed with the tools of behavioral genetics, we can actually *quantify* how much of the mind is innate and how much is determined by experience.[16] But the very picture of *genes* on which this thesis rests has been vigorously challenged in recent years by dynamic systems theorists.[17]

Despite the endless attempts to move beyond this polarizing framework, the basic issue remains as compelling today as it ever was.[18] It surfaces in psychology as the nature/nurture debate; in biology as the clash between genetic determinism and dynamic systems theory; in linguistics as the nativist/usage debate; in history as the determinist/pluralist debate; in political theory as the Marxist/libertarian debate. In this section we will explore the ways our developmental model can resolve some of the dilemmas arising from these polarities in anthropology and the social sciences.

In anthropology, the problem emerges in the attempt to account for the seemingly endless cultural diversity observed in different societies while at the same time constructing a unifying framework through which these cultural variations can be understood. It is not just a question of accounting for the different customs and conventions that have been observed, or for the different kinds of kinship systems, rituals, myths, social institutions, and cosmologies. At a deeper level, anthropologists have struggled with the question of whether, underneath the apparently endless diversity, there are certain universals that are a function of the innate structure of the human mind, or whether the very manner in which an individual thinks and feels as well as acts and speaks is a function of the society into which that individual is born. In other words, the great challenge for anthropology remains that of determining whether human nature is in part innately determined or whether human nature, including our biology as well as our minds, is ineluctably enculturated.

Franz Boas stands out as the father of modern cultural relativism in anthropology. Few figures in the history of anthropology have had as great an influence as he has had, and not just on their own field but on much broader intellectual currents. Boas, who trained some of the most influential of early twentieth-century American anthropologists (including Ruth Benedict, Alfred Kroeber, Margaret Mead, and Ashley Montagu), is most famous for his claim that cultures differ in indefinitely many ways and that the variations between different groups are the result of historical, social, and geographic conditions. He came to this view as a result of his early research with the Inuit on Baffin Island. Boas was struck by how strongly the Inuit were bound together by a shared sense of destiny and reality. Far from seeing them as a "primitive" culture, he insisted that they displayed a complete and quite advanced culture that was admirably suited to their conditions.

Boas was not just questioning the racist ideas dating back to the so-called cultural evolutionists, such as the English naturalist Charles White whose *Account of the Regular Gradation in Man, and in Different Animals and Vegetables; and from the Former to the Latter* maintained that "the white European, who being most removed from the brute creation, may, on that account, be considered as the most beautiful of the human race."[19] Boas was making the point that the ways in which human beings actually *think* are, like their social customs, endlessly diverse, and that no mode of thinking can be said to be more "primitive" than another. Indeed, according to Boas, even the emotions that human beings experience or the values that humans embrace are forged by their culture.

The great problem for Boas's theory, however, is that he had no coherent model of the processes whereby a culture influences how a child behaves, thinks, and feels; that is, an account, such as we have outlined above, of how, through the near infinite variations in the ways infants, young children, and their caregivers negotiate the basic six stages of f/e development described earlier, the caregivers and adults of a group shape a child's development. Because Boas lacked a model of the processes whereby a child becomes enculturated, his theory of cultural relativism failed to persuade the legion of social scientists who remained convinced that, because certain basic traits are present in virtually every society studied, there must indeed be human "universals" in the realms of social behavior, cognition, emotion, and language, and that these common human traits are universals precisely because they are innate biological phenomena.

Under the influence of Boas's teachings, Benedict and Mead traveled to indigenous cultures and began to incorporate observations of human emotions, attitudes, and other psychological states into their fieldwork.

Here, according to Mead, is where anthropology steps in; for the anthropologist can not only provide cross-cultural data, but can also test current developmental theories about, for example, the relationship between cognitive development and family or social conditions in the template of an indigenous culture. Thus, it is through the union of anthropology and psychology that we can hope to discover the *physis* of human nature.

The problem with this argument, however, is that it renders the psychological anthropologist hostage to the developmental models available; and those developmental models remained woefully lacking throughout the twentieth century. One was forced to choose between a series of mechanist models, beginning with early behaviorist theories of classical conditioning, and then progressing through mechanist theories grounded in cybernetics and automata theory, followed by the sophistical algorithms of Artificial Intelligence, and culminating in the genetic determinist hypothesis that the development of the child's higher capacities is not so much a matter of *development* as of *biological maturation.* Thus, without a proper developmental framework to work with, Mead's program for a psychological anthropology was doomed to remain a utopian vision (which, ironically, is the charge that was leveled against Mead by later cultural anthropologists).

The attempt to discover the underlying *physis* of human nature by wedding anthropology to psychology enjoyed one of its greatest advances in the work of the Harvard anthropologist Clyde Kluckhohn. According to Kluckhohn, the first task of the anthropologist is to record the variations that occur in human forms of life, which encompasses the differences seen in body types, languages, and the various sorts of things that people make and do. Only then can we identify what humans have in common—that is, glimpse the essence of human nature. The fact that all human beings undergo the same major life experiences, such as birth, helplessness, illness, old age, and death, accounts for some of the regularities seen in human societies: the differences between sexes, persons of different ages, the varying physical states of individuals. But there are many ways in which social activities, such as daily habits of food intake, change biological processes. Thus, a field worker must record such diverse things as head shape, health practices, diet, motor habits, agriculture, animal husbandry, music, and language.

Culture, in Kluckhohn's view, is fundamentally different from society. The term "society" refers to a group of people who interact more with each other than with other individuals and cooperate with each other in the attainment of certain goals. But "culture" refers to the distinctive *ways of being* displayed by a group of people. Perhaps most significant of all, "culture" refers to the manner in which the members of a society communicate with one another, nonverbally and verbally, and the things that are understood intuitively, or implicitly.

Here we can begin to see a definite growing awareness of the importance of preverbal communicative processes in the structure and functioning of a culture. The developmental forces at work are not simply a matter of socialization, that is, of adapting an innate cognitive, linguistic, and emotional template to the particular contingencies of the society into which one is born. Rather, what we have seen throughout Part III is how the process of co-regulated emotional signaling whereby the child develops her cognitive, linguistic, and emotional abilities is the very same as the process whereby the child learns which actions, values, and attitudes are valued by her society and which are discouraged or taboo. That is, the two processes involved here—the growth of the mind and the child's enculturation—are internally, not externally related; they are two different sides of the same developmental coin.

In practical terms, the great problem raised by the approaches taken by Mead and Kluckhohn is that the ways of life being observed are completely foreign and can be interpreted only through one's own system of values. Accordingly, anthropologists have to learn not just the language but also the customs and conventions of that society if they are to understand a child's development in that society; and as philosophers have shown, to learn a "form of life" by being immersed in it is an enormously complicated matter. Indeed, philosophers have debated whether it is ever possible to understand completely a foreign way of life or whether our own enculturated modes of thinking and feeling constrain our capacity to understand a foreign mindset.[20] Thus, it was imperative that cultural anthropologists refine the methodological principles developed by the founder of "functionalism," Branislow Malinowski.

THE RISE OF FUNCTIONALISM

Early in the twentieth century it became clear that cultural anthropology desperately needed a formal methodology for conducting ethnographic research. Such a methodology was developed by Bronislaw Malinowski, the so-called William the Conqueror of British Social

Anthropology, as Ernest Gellner so memorably described him.[21] Over the course of his lifetime, Malinowski constructed general rules based on his considerable fieldwork for conducting "participant observation." That is, anthropologists don't just *observe* but actually *enter* the society they are studying; they must completely immerse themselves in an indigenous culture so that, over several years, they can learn how that community "functions" as a social system.

The basic principle of such a "functionalist" framework is that one looks at the role that particular customs, beliefs, and institutions play within the social system as a whole rather than viewing them as isolated elements, for the function of any one of these items consists in the contribution that it makes to the survival or maintenance of the system of which it is a part. Hence the anthropologist must identify the needs of the group to show how its particular customs, beliefs, and institutions contribute to the satisfaction of those needs. In other words, when anthropology studies a culture, it has to approach it as an integral composed of various institutions, values, and principles. A culture enjoys completeness and self-sufficiency when it satisfies the range of basic instrumental and integrative needs that confront that society.

Challenges to this functionalist framework came both from the structuralist school of anthropology inspired by the work of Claude Levi-Strauss and from the hermeneutic tradition of social explanation, both of which we will examine below. A third challenge came from within the "functionalist" school itself. The great problem with this approach was that one had to show *how* these cultural items contributed to the functioning of the society before one could explain their purpose. After all, some particular custom or ritual might be very prominent and yet not play a role in the maintenance of that society.

THE SYSTEMS VIEW OF CULTURE

In *Steps to an Ecology of Mind*, the English-born polymath Gregory Bateson argued that anthropologists should avoid constructing systems of categories to analyze a society until the actual problems that any such system is designed to elucidate have been clearly formulated. According to Bateson, familiar categories such as "religion" or "economic factors" are really ones that anthropologists make for their own purposes when they set out to describe cultures. In place of these familiar constructs

Bateson proposed five types of unities that, he argued, afford a much clearer picture of the types of mechanisms holding a society together:

1. The structural aspect of unity: The behavior of any individual in any context is consistent with the behavior of other individuals in other contexts.
2. Affective aspects of unity: Behavior is seen as a mechanism oriented towards the emotional needs of individuals.
3. Economic unity: Behavior is oriented towards the production and distribution of material objects.
4. Chronological and spatial unity: Behavior is schematically ordered according to time and place.
5. Sociological unity: Behavior is oriented towards the integration of a major unit, the group as a whole.

Bateson's systems theory introduced not just a much deeper level of complexity into cultural anthropology but, more fundamentally, a different manner of viewing the development of a child's cognitive, communicative, and social skills; for in his model, learning and teaching involve joint participation through social interaction. That is, Bateson postulated that a society coheres around fundamental dimensions that a child absorbs through her emotional interactions with her caregivers. What he was unable to explain, however, was precisely how this process of absorption occurs.

Each of the unities that Bateson postulates demands exactly the sort of explanation afforded by our developmental approach to group behavior. It serves to establish an underlying unity in Bateson's theory by clarifying the formative role that specific types of affective interactions play in negotiating each of the six early functional emotional stages. In turn, the mastery of these stages provides a developmental road map for each of Bateson's unities.

For example, the commonality of behavior across contexts comes from the relative mastery of Level 4, co-regulated affect gesturing, leading to pattern recognition and implicit cultural practices that define the "cultural group." Economic concepts involving the relative value of material objects, as well as time and space constructs, emerge from all six levels as affective (basic needs) patterns and are formed and symbolized in quantitative spatial and temporal contexts. Similarly, the social group and its practices emerge from the development of engagement, affective interaction, and social problem-solving.

THE STRUCTURALIST SEARCH FOR CULTURAL UNIVERSALS

In the search for universals, one figure who most stands out is the great French structuralist, Claude Levi-Strauss. Levi-Strauss argued that culture is made up of rules and systems the function of which is rarely understood by the people who follow them. Some of these conventions are the residues of traditions acquired from earlier social structures. Others have been unconsciously accepted or modified for the sake of a specific goal.

In this approach, the anthropologist endeavors to formulate systematic laws in several aspects of human thought and action. Such laws, which Levi-Strauss regarded as unvarying throughout eras and cultures, enable the anthropologist to surmount the apparent anomaly between the universality of the human condition and the apparently inexhaustible diversity of cultural forms. Thus, anthropology is both a structural and an empirical science, for each culture has to be observed in order to reveal the traits that vary from one culture to another as well as those that are constant.

Levi-Strauss's view of methodology is almost the complete opposite from Malinowski's. During his field studies in Brazil, Levi-Strauss never stayed in one village for more than a couple of weeks; never mastered the language or customs of a culture; and relied heavily on interpreters for his information. But for Levi-Strauss, that was the whole point of the exercise: What one is trying to record, he felt, is precisely one's first, initial impression, untainted by the penumbra of anthropological theory. In fact, in an interesting twist on Mead's argument, Levi-Strauss argued that Western psychology is faced with a fundamental obstacle in its quest to disclose the innate principles of the mind, for the mind, according to Levi-Strauss, is innately governed by a primitive logic of thought. But the Western mind is so thoroughly grounded in logical modes of thinking that it is virtually impossible to disclose the universal primitive logic of the mind. Hence one seeks to study primitive societies that have been as unaffected as possible by Western influences.

The key to Levi-Strauss's theory is that it is based on what is known as the "inferential model of perception." According to this theory, what we really "see" are sense data that the mind must organize according to certain principles that are hardwired into the human brain. In Levi-Strauss's anthropological version of this inferential model of perception,[22] we can discover the kinds of principles that the brain uses to organize experience by examining kinship systems, mythologies, and primitive classification systems, for these cultural phenomena are said to display certain structural patterns that are a reflection of how the brain organizes experience.

For example, the color spectrum is a naturally occurring continuum. But the brain organizes this phenomenon in discrete terms: It sees color as binary opposites (e.g., black and white, red and green). The brain then searches for intermediary positions to bridge the gap between these binary opposites (gray, yellow). There are various cultural phenomena that mirror the structural pattern of this experiential phenomenon, for example, the traffic light system, which represents binary actions (stop and go) with color opposites (red and green) and an intermediary action (proceed with caution) with the corresponding intermediary color (yellow). It is in this sense that the structural patterns observed by anthropologists are said to be a reflection of underlying mental universals.[23]

Thus, Levi-Strauss combines a realist view of nature with an inferential theory of perception to discover underlying structural patterns that can be observed in all cultures. These universal patterns, Levi-Strauss insists, are not environmentally or socially determined phenomena, but, rather, they reflect the basic manner in which the mind organizes experience. These organizing principles are said to be prior to rather than a product of enculturation; that is, they are basic cognitive principles that make social existence possible.[24]

To be sure, the "surface" differences between cultures can be so striking that they arrest our attention to the point that we cease to ask ourselves about the underlying or "deep" universal features of experience. So even though different cultures may have what on the surface appear to be vastly different kinds of kinship systems, these are of no concern to Levi-Strauss; all he is interested in are the deep, structural regularities. That is, no matter how different kinship systems might appear, we will discover structural universals running throughout this diversity. And these structural universals, according to Levi-Strauss, are a function of the innate structure of the human mind.

It is not surprising, given the nativist foundation underpinning Levi-Strauss's argument, that in recent years the search for cultural universals has turned its attention to evolutionary psychology. The paradigm for this argument is Chomsky's generativist view of language (although, it must be stressed, Chomsky distanced himself from Levi-Strauss's use of language as an analogy for culture). But what appeals to contemporary cultural universalists is Chomsky's distinction between the "depth structure" of language and its "surface structure," where the former is the underlying abstract structure that is a reflection of the "forms of thought."[25] Such a distinction offers an inviting paradigm for a corresponding distinction between the depth structure of cultures and their surface variations. That is, every culture might adopt vastly different social conventions for what is regarded as appropriate behavior towards different categories of agent within the society, but what is universal is that each culture has some form of convention for regulating social behavior. Furthermore, as we saw in Chapter 9, Chomsky's argument is grounded in the genotypic hypothesis that sometime during the Pleistocene a language gene *must* have been selected. Given the enormous success that his theory enjoyed in the late twentieth century, it is not surprising that structuralists and evolutionary psychologists appealed to Chomsky's "language gene" argument to support their adoption of the hypothesis that cultural universals were also genetically selected during the Pleistocene.[26] Indeed, such an argument becomes self-reinforcing insofar as these cultural universals must encompass language.

We can take a simpler example to illustrate how this argument works and why it resorts to genetic determinism to support the existence of cultural universals. Consider the fact that every human culture has its own unique system of gestures. To be sure, some cultures gesture far more than others; but to a greater or lesser extent, every culture gestures. But despite extensive efforts to construct a universal "dictionary" of gestures, it turns out that even the simplest of gestures, such as those expressing consent and negation, can vary in quite significant ways between cultures.[27] Rather, what is universal are certain more general properties; for example, some gestures are purely indicative or declarative, some are mimetic, and so on.[28] Thus, the evolutionary psychologist concludes that there must be some sort of genetic template for gesturing that is modified by social convention from one society to the next. Different cultures may have different conventions for expressing consent gesturally; but what every culture has in common is some gestural method of expressing consent. And the hypothesis that gesturing must be hardwired into the human brain can be further supported by the fact that we gesture unconsciously while we speak; we even gesture, completely oblivious of doing so, when there is no need to gesture, such as while speaking on the telephone.

We have already seen, however, how the child's gestural capacities develop through long sustained bouts of co-regulated emotional signaling, and indeed, how the great apes go through similar stages of gestural development (see Chapters 3 and 4). One of the most vivid examples of the developmental processes that are at work here can be found in deprivation studies. In a particularly famous experiment by Edward Tronick, caregivers were told to look at their four-month-olds with a completely impassive face and to resist all temptation to smile or coo.[29] The babies responded by making an extra effort, through smiling and vocalizations, to coax their caregivers to reciprocate. When these actions met with no response, the babies soon became irritated and their gestures became increasingly disorganized and purposeless. At last they became withdrawn and listless, and when the caregivers were told to resume their normal interactive behavior, it took some time to reengage the child, despite the caregiver's energetic attempts.

In the middle of the twentieth century, Bowlby described (and later James Robertson illustrated with remarkable films) what happens when a young child goes to a hospital and is separated from her primary caregivers. Protest and anger give way to despondency and withdrawal and a loss of capacity for affect gesturing. Similarly, Spitz described and filmed the self-absorption and lack of affective gesturing that occurs in institutional settings where infants are cared for physically, but not engaged in affective interactions. Harlow and Soumi have described similar processes.[30] Most important, we have found that deprived children can learn affect gesturing in a special therapeutic program.[31] In addition, in our work with children with autistic spectrum disorders, who also have limitations in this system, but for different reasons (biological ones), we have found it is possible also to help them learn affect gesturing.

There is thus considerable evidence that affective gesturing is learned—not hardwired—and must progress through the stages we have outlined.

LINKING THE UNIVERSALS OF HUMAN NATURE WITH CULTURAL DIVERSITY

From the point of view of the cultural relativist, our model of group development may seem to be too firmly ensconced in the structuralist family of anthropological theories; and from the structuralist vantage-point, the argument might seem to be far too imbued with relativist ideas. The f/e developmental model does indeed have elements in common with both schools of thought. On the one hand, our model offers a way of understanding developmental processes that accounts for the endless cultural variations that ethnographers have observed and, on the other, a way of looking at the developmental steps that lead to relatively universal human processes that underlie the diversity. But, as we have shown, the relatively universal characteristics such as the capacity to communicate with gestures in general (rather than any specific gesture or "content") often has a learned, developmental sequence and is, therefore, not hard-wired as often alleged.

For example, consider the incest taboo. According to Levi-Strauss, the incest taboo is the defining feature of culture: It makes us human by reaffirming the separation of humanity and culture from nature.[32] But although most human societies have some form of incest taboo, not all these societies have the same incest taboo. Anthropologists have explored three basic kinds of explanation for this phenomenon: psychological (e.g., Freud's hypothesis that incest would disrupt the hierarchical authority structure of a family and lead to rivalries and violence within small groups);[33] sociological (e.g., Malinowski's argument that incest would impair the socialization of a child);[34] and biological (e.g., the argument that incest avoidance serves to lower the incidence of congenital birth defects, and thus was naturally selected). Arguments for and against each of these hypotheses have been widely discussed and it is not our intention to review here the pros and cons of each position. Rather, our concern is with how our developmental model suggests an alternative explanation of the genesis of the seemingly near universal human aversion to incest and the mechanism involved in the transmission of each culture's version of the incest taboo.

We have already seen how, to get to the fourth stage of f/e development, evolution favored those practices that enabled hominid infants to enter into long chains of co-regulated emotional signaling with their caregivers, and that for such an ability to develop, it was critical for infants to learn how to tame their catastrophic emotions. In our discussion of this process in Part II, we naturally focused on the sorts of practices that hominids might have adopted to facilitate this development; but the converse side of this argument is to consider what sorts of practices early humans would have avoided to prevent provoking catastrophic reactions in their infants. And the overstimulation of an infant's genitalia would naturally fall under the last category.

Thus, we might see the incest taboo as part of the general process in human evolution of developing caregiving practices that would promote a child's capacity to regulate her mood and behavior, which, as we saw in

Chapter 2, is a critical step on the path towards long chains of co-regulated affect gesturing that, in turn, leads to developing symbolic and language abilities, including the basic capacity to produce finely regulated motor-vocal patterns. Furthermore, the incest taboo would have been transmitted through these very same caregiving practices. Children would have acquired their culture's attitudes toward incest through subtle emotional signals in which parents convey that one's private parts are different from the other parts of one's body and that they can't be brought into pleasurable interaction (e.g., tickling the tummy). That is, parents would have responded differently—through facial expressions, vocalizations, even aversive behaviors—both to touching their child's genitalia and to having their own genitalia touched. Many parents convey their special feelings about the genitals by excluding them from routine interactive games that involve rubbing the hands, legs, or tummy.

We can see precursors of the incest taboo in the avoidance of inbreeding amongst different animal species. But what is particularly interesting here is that, although we see mother-son avoidance in, for example, macaques and baboons, in chimpanzees and bonobos we actually see cultural variations in inbreeding behavior. Some groups of chimpanzees avoid inbreeding, others do not. A moderate amount of inbreeding has been observed in the chimpanzees at Gombe, perhaps because the chimps living there are so isolated that there is relatively little opportunity for the sort of female migration that is normally seen amongst chimpanzees.[35] Thus, whereas the avoidance of inbreeding appears to be a hardwired behavior in the lower primates in which this occurs, one can discern flexibility concerning this practice in the great apes, suggesting the influence of culture and learning.

What our developmental model draws attention to is that only in a species that has begun to master the fourth level of f/e development could one expect to see such cultural variation. For only such a species possesses the ability, through its more subtle affective signaling, to change attitudes towards inbreeding or to adopt a variety of them. Thus, far from being forced to postulate some sort of fail-safe genetic mechanism to explain how it is that these apes are able to respond in such a manner to social isolation, we can see how the group is able to redefine its attitude towards inbreeding.

Furthermore, our developmental model suggests that the incest taboo, as it is known, is specific to species that have negotiated the fifth stage, mastering the creative use of symbols and words. Bonobo chimps, as indicated in Part II, only partially master Level 4 and, therefore, are limited in their symbolic capacities. Humans can engage in much longer series of co-regulated affective exchanges and, therefore, can "enculturate" new offspring with far more complex social capacities and "norms."

A similar approach can be applied to any of the basic universals and their cultural variations studied by anthropologists. Our developmental model provides a way of reinterpreting the observations that have led to the different schools of thought within anthropology and the social sciences. It offers a new model for understanding the full complexity in the universal and relativistic features of human functioning in individual, group, and cultural contexts. As we have seen, the ability to form and give content to symbols, or to share in the collective stories that characterize a culture, depends on a series of developmental steps that both the individual and the society need to master. Family, kinship systems, and community structures—that is, the very ability to form relationships of any kind—also depend on how individuals and groups meet these basic developmental needs.

As we apply our developmental lens to different groups, it is important to emphasize that a group may negotiate different f/e developmental levels for different aspects of its experience. Dependency and nurturing care may be at a different level than dealing with aggression. The f/e developmental model can, therefore, help us understand the ways in which different groups deal with certain themes, such as dependency or aggression. Are they reflective with that theme? Or do they tend to use impulsive actions with that theme? Do they integrate many different subtleties around that theme, such as the many nuances of intimacy, warmth, and love, or are they polarized ("You either love me or hate me")? In and of itself, however, this developmental perspective does not provide a value judgment. Rather, it provides a profile that helps to describe the different ways in which different experiences are handled in different groups.

THE HERMENEUTIC TRADITION

Another of the great challenges to functionalist thinking came from within the ranks of cultural relativists and stemmed from the hermeneutic tradition in the human sciences. We begin here with the thought of Emil Durkheim (1858–1917), the great nineteenth-century French sociologist. Durkheim set out to establish sociology as a science that as such studies a unique realm of human phenomena, different in kind both from the physical world (the province of the life sciences) and the mental (the province of psychology). Sociology is concerned with the analysis of "social facts"—laws, conventions, automatic actions—that can be treated as "things" just as much as material objects, even though they belong to a different logical category. "Social facts" are "ideas" (representations) in the "collective mind" (the *âme collective,* or conscience collective) of a society; that is, something that exists over and above the individual members of the society and that has concrete effects on the behavior of the members of that society. For example, we may not be conscious of the rules proscribing how to dress but we nonetheless follow them. Similarly, different societies have different rules about how close to stand to one another in different circumstances. We all follow these rules quite unconsciously, unaware of how carefully we regulate the distance we put between one another, for example, when standing in a crowded elevator, standing in a line, or having an ordinary conversation on the street.

Durkheim built up a Darwinian scenario about the evolution of societies based on this notion of the "collective mind." The simplest form of social organization is said to be the "horde," in which individuals are attached to the group by their adherence to the same powerful set of collective representations. The more complex a society becomes, the weaker is said to be its "collective conscience" and the more that society is bound together by the complex interdependence of small groups to which individuals are bound by their shared "collective representations." In these assertions, Durkheim is suggesting developmental steps leading to cohesion at different levels very similar to the ones we have proposed (earlier this chapter).

What is missing in Durkheim's model is an account of how these "collective representations" are imprinted on the mind of the individual, a matter that Durkheim believed would eventually be resolved by psychology. The same assumption carried through in the work of Ferdinand de Saussure (1857–1913), who in turn had a profound influence on Levi-Strauss. Saussure was heavily influenced by Durkheim's conception of social facts. According to Saussure, the linguist must distinguish between two different sorts of things: *langue* and *parole.* The former is the abstract system that, as social fact, exists over and above the individual speaker. *Parole* consists of the tangible speech sounds that the linguist can observe. If two people speak the same language, then they *must* understand each other. If two people do not understand each other, then they *cannot* be speaking the same language. When a language speaker speaks she is simply using the *langue* that is imprinted on her brain (and the brain of anyone else in her community whom she speaks with). *Langue* is thus a "social fact" in the sense spelled out by Durkheim; that is, something that exists over and above the individual members of the society.

How *langue* comes to be imprinted on the mind of the individual was, Saussure believed, a matter for psychologists to address; the linguist's task was solely to discover the rules the language speaker follows, even though the language speaker is not conscious of following those rules. The f/e framework provides us with precisely the sort of psychological explanation that Durkheim and Saussure anticipated. We have already seen how the ability to use symbols to sequence and organize the world and form a higher level of pattern construction rests on a continuous flow of affective interaction that enables a child to separate perceptions from actions and thereby form symbols (see Chapter 2). A baby is able to recognize patterns in all the sensory areas of her experience: visual, spatial, auditory, tactile, motor. Usage linguists have shown that these patterns can be grammatical; for when humans communicate linguistically, they string symbols together into sequences. Patterns of use emerge from these co-regulated symbolic interactions and through constant repetition become stable features of linguistic communication: a process that usage linguists refer to as "grammaticalization." In fact, as we explored in Chapter 7 on language, the f/e developmental model emphasizes the role of affect and intent in creating operational grammar. As we described, to convey affective intent, which is intrinsic to getting needs met through co-regulated, affective interactions, the child needs to combine verbs and nouns together in patterns—"I want juice." As needs and goals become more complex, so do affect gesturing and grammatical structures—"I want juice now" and, in adolescence, "I want more juice now, you turkey!"

We have already seen such a phenomenon in Chapter 4 in our discussion of the bonobo Kanzi's remarkable linguistic achievements. It was by being engaged countless times in the day-to-day activities of life at the Language Research Center, for example, fetching a bottle of juice from the fridge or putting his toys back into the box, that Kanzi was able to understand grammatical sentences such as "Get the juice that's in the fridge" and "Put the toys into the box." At first this extraordinary breakthrough in ape language research aroused the concern that the sorts of simple grammatical patterns that Kanzi was mastering could not explain the more complex kinds of linguistic constructions that young children master. But in recent years, usage theorists have shown that exactly the same picture applies to human language acquisition; that is, by constant exposure to grammatical patterns in the language that envelops her day-to-day activities with her caregivers, the child does indeed develop the sort of implicit knowledge of grammatical structure that Chomskeans thought could never be *acquired* but must rather be innately present at birth.[36]

One of the great advances in the development of sociology was due to the thought of Max Weber (1864–1920). Weber's basic idea—which he developed on the basis of Dilthey's argument about the nature of the "human sciences" as an interpretive endeavor—was that sociology must wed causal explanation with interpretive understanding. According to Weber, what makes an action *social* is the meaning attached to it by the actor. That is, just having an effect on another actor does not render an action social. For example, if I trip on a crowded sidewalk, thereby causing someone else to topple over, my action is not in itself "social." However, in one of Weber's examples, if two cyclists see that they are about to crash into each other and try to avoid this, then their actions have a social character.

The meaning that we assign to the actions of others depends on our knowledge of the norms and standards regulating their behavior in such-and-such a context. So, for example, my knowledge that the driver ahead of me intends to turn right at the next intersection is grounded in my knowledge of the rules of driving (the use of turn indicators). Thus, when explaining social patterns, or regularities, the sociologist must explain the subjectively intended "meaning" of the actions for the individuals concerned in order to arrive at a causal explanation of the causes and effects of actions. That is, in understanding the motive for an action, one understands it both as a cause of the action and its meaning.

Weber's argument raises a profound question, which was widely discussed by philosophers in the 1960s and '70s, about the relationship of motives, wishes, and intentions to actions. Weber construed this relationship as causal, on the paradigm of physical causality; but much careful philosophical elucidation has explained the categorical distinction between physical causation and the sociological explanation of actions in which motives, wishes, and intentions are construed as the *reasons as opposed to the causes* of actions.[37] But such an argument does not obviate the problem that Weber was highlighting here, namely, that the very nature of human functioning challenges us to use hermeneutic tools as well as methodologies developed by the natural sciences.

BENEATH THE SURFACE OF GROUPS: CONTRIBUTIONS FROM DEPTH PSYCHOLOGY

In addition to his work on individuals, Sigmund Freud also looked beneath the surface describing irrational forces operating in groups. In the early 1900s, when Freud was formulating his basic themes, Wilfred Trotter was writing about the "herd instinct." He hypothesized a basic instinct for individuals to be part of a larger entity, akin to a multicellular organism. Gustav Le Bon was exploring the phenomena of crowds and observed that these groups operated irrationally with "contagion" and "suggestibility." William McDougal, in the 1920s, observed that in groups, the emotions of one group member tended to synergize the emotions of another, leading to intense irrational actions.

Freud sought to clarify these hypotheses with a more comprehensive theory. He hypothesized that individuals projected components of their intrapsychic lives onto a leader (namely, their primitive ego ideal) and that this process, in which the standards of the leader replace those of society, provided the group with its force, its seeming immorality, its lack of reflectiveness, and its irrationality. These projected intrapsychic components were based on early primitive irrational processes.

The primal horde with its feared, powerful leader was the model for Freud's concept of the group: The group's members work together through fear of aggression and destruction that destroys the leader and threatens the members. Suggestibility, identification, and other related phenomena were operative processes.

Since Freud, psychoanalytic observers have further developed these insights describing elements in a group's irrational infrastructure that relate to early developmental processes. For example, Wilfred Bion developed methods of exploring group processes and described the irrational interactive processes in groups that must be dealt with for rational problem solving to occur. He described how group members become dependent on the leader, develop fight/flight patterns with an imagined outside danger or foe, and form pairings with magical, grandiose expectations.

Eliot Jaques saw the group as a social system to deal with potential irrational persecutory feelings. Otto Kernberg explored the group process in relationship to the projection of early primitive aggression and the projection of related intrapsychic components (e.g., the super ego). Didier Anzieu and Janine Chassquet-Smirgel explored the processes of fusion as well as the projection of primitive intrapsychic contents and the concomitant creation of illusions and ideologies.

The developmental framework presented in this work looks at the irrational elements in groups systematically in terms of contributions from different developmental levels. This framework integrates our understanding of irrational and rational elements in group processes as well as the contributions from other disciplines. Unlike much of the work just mentioned, it also looks at motives for group cohesion that do not derive from fear of aggression but emanate from the capacities for relating to others and empathy.

UNITING THE DESCRIPTIVE (HUMANISTIC) AND NATURAL SCIENCE TRADITIONS

Traditionally, humanistic approaches that focus on subjective meaning, and natural science, which focuses on replicable experiments of measurable phenomena, have existed in independent realms. The former is part of literature, philosophy, and, more broadly, the humanities, whereas the latter has come to dominate the "hard" sciences. Depth psychology (e.g., psychoanalysis), cultural anthropology, and sociology, however, appear to need both approaches. To illustrate this principle, we might return to the grammaticalization example discussed above.

We saw how usage linguists have shown that a child develops her implicit knowledge of grammatical structure through constant exposure to the grammatical patterns embedded in presymbolic and symbolic communications. What we also need to bear in mind here is the point, discussed in Chapter 9, that recent evidence has shown that when the human nervous system is engaged in this type of affectively mediated growth it is also likely laying down new pathways; these are not the result of a genetic blueprint but of some fundamental capacities that are now being defined and redefined and organized and reorganized at new levels through interactions with the affective world. Such an argument would apply to the formation of the so-called language centers in the left perisylvan region of the brain. Thus, the f/e developmental model also accounts for the sorts of findings in aphasiology that further encouraged Chomskeans to postulate that the child must be born with a genetic blueprint containing the plans for her language centers. Also relevant to this blending of the subjective and the quantifiable is Terrence Deacon's interesting argument that the very structure of language is a reflection of the structure of the human brain—that is, natural languages evolved in a way that conforms to the natural pattern-recognition capacities of the human brain[38]—and Sonia Ragir's research showing that language genesis is constrained at three epigenetic levels: between cortical and subcortical structures, between the ontogeny of the brain and the activity of the individual, and between the development of the individual and the adaptation of the group to its environment.[39]

One of the primary figures in the development of the view that anthropology should not, as was argued by Alfred Radcliffe-Brown, be seen as a natural science but rather as a member of the *Geisteswissenschaften* was the Oxford social anthropologist E. E. Evans-Pritchard. Where Radcliffe-Brown was interested in the role that institutions played in maintaining social structure and solidarity and in determining an individual's emotions, Evans-Pritchard sought to show how anthropology should be seen foremost as an interpretative discipline.

According to Evans-Pritchard, there must be uniformity and regularities in social life that serve to establish some sort of order; otherwise, a group of diverse individuals could not live together. It is only because people know what kind of behavior is expected of them and what to expect from others in the various situations of social life that they can coordinate their actions under the guidance of various rules and informed by the values of their culture. They can make predictions, anticipate events, and lead their lives in harmony with their fellows because every society has a moral form or pattern. Thus, the anthropologist is engaged in the study of *normative,* not natural, phenomena, which as such must be interpreted to be understood.

But then, *how* do individuals come to know what kind of behavior is expected of them and what to expect from others? How do they come to make the predictions and to anticipate events that, as Evans-Pritchard so rightly stressed, enable them to live in harmony with their fellows? It has long been clear to social theorists of all persuasions that such knowledge is rarely taught, and, indeed, that such knowledge is rarely explicit. When an earlier generation of cognitive scientists sought to answer this question they assumed that, since such knowledge was not conscious, it must be preconscious: a construct that was heavily influenced by Artificial Intelligence models of the kinds of processes that the brain as computational organ must go through while processing information. On this line of thinking, the brain must be primed to make predictions, be this of physical phenomena or social behavior. That is, the brain is like a powerful computer, programmed with some very general processing strategies, just waiting to be fed information.

In contrast, in our developmental model, the ability to make predictions, which begins to emerge at our third level of development in terms of pattern expectation, continues to develop throughout the following stages, and then to develop at a high symbolic level at the ninth through sixteenth stages in terms of affectively valuing and even quantifying future possibilities, does not arise until the child is able to attend and self-regulate and to engage and relate with her caregivers. Furthermore, the child's progression through these stages is not some sort of predetermined maturational phenomenon but rather, as we saw in Chapter 2, depends vitally on the sorts of practices in which caregivers engage to promote the development of these capacities.

Such an argument contributes a critical foundation to Evans-Pritchard's basic hypothesis that anthropology studies the sorts of practices and techniques that the members of a society must master to manage social relations. The significance of such practices, Evans-Pritchard pointed out, cannot be explained without understanding what they mean to the participants in the society. But then, the participants in a society may themselves be unable to explain the meaning of such practices. Hence, social anthropology, as an interpretative discipline, must account for the different types of procedural and implicit knowledge that underpin a society and for how this knowledge is acquired.

Furthermore, Evans-Pritchard's argument exemplifies how universals provide us with frames of reference within which there can be endless variations. For that is precisely the point of his "levels of abstraction." Along these lines, our developmental model suggests that each stage of f/e development is a new, relatively universal feature of human functioning. With each new stage and new "universal," there is a new lens through which to view the near-infinite variations that occur in social behavior. The very same emotional signaling that underpins these ubiquitous social patterns allows for endless variations within those patterns, in much the same way that mixing pigments together enables us to paint, but the paintings we can make are infinitely diverse. The lens of "thinking off an internal sense of self and internal standard" affords near infinite variation in "content."

The same issue arises in regards to the structural-functional theory developed by the Harvard sociologist Talcott Parsons, who did so much to expound and develop the views of Durkheim and Weber. Parsons saw societies as open systems that are constantly responding and adjusting to changing environmental and political contingencies. Hence, the fundamental problem faced by any society is what he called "boundary maintenance": establishing a set of criteria by which individuals identify themselves as members of that society. This is precisely the point that we looked at earlier in this chapter when we considered the different practices that Japanese and American caregivers engage in with their infants.

According to Parsons, societies are complex systems composed of many interdependent subsystems—cultural, political, economic—each of which is also bound together by a common set of values. Hence, "society" as such is the highest order of the collective as defined by a set of values that enable its members to identify with each other as members of that "distinctive institutionalized culture."[40] But conspicuously missing in his theory is an account of the developmental processes whereby the individuals in a collective come to develop the shared sense of reality—what he referred to as a group's distinct mode of being-in-the-world—which, according to his theory, is the glue that holds society together.

The same point applies to the work of one of the most important contemporary social scientists, the French sociologist Pierre Bourdieu. Bourdieu argues that modern society employs two primary systems of stratification: economic (one's status is determined by one's capital) and cultural (one's status is determined by one's "symbolic capital"). All human action, according to Bourdieu, takes place within social fields in which individuals, groups, and institutions compete for limited resources.[41]

Bourdieu is perhaps most famous for developing the concept of habitus: the idea that, through upbringing and education, an individual acquires a system of dispositions that serve as categories of perception and organizing principles of action. This, of course, again leaves the question of *how* the individual acquires this "system of dispositions."

Indeed, it is this basic problem of how dispositions or other traits are acquired that, as we have seen in the foregoing sections, has dogged anthropology and the social sciences from their inception. And it is precisely this lacuna that our developmental model fills. The glue that holds a society together is not some set of concepts or an abstract ideology; rather, it consists of the basic types of interactions that are embedded in the functional emotional developmental stages described earlier. This preverbal communication becomes a core part of the individual's personality and thence the basis for the "distinctive institutionalized culture" and the individual's "system of dispositions."

A DEVELOPMENTAL MODEL FOR MENTAL HEALTH AND MENTAL DISORDERS

The developmental model of groups that we have been proposing also provides a useful lens for looking at mental health and mental illness. According to this model, mental health is indicated by the levels of f/e development achieved by an individual vis-à-vis his society, and mental illness as deficits in these core capacities. For example, an individual may have a deficit in his ability to connect gestures together into patterns (Level 4) and then, later, to connect symbols or ideas together with logical bridges (Level 6). If one or both of these f/e levels are not achieved, an individual in most cultures will almost invariably have difficulty functioning in that culture; for such an individual is unable to read or respond appropriately to preverbal emotional and social signals or to participate functionally in a range of group activities, from the very simple, such as eating, to the complex, such as rituals.

There is an important difference between an individual who "hears voices" as part of a religious experience or in a culturally sanctioned role but is otherwise capable of logical thinking and of following the rules of the group, and an individual who has difficulty connecting ideas together. In other words, in assessing mental health and mental illness, our developmental approach looks at the individual's ability to relate to others, to communicate presymbolically, and to think and speak logically.

Near infinite variations occur within these abilities, and many different perceptual patterns are possible that may appear to represent significant deficits in these fundamentals but in fact do not. But the individual who cannot master these fundamental abilities will usually evidence aberrant behavior, which is identified by that culture as not only different but in some respects as defective. An individual who does not successfully negotiate these stages experiences a personal deficit and a deficit in his or her ability to function at the level that the group has achieved; indeed, the two aspects of development are inextricably connected. Thus, the individual who is incapable of connecting symbols together logically will be unable to participate in the most basic social patterns that the group organizes to define its cultural uniqueness.

This model of mental health and illness also helps us understand how biology and experience come together. There is an equilibrium at each of our developmental stages between biology and experience that underpins the individual's ability to operate at that level, be it relating, perceiving patterns, or creating and using symbols. Each level is an intermediary organization (a series of steps) in the developmental pathways towards mental health or mental illness. What we have, therefore, is a dynamic developmental model with a series of levels that brings together biology and experience leading to specific behavior or abilities.

This type of developmental bio-psycho-social model enables us to look at human behavior in terms of its relatively universal features and its infinite variations, and to see the way in which biology and experience are inextricably interconnected as the building blocks and pathway towards mental health and mental illness.

CONSTANCY AS THE SOURCE OF VARIATION

The developmental model that we have outlined in this chapter is based on three basic principles:

- The universals that exist in human societies are not somehow "encoded" in our biology but are embedded in yet more basic developmental processes.

The most basic constants of human society—universals such as the fact that the members of every human society are able to attend to one another, or to engage in preverbal communication—are universals not because they are part of our biological makeup but because they are learned abilities that predate the earliest hominid societies. That is, the caregiving practices that promote the development of these universals have been passed down and thus learned anew from one generation to the next, and even from one species to another, over the course of millions of years. Thus, as we have seen throughout this chapter, postulating the presence of certain individual or social universals does not compel one to adopt a biological framework to explain their existence. Even though basic social behaviors such as mating and caregiving are found everywhere, this does not mean that these behaviors are hardwired. Rather, it suggests that these behaviors, which, like all our behaviors, are culturally mediated, occurred very early in our evolutionary pathways. To be sure, behaviors such as pair bonding or caring for our infants are such a standard feature of human functioning that individuals who do not engage in them have commonly been seen as dysfunctional. But the fact that these behaviors are ubiquitous does not entail that they must in some sense be "instincts" (regulated by our genes); rather, it indicates that these behaviors proved to be fundamental to the development of our higher reflective capacities, and for that reason were culturally transmitted over aeons. But then, the very fact that these fundamental behaviors are all learned entails that they can all be unlearned.

- Every universal affords endless cultural variations.

The nature of these caregiving practices is such that it provides the means for near infinite cultural variations. Different cultures engage in different kinds of preverbal patterns of regulation, engagement, and affect gesturing. There is, for example, different signaling around different customs, beliefs, and values. These attitudes of the group are experienced from earliest infancy and form the psyche of the individual. Throughout her life the individual conveys to others, through the same sorts of affective regulation, patterns of engagement, gestures and head nods and vocalizations that promoted her own development, her culture's distinctive attitudes towards different customs, beliefs, values. Thus, both the universals and the variations come from culture. The universals—the highly stable features of human functioning—derive from very early developments in the cultural transmission process and provide the foundations for cultural variations—that is, the abilities to engage in co-regulated emotional communication, to form symbols, and to think logically and creatively all reflect interactive learning mediated through culture over millions of years, but the *contents* of these processes vary from group to group.

- The very nature of human beings and human functioning challenges us to use tools developed by both the natural sciences and humanities.

As we saw in Parts I and II, the techniques used to study the developmental processes that result in the universals of human functioning and human societies are grounded in the traditions of the natural sciences. But the techniques that are used to study the near-infinite variations in human functioning and human societies are grounded in the great hermeneutic traditions of the human sciences, including literature, philosophy, and psychoanalysis, as well as in traditions developed by the natural sciences, including the experimental method. The descriptive tools of the humanities are essential because the variations in human functioning include subtle differences in behavior, thoughts, beliefs, feelings, and relationship patterns.

According to these basic principles, we can think of universals as the lens through which one looks at a large group or society. Each stable or structural feature that we highlight opens a door to understanding the near infinite variations that occur in human societies. In this chapter we have explored a new set of relative universals and a new set of near-infinite variations in terms of the developmental levels and needs of social groups. Each basic developmental need is at once a new universal and a new lens through which to see infinite variations. For example:

1. The first basic developmental need that a group must negotiate is its security. Some groups are able to negotiate their security with organizations and practices that maintain order without impinging on the rights and freedoms of the individual, whereas other groups can maintain security only through practices that infringe significantly on personal freedoms.
2. The next developmental need that groups must negotiate is cohesion and shared allegiance. Some groups are able to cohere around reflective institutions and processes that promote a strong sense of shared identity,

whereas other groups can cohere around superficial traits only, such as skin color or ethnicity.
3. Next, groups must develop a shared sense of reality at the presymbolic level. Some groups are able to do this by embracing practices that enable them to develop, through frequent interactions with a wide range of others, an implicit understanding of a broad range of differences and themes. Other groups may fragment into small subgroups that disengage from each other and become more and more isolated and self-absorbed.
4. After this, groups must develop a shared sense of reality at the symbolic level. Some groups establish institutions and practices that promote a shared understanding amongst its various subgroups of each others' symbols, but other groups fail to establish the sorts of institutions and practices that promote mutual understanding, and thus, have subgroups that tend to dismiss each others' symbols as irrational.
5. Beyond this, some groups go on to develop reflective institutions and practices that support logical reality-based problem solving, maintain stability through change, and promote the development of highly reflective individuals, but other groups tend to relinquish their capacity for reflective thinking under the pressure of strong feelings, and are unable to maintain their institutions and practices at the same level of reflectiveness during periods of acute stress.

All human societies have progressed through some way of negotiating the first six levels of our model of development, and many societies have progressed quite far into the higher levels as well. These core developmental processes go back millions of years. It is for that reason that they are found in every human culture—not because they are biological but because they predate human culture and were a critical part of the caregiving practices that generated human culture.

Once we understand the developmental processes that underpin the various universals and variations discovered by anthropologists and social scientists—processes that extend back millions of years, long before the emergence of human societies—we can transcend the conflict between cultural relativists and structuralists that has hitherto dominated these fields. With the specter of biological determinism removed from the scene, social theorists can build on the search for developmental processes that provide a frame of reference within which there can be endless variations in cultural and social patterns.

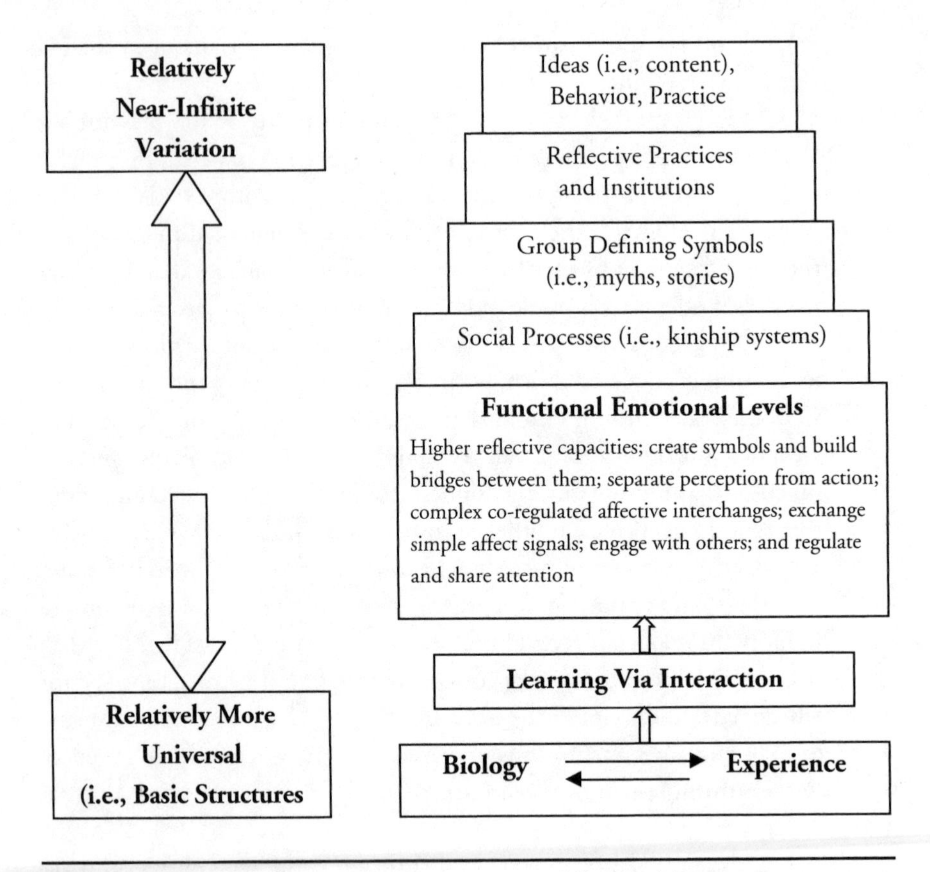

FIGURE 13.3 Developmental Processes and Cultural Patterns

—14—

A New History of *History*

IN PART II WE presented a view of human evolution as a dynamic developmental process in which early prehuman, and even before that, nonhuman species underwent the stages of functional/emotional development outlined in Chapters 1 and 2 and communicated these advances to succeeding generations. This cumulative social process was the result of cultural practices that promoted increasingly elaborate emotional gesturing, passed down from one generation to the next, rather than genetic mutations that were biologically encoded and transmitted. These practices were taught over and over again. Sometimes they were expanded and embellished, leading to further cognitive, communicative, and social advances; sometimes they lapsed and were forgotten and the very survival of the species was put into jeopardy.

To borrow a mathematical figure, the human evolutionary trajectory can be viewed as a square-wave, not a continuous function, in which regressions as well as developmental leaps occurred.[1] Although random genetic mutations were surely instrumental in many cases of extinction and adaptation—especially for lower organisms, where, for instance, a significant climatic change proved too taxing for the capacities of one organism yet spurred another to develop in some new direction—biology alone is not sufficient to explain the survival of higher social organisms that require flexible problem solving abilities in order to adapt to changing environments. The driving mechanism of evolutionary change for complex social organisms, such as the great apes and hominids, was primarily cultural.

The predominant tendency in evolutionary psychology has thus been to view the human trajectory as a series of discontinuous advances in the cognitive abilities of each successive species, as if a skill were somehow randomly generated and then naturally selected because of its adaptive benefits.[2] How else, according to this way of thinking, is one to explain why some new hominid species suddenly started to demonstrate a technological advance, in, say, tool manufacture? But consider the numerous accounts by famous scientists of the sudden "moment of insight" they experienced. Once examined, these experiences from "out of the blue" are anything but; they come after years spent developing skills and thinking about a problem, not to mention the social milieu that nurtured those skills or the temperament that drove the scientist in question. Likewise, there were no doubt countless "moments of insight" in human evolution; but that does not call for the out-of-the-blue mutations embraced by genetic determinists. Rather, it demands that we look at the sorts of practices that made such technological advances possible.

But what about actual human history, from the end of the last ice age to the present? How is one to explain the extraordinary march of technology, or the rise and fall of human civilizations, or the unparalleled advances in science, medicine, and the arts? How did we go so quickly from hunting mammoths with spears to talking about cloning a mammoth from the residues of DNA found preserved in the ice? On the standard evolutionary map, human history is really something of an appendage. All the biological changes that produced the cognitive, communicative, emotional, and social capacities of humans are considered complete by the end of the late Paleolithic period, and after that it was just a matter of human history unfolding, given this mental platform.[3] In our view, evolutionary processes have continued to operate over the past 12,000 years, and indeed, are continuing today.

In what follows, we will see how our developmental model presents us with a framework for conceptualizing the dynamics of history. Our goal here is to present only a sampling, not a comprehensive application of our approach, to illustrate the kind of new interpretation of historical events afforded by the f/e developmental model. Obviously, given the sheer enormity of the material involved here, we cannot hope to cover this topic in a comprehensive manner. Rather, to illustrate the point that throughout human history the mind was continuing to evolve, and indeed, that the human mind continues to evolve, we will focus on a few selected periods.

THE DISTINCTION BETWEEN THE STRUCTURE OF SOCIETY AND THE CONTENT OF ITS BELIEF SYSTEMS

What gets passed down in history—in addition to formative cultural practices, which, for better or for worse, are the cornerstone of human existence—are all manner of economic and political structures, scientific and folk theories, customs and beliefs, internal and external relationships, and so on. Many historians have investigated how these factors relate to a society's achievements and/or failures, but their views have been highly influenced by the determinist paradigm that has largely governed thinking about history as well as evolution. We attempt here to break free from this grip of determinist thinking by looking at the evolving f/e levels of societies as a whole and how these relate to each society's fortunes and to its own view of history.

The key to this approach to the study of history is to consider the relationship between the *structure* of a society and the *content* of its knowledge and belief systems—that is, its views about human functioning and society itself. To understand how this relationship applies to societies as a whole, first consider the CEO of a large multinational concern whose thinking has advanced to a very high level of f/e development. He regularly makes multicausal gray-area decisions in the day-to-day management of the company; is able to balance risks against rewards and plan the company's future investments accordingly; and he frequently expresses judgments in the realm of politics that reflect a strong internal standard. Thus, he is never reluctant to explain why he disagrees with the administration's policies and what he would do differently, and why his policies and not the administration's are more consistent with the long-term needs of the country. In showing this ability to think off an internal standard and to make informed judgments about international affairs, he evidences high levels of reflective skills, including the ability to apply this level of reflective thinking to a broader cultural context and to think in probabilistic terms about the future.

In the realm of human relations, however, he operates at a much lower level of f/e development, for he finds it very difficult to understand other people's emotions and virtually impossible to deal with other people's emotional problems. In fact, he believes that all individuals are motivated by fear and greed and that the best way to manage a large corporation is through reward and intimidation. As it happens, this narrow view of human functioning has proved highly adaptive for his ascent through the corporate ranks. For where others might worry about offending coworkers' sensibilities, he bulldozes his way through delicate negotiations and forces others to acquiesce to his will.

These rather undeveloped views of human functioning are particularly evident in his family life. For example, should his spouse be "too busy to attend to his needs," he immediately assumes that she is out to control him, rather than considering the possibility that she may be exhausted from taking care of their three children who have been making demands on her all day. In other words, in his understanding of human behavior and emotions, he is not able to embrace multicausal and gray-area thinking and often regresses below even reality-based thinking as he grabs onto unproven assumptions fueled by his fears and anxieties.

As it happens, this individual enjoyed a very successful tenure at the helm of his company for several years while the economy was going through a period of robust growth. Year after year, the company reported a 20 percent rise in earnings, and everyone involved, from the lowest office worker to the highest executive, was content to work in the company's inhospitable climate. All this began to change dramatically, however, as the economy reached the end of the business cycle. With a serious recession looming, it was clear that significant retrenchments would have to be made, but for this to happen the company needed someone who could negotiate effectively with the unions. "Ruling with an iron fist" no longer had the same effect it once did; now it seemed only to stir up internal dissension and threatened to provoke the workers into crippling actions. Faced with an escalating crisis, the board of directors had no choice but to lay off the CEO, while paying tribute, of course, to his inspired leadership through the golden years of growth.

When we study history, it is equally illuminating to compare the developmental level of the institutions and decision-making processes of a society with its views about human functioning and the nature of society. What we see is that the more a society's institutions and decision-making processes reflect high levels of f/e development, the more stable that society, for they are better able to deal with regressive beliefs and behavior in encapsulated or narrow areas. Moreover, the greater the number of individuals in a society who, because of their high levels of f/e development, can support these institutions, the better those societies can withstand internal and external stresses: economic crises brought about by war, environmental catastrophes, serious declines in the business cycle. But the converse is also true, as was seen time and again in the twentieth century when attempts were made to impose democratic structures on nations whose citizenry had not yet developed the capacity to support them. A society's ability to sustain its advanced institutions and decision-making processes is always at risk, should a significant proportion of the population cease to develop these higher reflective capacities.

Hence, the historical process is not unidirectional; rather, history is marked by countless periods of stasis and regression. In place of determinist models of this dynamic, however, our developmental view of history illuminates how the structure/content relationship shapes a society's capacity to deal with the internal stresses that are inherent to social existence and the external stresses with which all human collectives must contend. The basic principle here is that the more advanced a society's views of human functioning and the nature of society, as reflected in its science, technology, philosophy, art, and literature, the more these views support the structures that a society creates to govern itself. Thus, as we trace levels of f/e development in selected periods of history, we will look at the changing content in all these areas as it shapes views of human behavior and society, and their influence on the structure of governance and related economic and political institutions. Our basic thesis is that the breadth and stability of the structures that determine the stability of societies are proportional to the degree to which higher developmental levels characterize their knowledge and belief systems, especially concepts of human nature.

As we look below at some of the more striking features of a few of the major periods in history, our emphasis is not just on the growth of knowledge but on a society's concepts and belief systems in regard to human functioning and group behavior. In our view, *a society's fundamental assumptions about human behavior underpin its institutional structures*; these assumptions can have frightening effects, as seen in totalitarian regimes, or inspiring results, as can happen in democratic societies. Advances in understanding human nature come to be reflected in institutions that contribute to the structure of the society. For example, as we shall see below, advances in Enlightenment philosophers' view of human beings came to play a critical role in the governmental and decision-making structures that were established during the Enlightenment and were grounded in Enlightenment views of free will and human autonomy. From the proliferation of nineteenth-century Learned Societies, which were supported by the government and called upon to advise in all manner of policy decisions, to today's reliance on scientists for decisions pertaining to health, criminal justice, the environment, and the economy, the concepts developed in biological and social sciences influence—for better or for worse—government institutions and policies.

ACCOUNTING FOR HUMAN PROGRESS

Although on an evolutionary scale the past 12,000 years amount to just the blink of an eye, in technological and cultural accomplishments they represent an enormous period in human development. How is one to account for this explosion of human knowledge and technology? From the perspective of our developmental approach, the answer lies in the fact that, at some point near the start of recorded human history, large numbers of humans began to master abstract reasoning, multicausal thinking, gray-area differential thinking, and thinking according to an internal standard (Levels 6 to 9, as outlined in Chapter 2). The reason this is so critical can be observed in the study of child development, where it is clear that these higher reflective capacities constitute a critical threshold in a human being's cognitive, social, and moral development.

A similar process took place over the past 12,000 years. Once humans began to think logically about the problems of survival and the problems of social existence, and, contingent on this, became able to frame multiple hypotheses, to think in matters of degree and make relativistic comparisons, and to make judgments according to an internal sense of self and internalized standards of behavior, they started to make remarkable advances. Not only did abstract thinking in technology and science flourish but so did much more complex social organizations, and there were striking developments in representational art, writing, literature, philosophy, and so on.

These advances, however, cannot be charted as some sort of Panglossian linear ascent, a steady progression in which successive generations became more flexible and abstract thinkers. Rather, when we look at the various periods of human history, the picture that emerges is much more complex. For example, we see elements of very high levels of f/e development in some areas, such as science and technology, alongside much lower levels of f/e thinking in regards, for example, to the persistence of superstitious and magical thinking. Or we see some small sector of a society—a governing class, for example—advancing in their f/e capacities while the vast majority of the society operated at a much lower level of development. Or we see a few individuals reaching quite high levels in a particular area of abstract reasoning—mathematics, science, philosophy—yet their social and emotional development remained constricted.

Furthermore, as we noted above, human history is marked by a constant see-sawing between cultural advances and regressions. This is manifested in all sorts of ways: by the rise and fall of civilizations; or, within civilizations, the rise and fall of enlightened social institutions, such as the first stirrings of democracy and the rule of an impartial legal code followed by periods of tyranny and lawlessness (as in the decline and fall of Rome). There have been, in fact, not just one, but many "dark ages," so-called because of the virtual disappearance of written historical records, sometimes brought about by deliberate acts (the burning of the library at Alexandria, for example) and sometimes by a collapse into anarchy.

Nevertheless, when we look at the grand sweep of human history, we see a gradual progression from lower to higher levels of f/e development. Indeed, one might argue that, although they certainly did not think in these terms, this is the reason why so many great thinkers—Hegel, Marx, Spencer—were drawn to a teleological view of human history. For that matter, even though classical evolutionary theory was first and foremost supposed to be a nonteleological theory, it has often slipped back into this way of thinking.[4] In its own way, genetic determinism often seems to be an attempt to account for this sense of teleological or purposive evolution.[5]

There is really a *dual problem* in human history, therefore, which needs to be explored. On the one hand, there is the extraordinary technological and scientific ascent of humans over the past 12,000 years, yet nonhuman primates, such as the great apes, have remained at the same level of technological development for millions of years;[6] but on the other hand, human history is characterized by constant regressions as well as advances. Some teleological theories (Hegel and Marx) attempted to account for the latter, but for the most part theorists have tended to focus almost exclusively on the reasons for the advance and ignored the phenomenon of regressions. But virtually all teleological theories have tried to explain the dynamics of human progress deterministically, either in terms of a theological mechanism (e.g., Hegel's Spirit that "determines history absolutely"); a social mechanism (e.g., Marx's modes of production); a biological mechanism (e.g., the selection of certain genes, such as a "language gene"); or a psychological mechanism (e.g., the development of certain modules, or a "mindreading" mechanism).[7]

Our developmental approach to human history instead attempts to explain the progressive nature of human history in nondeterministic, nonlinear terms. Any such theory must at the same time explain the reasons for the remarkable technological and social ascent of humans over the past 12,000 years and the constant swings between advances and regressions. Rather than postulating a deistic hypothesis that some special mechanism must have begun to operate, miraculously, at the dawn of human history and has been steadily ticking away ever since, a convincing explanation must view human evolution as an *ongoing* rather than a completed process, a process always vulnerable to regression, rather than a one-directional progression towards some predetermined utopia (such as Marx's socialist state).

Over the broad expanse of history, individuals were developing their higher levels of thinking in an expanding range—for example, emotional, social, and linguistic development as well as technology, science, literature, and the arts—though not always at the same pace. We see a similar phenomenon here to what Piaget attempted to capture with his notion of *décalage*; that is, some particular group may have been at a fairly low level of f/e development in regards to their emotions or their shared sense of reality yet have reached a fairly high technological level, or vice versa. This is a point that the great anthropologists stressed at the dawn of the twentieth century (see Chapter 12). The engine underlying the development of these complex mental processes over the past 12,000 years was the same as the engine underlying the development of the more fundamental processes (those that we saw in Chapters 3 to 5). This engine has been operating for a million years, and dates back to the Common Ancestor and beyond.

The reason human history has the teleological appearance that it does, therefore, is because, over the course of human history, there has been

A. a gradual development in the kinds of complex emotional exchanges that underpin the growth of the mind's higher reflective capacities; and
B. a gradual broadening in the number of children who experience these nurturing exchanges, and thus, a growth in both the proportion of society and the number of societies operating at higher stages of f/e development.

In other words, the fundamental dynamic operating in human history is neither biological nor material: It is the cultural transmission of caregiving practices that support the development of higher reflective capacities in individuals and groups. Precisely because this is a cultural phenomenon, it is not predetermined and is highly vulnerable to regression.

Our basic hypothesis, therefore, is that societies flourish and advance when they support the sorts of caregiving practices that we have been examining for the vast majority of their children; and societies decline and regress when vast numbers of children are deprived of these co-regulated emotional interactions (or, of course, exposed to various kinds of environmental toxins or nutritional deficiencies that compromise their nervous systems). This hypothesis differs in a fundamental respect from—but complements—the classic view of human development championed by John Stuart Mill and Isaiah Berlin; for it holds that broad cultural advances are not simply the result of enshrining fundamental rights, nor of providing individuals with the means and opportunity to develop their minds. The greater the number of the members of a society that are capable of engaging in reflective thinking, the more stable the institutions in that society will be. But the development of such thoughtful individuals is a result of early interactive experience: Rights and opportunities are not sufficient. Hence the liberal-democratic vision put forth by Mill and Berlin needs to be grounded in a *developmental perspective.*

THE MAGDELENIAN PERIOD

The Magdelenian period dates to between 12,000 and 8,000 years ago. The area we are particularly interested in is Europe, where temperatures hovered around minus 22 degrees Fahrenheit in winter. In addition to the cold and fierce winds, there was little sunlight during the winter and game was scarce. Yet the peoples who lived there not only survived but actually thrived. Indeed, there is no evidence of widespread disease, which did not appear until humans started to live in large settlements.

These were a seminomadic people who lived in small tribes. Apparently they were trading with one another (as is indicated by the presence of salt, bitumen, and nonnative plants). Mostly they survived by hunting large animals, such as caribou and mammoths, and by fishing. They likely spent the year tracking game according to seasonal patterns. They lived in tents that were conical in shape, apparently to prevent snow from collecting, and easily assembled.

Every part of the animals that they hunted was used for a different purpose. They wore furs to keep warm, using different kinds of skins for the different parts of clothing (which included mittens, anoraks, and fur boots). By curing the skins they were able to create pots in which they cooked soups that were probably a mixture of meat and any vegetables they could find. Animal skins were also used to make leather containers in which they collected drinking water. Animal fat provided them with fuel for light and and heat lamps. And bones were used to make weapons, sewing needles, and even flutes and whistles.

Particularly interesting is how they treated their dead. Bodies were placed in graves, sometimes in groups of two or three. They were placed on their sides, the legs slightly bent. In a camp on the shores of Lake Ushki, Siberia, a burial pit has been discovered covered with red ochre. Paleoanthropologists have speculated that this may have been because they used the ochre to simulate blood, or to provide the corpse with a life-like aspect. In addition, tools and ornaments were placed in the grave in a manner that reflects gender and social status. In a grave from this period in Monte Verde, Chile, plants with special medicinal properties have been found.

To judge from these cultural advances the stage of f/e development reached by the people of the Magdalenian period obviously involves making speculative assumptions, given the limited archaeological information available. The standard way to approach such a problem in anthropology is to look for a comparable culture for which we have substantial information. When looking for a model for the people of the Magdalenian period, the important factor to be considered is just how harsh the living conditions were. Survival would have been a constant challenge. They had to find ways to keep warm and healthy and to obtain sufficient food and water. And they had to construct shelters and protective clothing from very limited resources.

One of the best models we have for a culture that has survived under similar conditions is that of the Inuit, the seminomadic peoples of the Arctic. Indeed, the Inuit were commonly described as still having a prehistoric culture. To be sure, one significant difference between the two cultures is that the Inuit were largely a maritime people (although there were also inland Inuit communities), whereas the people of the Magdelenian period were living in the glaciated regions of Europe. But the traditional Inuit way of life probably bears a striking resemblance to that of the Magdelenians. There have, of course, been great changes in Inuit culture over the past few decades as a result of the introduction of Western technology and the rapid urbanization that has taken place. But we have detailed historical records by late nineteenth-century and early twentieth-century explorers and anthropologists that may provide us with a penetrating insight into the lifestyle of the Magdelenian peoples.

The Inuit lived on the tundra, which consists of low, flat, treeless plains with permanently frozen ground (permafrost). The primary social unit in the traditional Inuit way of life was the extended family. For much of the autumn, winter, and spring these families would hunt on their own, but during the late spring and summer months they would join together in base camps that were positioned close to rich feeding grounds. Historical accounts provide us with a vivid insight of what it was like to face each day in frigid conditions with the task of finding enough food to survive that day. The primary source of food for the Inuit was fish, eaten frozen or raw (hence the name Eskimo, which in one of the Northern Indian languages means "eater of raw fish"). The most prized sources of food were caribou, seal, walrus, and whale. Seal was perhaps their most important source of food. The meat was eaten raw, frozen, boiled, dried, and fermented. The fat was regarded as a delicacy and also used as a fuel. The skins were used to make boots and outer clothing, and also to make umiaks, kayaks, tents, water floats, and cooking pots.

The manner in which the Inuit capitalized on seals for so many purposes is an example of fairly sophisticated multicausal thinking. They were experimenting with different ways of eating the flesh appropriate for different circumstances and mastering the techniques required for these different methods of food preparation. They were also figuring out such things as the way in which fat could be used as a dip, a source of fuel and lighting, and even as a medicine. These developments reflect the sorts of abstract logical thinking that we see at Level 6 of our scales, as well as the communication skills necessary to impart these techniques to the next generation.

For example, to figure out that seal fat could be used as a medicine would have involved identifying a relationship between certain forms of illness and the effects of consuming seal fat. The Inuit had to figure out that it was the fat and not the meat that was producing the desired results; they also had to know how much fat to ingest, and when, and by whom, and for how long. Interestingly, this traditional knowledge about the medicinal benefits of seal fat has recently sparked off scientific studies worldwide. We now know that seal fat is extremely rich in selenium, which is an important antioxidant and is vital for a healthy immune system and thyroid. We also know that selenium is an effective agent in warding off arteriosclerosis. Selenium deficiency has been associated with numerous disorders, such as a compromised immune system, Keshan disease (enlarged heart and poor heart function), and arthritis.

Although they certainly didn't have this level of explicit, technical knowledge, the Inuit had figured out the basic principles involved here. Similarly, the Inuit were figuring out the advantages of seal skin over other sorts of skins for various purposes. For example, they figured out that seal skin is ideal for making boots and outer clothing because it is water repellant yet porous, which allows for body humidity to escape, and it is lighter than caribou skin. They also figured out that the intestines were ideal for making parkas. Thus, the Inuit saw the seal as a complex natural resource. Not only does this use of the seal involve multicausal thinking—for example, figuring out that one particular thing, such as seal fat, could serve many different functions—but it even broaches gray-area thinking: figuring out that such-and-such amount of seal fat was needed in such-and-such circumstances, or that seal skin was better than caribou for outerwear but not for innerwear.

We see planning and problem solving in their hunting. For example, the techniques used in seal hunting depended on the season; the hunters figured out different techniques depending on the weather, which, of course, was a constantly changing variable. Seals were hardest to hunt in the summer when they could swim out to open sea. Fall hunting was the most important of all the seasons because enough seals would have to be killed to last humans and dogs throughout the winter. To survive in the ice-bound water, the seals would develop "breathing holes," which they would carefully keep open throughout the winter. The hunters would track these breathing holes and, when they discovered one, stand over it, completely motionless, their spears poised, for hours on end as they waited for the seal to show its head when it needed to breathe. This was a solitary and remarkably disciplined pastime. A young boy would learn how to hunt seal by being taught by his father; first, he learned how to forge his weapons; then, how to track seals, and figure out where a breathing hole was most likely to be found; and, literally, how to stand and hold one's spear, how to capture the seal, and then what to do with it.

A more illustrative example of their complex planning and organization capacities can be seen in their whale hunting. Whales were perhaps even more highly prized by the Inuit, for their fat as much as their meat, which not only provided them with an essential nutrient, but also a critical source of fuel for cooking, heating, and illuminating their igloos. Whale hunting was a collaborative activity, and much more perilous. The hunters would camp out in the spring on promontories close to whales' migration routes. When a whale surfaced, they would jump into their umiaks and kayaks and wildly pursue it. The lead hunter would harpoon the animal. Seal-skin bags inflated with air were attached to a hunter's harpoon, which would slow the whale down and eventually exhaust it, at which point the hunters would all join in for the kill and then pull the carcass onto thick ice where it could be cut up and the pieces hauled back to the base camp.

Fairly advanced multicausal thinking was required to plan for and coordinate such a complex activity. Moreover, to function effectively, the hunting party had to adopt a hierarchical structure in which different hunters were assigned different roles. Given the perilous nature of the situation, they had to know exactly what each other was doing and how they would respond to various emergencies and to each other's actions. You might compare this to the sorts of lifeboat or surf life-saving drills that are still practiced. In dangerous situations, it is absolutely vital that each person involved knows at every moment what everyone else is doing, and that only one person is in command.

The other great source of food and skins for the Inuit was caribou. Caribou would migrate in huge herds along familiar routes. Inuit hunters would pursue these herds and divide up into beaters and another group of hunters who waited with spears and bows and arrows to attack the stampeding animals. When a herd was found, it would be followed for days and the hunters would collect large numbers of skins and as much meat as they could transport; but they would never kill more than they could use or carry. Here, too, we see complex problem-solving skills. They had to figure out the animal's behavior, for example, its migratory and feeding patterns and how it would respond when attacked; fashion appropriate weapons; assign different roles to different hunters; envision the problems that might arise and fashion appropriate strategies in advance; and be able to function effectively together in what, once again, could be a very perilous activity.

We know, from the hooks from bones and the nets that have been found, that the people of the Magdelenian period fished; we also know that they hunted for caribou. From the weapons that have been found it seems likely that they were using a similar hunting technique as that used by the Inuit. They were also using caribou skins to make the same sorts of clothing as the Inuit. To be sure, the people of the Magdelenian did not hunt for seal, walrus, or whale; but the manner in which they hunted mammoths was likely very similar to the manner in which the Inuit hunted whales insofar as it, too, would have been a complex, collaborative, and dangerous activity involving many families working together. And the Magdelenians' use of the mammoth appears to have had many things in common with the way that the Inuit used seals and whales. That is, the mammoth was a source of meat and fuel for them and they used its skin for their tents and other purposes.

Furthermore, like the Inuit, the people of the Magdalenian period were experimenting with the different kinds of skins available for different purposes—to make water containers and cooking pots, for example—and figuring out which skins were best for innerwear, for outerwear, for boots and hats; and how to prepare these skins so that they were supple and could be formed into clothing, as well as the techniques required to cut the pieces for the various parts of clothing, and how to prepare the sinew from the animals they hunted into a gut that could be used for sewing. When we look at the Inuit, we see just how much learning has to go into each of these skills. Just as the Inuit took great pride in decorating their clothes, it would seem from the archaeological evidence that the peoples of the Magdalenian were also decorating their clothing. So for all the reasons given above, it seems highly likely that the Magdelenians were mastering multicausal thinking and were even beginning to master the gray-area differentiated thinking that is required in any complex coordinative venture.

To imagine what life might have been like for the Magdelenians we might consider a typical winter day for the Inuit. The day would begin before dawn with the females fetching water and preparing the morning breakfast (typically, frozen fish). The males would be busy packing up and strapping their belongings to their sleds and checking the dog lines. Older children would be responsible for caring for the younger infants. After breakfast, they would break camp early, harnessing the dogs to the sled and then setting out shortly after daybreak. With so little sunlight, every available moment of daylight had to be spent in travel because it was far too treacherous to travel at night. There are all sorts of crevasses in the ice into which a team of dogs and sled could easily fall. Depending on the size of the family, there would be anywhere from a couple to several of these teams. They'd be careful to travel in sight of one another, for in a blizzard, which can happen quickly, it is incredibly easy to become disoriented and lost. When game was spotted, the males would quickly organize themselves into two groups: one to beat the animals towards another group waiting with bows and arrows and harpoons. To remain soundless, they would communicate with each other by arm and head movements.

The day would be spent searching for game. It would come to a close at dusk, when temporary shelters, the famous igloos, would be constructed. Early explorers and anthropologists all marveled at how quickly and efficiently the Inuit could construct an igloo. Everyone had an assigned job: One person would carve large blocks of ice from the ground; another would carry these to the chosen site; another would position the blocks for the igloo; another would pack all the cracks with snow; still another would flatten out the floor; another would prepare the "beds"—literally banks of ice blocks covered with thick caribou hides; meanwhile, others would be busy unloading the sleds, gathering and boiling water, preparing the evening meal. Finally, before anyone could eat, the dogs would all be carefully checked, fed, and then bedded down in the snow for the night.

The igloo represents an ideal solution for a seminomadic hunting people that had to construct a temporary shelter from the materials at hand as quickly and efficiently as possible. The domed shape of the igloo with its low narrow entrance was ideally suited to the Arctic because it could withstand the most severe of winter storms. Numerous explorers and anthropologists have described listening to such a storm raging about them, safe and snug in their igloos, and then digging out from several feet of snow the following morning. For more permanent structures, several igloos would be joined together so that the members of an extended family could enjoy a surprisingly spacious shared living space.

The Magdelenians were probably equally adept in their construction of shelters. Using the materials that were readily available, they figured out a solution that could have been transported easily and erected quickly, and that was able to withstand the storms that would have been constantly raging. As in the igloo, temperatures inside the tent would likely have hovered around the freezing mark; but these tents, heated, like igloos, with animal fuel lamps, would likely have felt snug and secure.

As we can see from the above, the Inuit would have had relatively few opportunities to exchange any sort of emotional signaling during the day: for one thing, young children were bundled up and strapped tightly to the sleds. The signaling that did occur between adults would primarily have related to coordinating travel and hunting. The opportunities for sustained exchanges would have occurred primarily at night, and, because of the scarcity of fuel oil, they would also have been limited. And there is every reason to believe that the same would have been true for the Magdelenians.

There have been few accounts of what sort of life the Inuit experienced during these intimate moments. There are some fascinating accounts (see Knud Rasmussen's famous *The People of the Polar North,* published in 1908, and Vilhjalmur Stefansson's *My Life with the Eskimo,* published in 1913).[8] But one of the most valuable sources that we have is Jane Briggs's *Never in Anger.*[9] Briggs had gone to study the Inuit in very much the capacity of an ethnographer, living in her own tent and staying as detached from them as possible. But the exigencies of their way of life proved too great and, somewhat against her will, she was forced to move in with the family of the headman, who adopted her as his honorary daughter. Because of this, she was able to participate in as well as observe their way of life in startling and intimate detail.

The picture that she presented went against many of the stereotypes that had long proliferated. The Inuit have long been famous for presenting a stoical and nonexpressive face to the world. This led many anthropologists to speculate that they do not experience the "basic" emotions, or at least not to the same degree as others; and that, unlike most of humanity, they do not speak in "motherese" to their infants. Briggs also shared many of these stereotypes in the beginning. But once she became a bona fide member of the family, she found that all the members were extraordinarily affectionate with one another. In fact, one could say that their primary activity once inside their dwelling was to engage in long chains of warm conversation with one another. Much time was spent cooking and eating while they joked happily with one another and gossiped about the other families.

One of the most surprising elements of Briggs's account is to read about the behavior of Inuttiag in the privacy of his igloo. The Inuit have a strong convention in regards to concealing their emotions when interacting with those outside the family. Much to Briggs's surprise, however, the chief who had adopted her as his honorary daughter was a completely different person when alone at night with his family. Because she had been made a member of the family, he was totally relaxed in her presence. Lounging about in his long underwear, he delighted in playing with his children, especially with his newest infant. He would play with the baby in a high sing-song voice, holding it gently, calling it various endearments, trying to make it laugh by tickling and caressing it and making funny faces and sounds, playing peek-a-boo, and in general taking great delight in the baby's antics. Indeed, it was clear that there was absolutely nothing the baby could do that would have provoked a strong rebuke. Although the same liberties were not extended to the older children, the chief was equally loving and gentle with them, and with his wife.

This loving behavior with young infants is a reflection of the Inuits' belief that a child younger than three years lacks *ihuma*—reason—and hence is not responsible for any of his actions. Once a child had acquired *ihuma,* even a child as young as four, he was now responsible for his actions and for contributing to the family welfare. This belief is particularly interesting because it signifies that the Inuit had a conception of childhood as a distinct stage in human development. To be sure, their view of when a child should assume the responsibilities of an adult is considerably earlier from ours today. But what is significant here is that their conception of *ihuma* demanded that infants receive the sorts of warm emotional experiences that Briggs observed in addition to the necessities of nourishment and protection. We can only speculate how far our model extends here and whether the people of the Magdelenian period had a similar conception of childhood—a conception, incidentally, that was quite widespread amongst North American indigenous peoples. If so, it would certainly be a potent factor to our argument about the sorts of changes that were taking place in caregiving practices.

The sorts of loving behavior that Briggs observed in the privacy of the igloo were a far cry from how the Inuits dealt with conflict. Rasmussen told a famous story in his account of the fifth Thule Expedition (from 1921 to 1924) of how he had visited a camp composed of fifteen families in which every adult male had been involved in some sort of killing. Rasmussen even produced a list detailing how

- Havgugaluk killed his wife in a fit of jealousy and was killed by her relative Makharaluk;

- Angulalik took part in a murderous assault on some neighbours, a vendetta being the reason;
- Uaquaq and Erfaluk killed Qutlaq in revenging their relative Qaitsaq, the father of Netsit, who had been killed by him;
- Norrahanaq stole a wife from Annernaq (there was a fight, but nobody was killed);
- Portoq also stole a wife, and he, too, later fought with the woman's husband—again no one was killed;
- Kivgaluk lost both his father and a brother through murder; he was all ready to take vengeance, but for the present was only restrained on account of the mounted police, who came on patrol every winter and took all killers for trial;
- Ingoreq in jealousy tried to kill both Arshuk and Orshoraq, but failed;
- Erfvana killed Kununasuaq and also took part in the murder of Qutlaq;
- Kingmerut shot Mangigshalialuk, and was also one of those who shot at Halggiuhialuk, who escaped death, however;
- Qaitalukaoq in anger struck down Maggararaq with a knife;
- Pangnaq, a boy of about twelve years, shot his father because he was cruel to his wife, the boy's mother;
- Maneraitsiaq fought a duel with Mangigshalialuk with bow and arrow, but neither was killed; and
- Tumaujoq stabbed Ailanaluk with a knife because he had killed the former's relative, Mahik.[10]

Thus, along with the sorts of warm affective gesturing that we see within the dynamics of the family, there is evidence that the Inuit would regress to a much earlier stage of polarized, all-or-nothing thinking when it came to dealing with interpersonal conflict. One could argue that they were regressing to presymbolic modes of behavior in which perception leads directly to action, rather as a toddler regresses to catastrophic behaviors when frustrated or angry.

If the Inuit model holds, it seems likely that, while the people of the Magdalenian period were mastering abstract reasoning and multicausal thinking, their reflective abilities in regards to interpersonal relations were probably at a much lower level of development. Skeletons have been found with massive head injuries that would appear to have been inflicted by humans. When we look at the types of cultural advances that the Magdelenians were making, the thing that stands out most is how the problems they were grappling with primarily related to their means of survival: food, shelter, clothing, medicine, cooking and cooking tools, hunting tools and techniques. Although these beginnings of human history are marked by some striking technological advances, reflective thinking in the domain of relationships and empathy almost certainly lagged far behind.

To be sure, pretty much the same pattern has played out in virtually every human culture ever since, and it continues to be a pressing concern in our own culture. As we shall see in the remainder of this chapter, the pattern in human history has essentially been one of a broadening in our higher-level thinking capacities, from areas largely concerned with survival to areas involving science, technology, philosophy, and the arts. Emotional and social development has, universally, lagged behind these other areas, which suggests that the emotional and interpersonal sphere must be one of the hardest domains in which to develop mature reflective capacities. What we see in the Magdelenians, therefore, is not just the first signs of this familiar pattern in human history, but, indeed, the foundation for this pattern in human history.

EVIDENCE OF DEVELOPING MINDS

Several kinds of evidence lead us to conclude that, in regard to their means of survival, the Magdelenians were mastering abstract logical reasoning and multicausal thinking (Levels 6 and 7 on our scale) and starting to master gray-area differentiated thinking (Level 8):

1. Some of the advances that we see here, such as figuring out how to adapt to their hostile environment with more complex kinds of shelters and more protective types of clothing, reveal a flexible form of abstract reasoning. For example, they must have experimented with different kinds of structures before they hit upon the advantages of a conical shape. They likely experimented with different materials and techniques before they figured out how to fashion cooking pots and protective clothing out of animal skins.
2. There are also signs of a methodical process of inductive reasoning. For example, they appear to have been starting to experiment with medicines as a means of enhancing their health, for skeletons have been found of what appear to be shamans buried with their medicine kits. Putting together such a medicine kit involves a long inductive process of gathering different

kinds of herbs and tubers and testing them to see what sorts of physical reactions they produced. This involves some systematic reasoning and indicates that knowledge acquired over generations was likely carefully passed down from elders to younger persons in a long line of shamans.

3. Evidence of collective problem solving can be seen in the hunting techniques they needed for capturing large and highly dangerous animals. A hierarchical set of relationships is needed for such endeavors. In carrying out such a kill, different individuals would have been assigned different roles and responsibilities and would have enjoyed differing stature within the group. These roles would have involved matters of degree and thus suggest differentiated, gray-area thinking.
4. Signs of early settlements at the end of this period suggest the development of further f/e development. As we will see in the following section, living in a stable society requires being able to think according to an internalized standard (Level 9). The harmonious existence of a settlement depends on the existence of shared social rules and conventions.

Thus, the Magdelenians were laying the foundation for the emergence of settlements during the Mesolithic and Neolithic periods not simply because they were making important technological advances but, more fundamentally, because they were mastering the sorts of complex thinking and internal standards that are absolutely critical for the gradual formation of social rules and conventions that sustain a settled community.

Such evidence supports a view of the dynamics of human history different from what one sees in standard teleological theories. In the standard view, humans started out with an established cognitive architecture and the rise of civilization was due to a "ratcheting effect" in which one generation after another capitalized on the technological accomplishments of the previous (e.g., how to harvest and store seeds, how to prepare the materials necessary to build permanent structures, and so on).[11] There can be no denying the importance of this "ratcheting effect" in human history. But to treat this phenomenon purely in terms of the storage and transmission of knowledge overlooks the manner in which a new generation takes in its stride concepts or technologies that its parents may have struggled with.

One need only consider how much smaller the world feels today for young children navigating the Internet. But the psychological mechanism operating here is not simply one of early exposure to new ideas or technologies, for it is not the ideas themselves, but more fundamentally, the manner in which they are communicated that determines how this knowledge is absorbed. The caregiver who communicates an idea with little or even negative affect is likely to constrict their child's interest in or facility with that concept. Conversely, the caregiver who communicates a concept with facial expressions and gestures that convey lively interest and excitement is likely to see the same response in the child.

Indeed, herein lies a critical dimension in the growth of science itself. The build-up of scientific knowledge is not at all comparable to the manner in which, say, a computer program builds up its "knowledge base" through multiple applications. Above all, science is a human endeavor and, as such, driven by human affects. To return to the example that we presented in the Introduction to Chapter 2, consider the reaction of Renaissance thinkers to Galileo's famous acceleration hypothesis. Galileo did not simply report an experiment showing that a falling body accelerates uniformly, which was so logically compelling that Renaissance thinkers had no choice but to add to their growing understanding of physics. Rather, Galileo presented this idea in *Two New Sciences* in the form of a dialogue in which the official Church position (Aristotelianism) is put forward by Simplicio (who, as his name implies, is profoundly simpleminded).

What Galileo was really arguing in *Two New Sciences* is that authority is irrelevant as far as concerns our knowledge of the physical world; that scientific assertions stand or fall by experimental test. And it was precisely because of its emotional overtones that Galileo's experiment had such a profound influence on the development of Renaissance culture; for, as we saw in Chapter 2, the generation of Renaissance thinkers who communicated this new knowledge to their children conveyed, through their facial expressions and gestures, those very values that came to inspire the basic principle of humanism that every individual has the ability to use his innate rational powers to discover the truth. That is, the foundations for the development of the toddler's creativity, pattern recognition, and problem-solving abilities would have been laid, long before he could speak, through the sorts of back-and-forth interaction patterns that we outlined in the introduction to Chapter 2, where we envisaged how a caregiver might have responded to her infant's refusal to eat a certain food by raising her arms in puzzlement with a matching facial expression and vocal tones that invited her toddler to communicate further his needs and wants while offering him various alternatives between which to choose.

In other words, the emergence of the humanist Weltanschaauung that marks the Renaissance, which led to the rise of skepticism and cultural relativism (as seen most notably in the writings of Montaigne), and even more fundamentally, to a growing challenge to the Church's authority to pronounce on matters pertaining to both the natural and the social world, would literally have been bred into the developing minds of Renaissance children. Such an argument thus provides the critical element that has hitherto been missing in historical accounts of the rise of humanism: namely, an account of how these new attitudes were inculcated. It is not simply that the Faculty of Reason, when presented with new truths, will immediately recognize and embrace these as such; there are far too many counterexamples to adopt such a rationalist oversimplification. Nor can one assert that there was some sort of contagion effect, without explaining why and how such a phenomenon should have occurred. These two theories, that reason automatically embraces a truth when it sees one, or that paradigm shifts occur as a sort of contagion, may both be true; but regardless of whether we believe in either—or neither—of these models, we still have to understand the mechanisms by which such a new ability can be learned on a mass scale: what common denominator could enable people to understand and embrace even a contagious idea. We need to understand what would lead to the appearance of an almost automatic embracing of a new truth, when that truth is a highly complex one. To answer these questions we need to understand the developmental processes that enable individuals to comprehend new ideas and become capable of a new and more complex basis of thinking.[12]

THE RISE OF CIVILIZATION

"Civilisation," Matthew Arnold remarked, "is the humanization of man in society." But the question we have been building up to in this chapter is how civilization itself was possible. How were the small farming and hunting communities of the late Magdelenian period able to create the complex social and political structures on which a civilization rests?

We would argue that over the course of millions of years humans attained a level of f/e development that enabled them to start living in settlements, and then a relatively rapid broadening of these abilities took place over the next 8,000 years that provided the foundation for the rise of civilizations. Although further technological advancements—say, grain storage or irrigation and writing systems or some other form of record keeping—were essential to the rise of civilization, such progress is not a sufficient explanation. Nor can Mesopotamia, Sumeria, or Egypt be explained by some hardwired biological instinct in human beings to congregate, as ants or large herd-dwelling creatures do, and then develop the modes of production necessary to sustain their large communities.

A "State" isn't simply a *larger* community than an agricultural village; a State represents a whole new level of communal existence. In Gregory Johnson and Henry Wright's classic definition, a *State* consists of a decision-making hierarchy with at least three levels, together with specialized administrators and some effective means of preserving records and other forms of information.[13] To understand what is meant by a "decision-making hierarchy," consider a simple agricultural village in which decisions about what crops to plant, how much of the harvest to store, who gets what share of the land, and even who marries whom, are made by a village headman. Such a structure represents a first-level decision-making hierarchy in which the headman directs the activities of all the other members of the group, at least in the areas that affect the community as a whole. A second level of decision-making hierarchy would exist if there were a small group of people, such as village elders, with whom the headman had to discuss the various issues confronting the life of the village and implement decisions. Suppose that, above the headman there was a tribal chieftain to whom all the village headmen had to report, binding all the villages together into a larger community; this would be a three-level decision-making hierarchy. And so on.

According to Johnson and Wright, the first of the great civilizations, the Mesopotamian State, had at least three such levels of specialized administrators. Furthermore, the State possessed the ability to process and store information relevant to overseeing the society's day-to-day operations. Numerous administrative records have been found. These documents, which were impressed on clay, consist of commodity seals, which were used to certify the contents of a container (such as a vessel, basket, bale, or storeroom), and message seals, which were used to store facts about goods and people. Writing as such did not arise until after the emergence of the Mesopotamian State; but these stamps and seals conveyed a great deal of information and are the direct precursors of later written scripts. Thus, for Johnson and Wright, a civilization is essentially a hierarchical society governed by some sort of administrative bureaucracy.

We need to go further than this classic view of what constitutes a State, however, and ask what forces could enable a disparate group of communities to establish and maintain such a burgeoning bureaucracy. When we look at the rise of the Mesopotamian State in the fourth millennium B.C., the feature that most stands out in the slow rise of urbanism that occurred is stratification: The larger the society became, the more stratified it became, and the more specialized each individual's role within that society. Everyone, from the king and priests down to the farmers, the soldiers, and the slaves at the bottom, acquired a specific social identity, and correspondingly, a highly circumscribed responsibility within the community. All were integrated with one another in the sense that all depended on one another to maintain their communal life. What held them all together? It can't have been simply necessity or self-interest, for this would not explain why some civilizations flourished and others failed to take hold, or why thriving civilizations would ever decline.

Sustaining this form of complex organization requires that the members of a community acquire a shared sense of reality: a common set of values and, indeed, a sense of identity as a member of that society, as well as a narrower sense of identity as a member of such-and-such a family or clan or community. Accordingly, a majority of individuals must be able to regulate their behavior according to the overall society's complex codes of behavior in addition to their family's or clan's or community's codes of behavior.

The point that we are making here is, of course, reminiscent of Durkheim's famous notion of the collective conscience of a society: a "collective mind" that exists over and above the individual members of the society. According to the social anthropologist Max Gluckman, the mind of the individual is molded from the moment of his birth by the ideas and behavior of his parents and peers. In this sense, the individual is said to *inherit* a "mind," and that this inheritance must be understood "in a social, not a physical sense."[14] As we saw in Chapter 13, because they did not understand the developmental processes involved, structuralist theorists attempted to construe this inheritance of a mind on a model that was very similar to that of a physically oriented geneticist explaining the inheritance of physical traits. What was not understood was the actual developmental pathways through which the culture and the social context within which children are born influence the development of their minds.

In Chapter 13, we looked at the developmental pathways through which a culture exerts its influence. In this way we saw how, long before the new mind acquires symbolic meaning, it acquires information from its culture and social context through the way in which presymbolic interactions, mutual regulation, engagement, emotional signaling, and co-regulated emotional exchanges associated with the social group are being negotiated. For example, the way in which such basics as closeness, dependency, aggression, sexuality, gender, dominance, and submission are negotiated is related to presymbolic interactions that are the features of learning in the first years of life, way before symbols themselves become a dominant factor. In other words, the near infinite number of emotional interactions that occur between babies and their caregivers create a structure for determining what is important, permissible, meaningful, and realistic—that is, for what will later become symbolically interpreted values of the collective mind and culture.

To establish a hierarchical society, its members must understand their roles in a truly relative sense. For example, there are different degrees of power or leadership, or different degrees of closeness and intimacy, or different levels of skills and responsibilities. The ability for this sort of relativistic thinking, which we have described as differentiated gray-area thinking, comes after basic causal thinking and multicausal levels of thinking.

For complex hierarchical societies to sustain themselves, a reasonable number of its members must have achieved this level of grey-area differentiated thinking. Otherwise, they will be unable to sustain flexible problem-solving strategies and group decisions that reflect compromise and reconciliation. In addition, to sustain a complex hierarchical society requires that its members share a common sense of reality: that they acquire a collective identity based on values rather than primitive beliefs. Achieving this level of social development requires that the individuals involved master the next level of development, namely, thinking off an internal standard and acquiring a more integrated sense of self. Here, too, there are different degrees. At a fairly concrete level, the shared sense of reality that binds together the members of a group may simply be based around common words and concepts. At a more abstract level, a shared sense of reality may be grounded in institutions and practices that support a shared system of values.

As we will see below, there is considerable evidence that, beginning around the fourth millennium B.C., societies were becoming increasingly based on these higher levels of development. In fact, we will see that social groups at this time were beginning to expand the application of these higher levels of reflective thinking by exploring such themes as sexuality, or by defining work and family roles. We will trace these developments in a summary way in the paragraphs that follow.

SUMERIAN CITY-STATES

Once a broad section of a society has developed the f/e abilities that are necessary to support a State, this provides the foundation for further dramatic technological advances to occur. This latter phenomenon can be seen in the rise of Sumerian civilization, which flourished between 3000 and 2350 B.C. We have a fairly detailed picture of what life was like in these Sumerian city-states because shortly after 3000 B.C. they began to develop a written language. The earliest known documents are the clay tablets and seals from early occupations at Uruk (c. 3400 B.C.). These documents include signs for carpenters, donkeys, boats, copper, and the word *en*—the title of the lord *unken*—which may have been a people's assembly. In all they had more than fifteen hundred symbols.

In 2900 B.C., the Sumerians introduced a phonetic system that served to represent distinct syllables and words of the spoken language. The pictographic elements slowly lost their iconic character and eventually Sumerian writing consisted primarily of these abstract symbols. But the Sumerians never developed an alphabetic system. The first truly alphabetic written languages appear to have developed around the middle of the second millennium B.C. among Semitic-speaking peoples in Palestine and northern Syria. Then, in the tenth or ninth centuries B.C., the ancient Greeks adopted and adapted the Syrian or Phoenician variant of these alphabets.

Children can begin to master reading and writing as early as the sixth stage of f/e development, when they start building bridges between ideas. One might argue that the precursors of writing, such as pictographs, originated at a comparable stage of development. But to create an alphabetic system of writing, the members of the society have to be capable of regulating their own and each other's behavior in various ways: how a sound should be represented; how a word should be spelled; how a sentence should be constructed; how different kinds of speech acts, such as asking a question or issuing a command, should be represented; how pauses and other so-called paralinguistic aspects of speech should be represented; and how different narrative forms should be constructed.[15] Furthermore, a society cannot develop an alphabetic system of writing unless there is some significant proportion of the populace who, on the one hand, are freed from the burden of meeting the daily needs for survival, and, on the other, from the burdens of governing. Thus, it is not surprising that writing should have originated in the sort of stratified society that we see in Sumerian civilization.

At the pinnacle of Sumerian society was a king who was assumed to be a descendent of the gods and who remained in contact with them. Beneath him was a group of nobles. There was also a class of wealthy businessmen who lived in the larger, better houses of the city. Beneath them, in terms of wealth and prestige, were the many artisans and farmers, including smiths, leather workers, fishermen, bricklayers, weavers, and potters. Scribes apparently held fairly important positions in the society because literacy was an admired accomplishment. At the bottom of society were the slaves, often war captives or dispossessed farmers.

Cities appear to have been divided into modern-like "wards," but we do not know on what organizing principles they were based. There is some evidence that people were still organized in clans and that kinship ties were instrumental in the formation of various social groups; for example, the members of military units appear to have been based around kinship ties and perhaps place of origin. Many Sumerian states probably had a public assembly in which the upper and middle classes of the society formulated policy and made decisions through consensus and even a form of simple democracy. Such political advances are an indicator that the Sumerians had developed refined reflective capacities and an expanded sense of self, for to reach consensus through public decision-making forums, individuals not only must be able to recognize that there are different degrees of satisfying their interests but also have to be capable of reconciling their personal wishes with the larger needs of a community that now includes members unrelated by blood or possibly even customs.

The Sumerian extended family appears to have been a strong and durable unit that was protected by laws governing such matters as rights of inheritance and recompense caused by injuries. This involvement of the State in the welfare of the family is yet a further indication of a widening sense of identity and internalized standard; for now individuals were starting to think of their families' rights, and what sorts of public actions were available to them to avenge transgressions. Similarly, even though it was a patriarchal society in which women were viewed as inferior members (as is evidenced, for example, by their divorce laws, which stated: "If a wife rejects her husband and says, 'You are not my husband,' they shall cast her into the river. If a husband says to his wife, 'You are not my wife,' he shall pay half a mina of silver"), nevertheless some laws were instituted to ensure that women were reasonably well treated. The basis for the introduction of such laws would have been the emergence, amongst a large section of the populace, of a growing concern that potentially vulnerable members of the society should be protected, if necessary, by the State.

Some aspects of the Sumerian way of life are still recognizable today. For example, there are records of children being shepherded to school against their will. Yet the Sumerians' view of reality is grounded in a magical way of thinking that is difficult for us to grasp fully today. In Sumerian mythology, the earth was viewed as a flat disk under a vaulted heaven. They believed that various gods controlled human destiny according to well laid out plans, and that the world would continue without end and with very little change. Mirroring society, the deities were hierarchically arranged in power and authority and were given to power struggles and many vices. Each god was in charge of something, such as the movements of the planets, or irrigation and brick making, and each one was immortal and inflexible. Almost every official Sumerian pronouncement and inscription invoked the favor of the gods who oversaw the city's fortunes.

When we look at the rise of Sumerian civilization from our developmental perspective, therefore, the feature that stands out is how radically divorced their technological advances were from their views of reality and human functioning. Judging, for example, from how difficult it is for children to learn how to write despite the sophisticated methods we now employ, one can appreciate just how advanced their abstract reflective skills had to be in order to *create* an alphabetic system of writing. What was involved here was a whole new way of looking at and analyzing speech: breaking it down into a small number of phonetic units that could be represented in a simple combinatorial system. But for the Sumerians, writing was far more than just a technological advance; indeed, they saw writing as a symbol of their civilization, which as such was bound up with mystical power. Thus, King Shulgi proclaims, in the poem that was recently discovered (c. 2100 B.C.):

"Now, I swear by the sun god Utu on this very day—and my younger brothers shall be witness of it in foreign lands where the sons of Sumer are not known, where people do not have the use of paved roads, where they have no access to the written word—that I, the firstborn son, am a fashioner of words, a composer of songs, a composer of words, and that they will recite my songs as heavenly writings, and that they will bow down before my words."

Despite their sophisticated technological advances, the Sumerians continued to believe in personalized gods who had a very direct involvement in human affairs, causing triumphs because of some personal investment in the actors involved or crises because of their own petty feuds. Thus, the Sumerians continued to subscribe to a version of henotheism, according to which there were a large number of gods, each of whom was personally connected to a clan or a locality. Because the ways of the gods were so unpredictable, one could seek to appease their anger or gain personal reward only by one's loyalty. But as primitive as the content of these beliefs may have been, the rise of Sumerian organized religion reflects advances in f/e development. On the one hand, we see a primitive belief system that in many respects was precausal; but on the other hand, the organizational structure for disseminating and sustaining those beliefs was remarkably sophisticated, demonstrating advanced levels of gray-area thinking and internalized standards. The sophisticated religious organization that they developed—the shrines and temples, the priestly hierarchy, the rituals and festivals—thus created a rational vehicle that seemingly legitimated regressions into primitive precausal ways of thinking.

ANCIENT GREECE

History as such really begins with the Ancient Greeks. We say this not for an anthropological reason but because of an argument made famous by Bishop Berkeley in his dictum *esse est percipi.* That is, for history to exist, a society must have a conception of history: of a past and a future and its place in the historical present. Herein lies one of the more interesting phenomena that separates humans from animals: We are the only species that has a history simply because we are the only species that has a conception of history. That is not to suggest, however, that to be human is to possess a concept of history; not all human societies have possessed a concept of history, and not all societies possess the same concept of history. Indeed, the concept of history, and thus, history itself, has evolved over the past 2,000 years and provides us with one of our most potent indicators of the evolution of our reflective abilities.

Before the Ancient Greeks, ancient civilizations, particularly the Sumerian and Egyptian, were keeping records of past events. But these ancient civilizations subscribed to cosmologies that were thoroughly grounded in magical thinking. Human existence was thought to have begun with an act of the gods, and, where it was believed that human existence would one day end, this would be Gotterdammerung-like, as the result of a cataclysmic conflict amongst the gods themselves. The lines between reality and subjectivity were blurred, and human existence was likened to a dream. As an inscription on an Egyptian tomb put it:

> Our people rest in [the land of eternity] since the earliest primordial time,
> and those who will be in infinite years,
> they all come to that place. . . .
> The time that one spends on earth is only a dream. But welcome safe and sound!
> One says to him who has reached the West.[16]

This picture of existence as infinite is not so much an expression of the view that time will never end—which is actually a somewhat abstract thought—as an indication that the beginning and end of time could not be conceptualized. According to these cosmologies, human existence is controlled by the gods at every point and in every respect. There is no making sense of human existence; the gods behave in arbitrary ways. All one can do is accept whatever happens and do one's best to appease the gods, which, in many early civilizations, included human sacrifice. Thus, as we saw in the preceding section, although these cultures were advancing well into the areas of gray-area thinking and beginning to form internalized standards in such areas as social organization and technology, their thinking in regards to reality remained at a fairly primitive, precausal level of development that stressed resignation and acquiescence to the supernatural forces thought to govern life.

The magical belief systems that we see in these early mythologies were transformed, in Ancient Greek culture, into a fatalistic view of history. Herodotus is commonly heralded as the father of history, even though he was preceded by a slightly older contemporary, Hecataeus, whose works seem to have had a historical element.[17] Herodotus, and following in his footsteps, Thucydides, set out not just to record factual events but to make sense of these events, even if much of this material was gathered from hearsay and the memories of soldiers and travelers. But the view of history that they introduced continues to display many of the features of the magical thinking of these earlier mythologies, as is epitomized, for example, in the *Iliad,* the *Odyssey,* and of course, the myth of Oedipus.

According to the myth of Oedipus, when Laius was the guest of King Pelops he was so captivated by Pelops's son Chrysoppos that he kidnapped the young boy. Outraged, Pelops uttered a curse that Laius, who was childless, would one day have a son and then be killed by his son's own hand. Zeus heard Pelops's curse and decided to grant it. Here, it should be noted, is the reason curses were taken so seriously: In uttering a curse, one was invoking the gods to take a punitive action on one's behalf. The danger posed by a curse lay not so much in the words themselves but in the possibility that the gods would listen and be moved to action by those words.

Because of the magical power invested in curses, it came to pass that Laius did indeed have a son, but, to avert his fate, he had this newborn child crucified to a tree, nails driven through his feet, and left him there to be eaten by wild animals. But a shepherd came across the infant and delivered him to his own king, Polybus, who himself was childless. Hearing a rumor when he was a young adult that he was adopted, Oedipus visited the Oracle at Delphi to discover his true roots. The oracle told him that he would kill his father, marry his mother, and sire a damned generation. Still believing that Polybus was his father, and wishing to avoid such a horrible fate, Oedipus fled Corinth and set out for Thebes. When he came to a crossroads, he was imperiously ordered by an old man to get out of the way. Infuriated, Oedipus got into a fight and ended up killing his true father, Laius.

In *King Oedipus,* Sophocles raises an intriguing question as to whether the source of Oedipus's fate lay in his character; for Oedipus regresses to catastrophic behavior, becoming far too angry far too quickly. Not only does he completely distort the significance of the incident, but in the process he commits the cardinal sin of assaulting and mortally wounding an elder. Thus we are left to ponder the sources of one's destiny in life: Does it lie in the machinations of the gods or in one's temperament?

Such a question marks a significant advance over the magical thinking of earlier, more primitive cultures. And yet, the Ancient Greeks still viewed one's character as a matter beyond one's control.

Sophocles never questions the fatalistic view of history. One might think that this is precisely what Sophocles wanted to challenge, insofar as the incident between Oedipus and Laius takes place at a crossroads; but the appearance of a crossroads is actually ironic, which perfectly encapsulates the Greeks' view of history. For Oedipus believes that he is controlling his own destiny, that he is traveling the road he has chosen of his own free will. But, of course, the audience knows that he is on the road he was predestined to choose and that it was carrying him unknowingly towards his predetermined union with his mother, Jocaste. That is not to say that the Ancient Greeks believed fate decreed everything that happened to one; they believed only that how we behave at critical moments in our lives, such as Oedipus when he was at the crossroads of his life, is predetermined by some agency that we can neither control nor understand.

The plot of Sophocles' *King Oedipus* is grounded in the Greeks' fatalistic view of human existence. And yet the play itself is quite complex, not just dramatically, but also in the institutional structure in which it appeared; the plays were performed during a religious festival in which the audience voted at the end for what they regarded as the best play. (Sophocles is said to have won eighteen times.) This element of competition is itself quite advanced insofar as it requires the members of the audience to make an aesthetic judgment. Viewing a play was regarded as a religious act rather than an entertainment (comparable, in some ways, to the medieval mystery plays). As such, the Greek plays represent a considerable advance over earlier primitive acts of obeisance to the gods. Yet the content of this practice still resonates with earlier views about the direct role that the gods play in human affairs, which is what demands such public acts of obeisance.

Perhaps the most important example of the f/e advances taking place in the structure of Greek thinking concerns their views about the nature of *psuchē*, or soul, which occurred in the fifth century B.C. Our understanding of the early Greek view of the mind is confined to what we can glean from the *Iliad* and the *Odyssey*, bearing in mind that one always has to allow for poetic license. Thus, for example, one might wonder whether the Ancient Greeks shared Homer's view of Dreaming as a god who stands next to the head of someone sleeping and instills a dream.

But, as Bruno Snell showed in his classic *The Discovery of the Mind,* the very language that Homer uses reveals important insights into how the ancient Greeks viewed the mind.[18]

One of the most striking aspects of their mental vocabulary is that their use of perceptual and cognitive terms was fundamentally bound up with affective overtones. For example, as Snell showed, there is no abstract verb corresponding to our use of "to see," that is, solely referring to the act of perceiving. Rather, Homer's characters have "a particular look in their eyes," or "look with a specific expression," or have "a stare that commands attention"; they might "glare intently" or "gaze forlornly"; look with "pride," "joy," "fear," or "apprehension." In other words, every perceptual act is fundamentally bound up with an affect. Similarly, Homer's terms for the emotional and cogitative aspects of the mind *(thymos* and *noos)* overlap to such an extent that one cannot draw the sort of fundamental distinction between the emotions and reason that was to become one of the defining features of the great "rationalist" revolution that occurred in Greece in the fifth century B.C.

It is not until the rationalist revolution, with its belief in a universal, uniform human mind compartmentalized into separate organs or faculties, that we encounter the use of perceptual verbs with our familiar sense of "seeing" as a purely visual act: an indication that the great changes taking place in the philosophy of mind, as found in Anaxogoras, Socrates, and Plato, were preceded by an important linguistic shift. Indeed, one might even argue that it were precisely because language speakers were starting to distinguish between body *(soma)* and soul *(psuchē)* that the nature of *psuchē* and its relationship to the body had become such a pressing philosophical issue.

From a f/e perspective, this shift from the Homeric to the rationalist language of the mind reflects an important advance. We saw in Chapter 1 how, in impulsive humans, perception is closely tied to action: A perception triggers an action. This is very much the picture of action that we see in Homer. For example, the *Iliad* begins with the great quarrel between Agamemnon and Achilles that precipitates the ensuing drama. Agamemnon is behaving in an unbelievably churlish manner (which itself is dictated by the gods), and one fully expects Achilles, who immediately draws his sword, to strike Agamemnon down; and yet Achilles refrains from doing so and places his sword back in its scabbard. He does so, however, not because of any fleeting concern about a possible civil war erupting between their two camps but because Athena suddenly appears to him (and him alone) in a vision, counseling him to curb his wrath. So right from the beginning we are presented with a picture in which perception would indeed immediately lead to (in this epic poem, calamitous) action were it not for the intercession of an external, higher force. Throughout the *Iliad* we see this pattern repeated: Perceptions trigger actions, and when characters do think reflectively, their "intentions," "motives," and "beliefs" are placed in them by external forces. With the rationalists, however, the *psuchē* takes over the role of making rational choices and decisions (as in Plato's "Myth of Er" in Book X of *The Republic*).

The shift that we see in the rationalist picture of the mind, therefore, with its bifurcation between reason and the emotions, is a reflection of what we described in Chapter 1 as the f/e advance in which the child or impulsive adult is able to separate perceptions and actions. This separation between perception and action is reflected not simply in philosophical discourse on the tasks of the rational part of the *psuchē* but on a much more subtle level in the introduction of mental terms that themselves separate perceptions, thoughts, and beliefs from affective actions. As we will see, this rationalist development was to undergo several important advances in Western thought, perhaps most notably in the great Cartesian revolution that took place in the seventeenth century and that laid the foundation for our modern science of the mind (see below).

Completing these advances in the rationalist view of *psuchē* is the creation of the *polis:* the city-state that was based on the idea that people should live in a community of individuals who were all conscious of their shared interests and common goals and who would manage collective concerns by debating possible action in a public setting. However, we can see a significant disparity between the Greeks' view of human functioning and the democratic institutions that they were creating. As we saw earlier, during times of acute national stress such as occur, for example, during times of war or natural disaster (famine, earthquakes), such disparities can undermine the stability of a society. Should internal stresses become too pronounced—should the thinking of large elements of the populace become too polarized—the society will no longer be able to sustain its more advanced institutions and processes.

To be sure, there are areas in which the content of ancient Greek thinking was advancing into the realms of future-based probabilistic thinking (our eleventh stage) and reflective thinking according to a stable, broad inner framework (the twelfth stage). For example, in Miletus in the sixth century B.C., thinkers such as Thales and Anaximander began to develop the first concept of science: the first attempt to explain natural phenomena empirically. These early scientific advances were later followed by remarkable advances in abstract thought, for example, in the development of metaphysics, logic, mathematics, and geometry, as well, of course, as the elaborate forms of ethical, aesthetic, and political theories found in Plato, the Phytagoreans, and Aristotle.

Yet, as sophisticated as was the Greek canon of philosophical argumentation, the great philosophers nonetheless believed that emotions are an irrational force. Indeed, Oedipus's belief in his own free will is represented by Sophocles as an act of hubris. Thus, as far as the Ancient Greeks were concerned, the causes of human events were irrational, or transcendental and ineffable: The things that humans do or that happen to humans are caused by forces over which we have no control. Thus an interesting question arises in regard to Greek history: Did the disparity between the evolved institutional structures and the primitive content of the society's beliefs about human functioning erode the former and render Greek institutions more and more vulnerable to the regressive content of those beliefs? Such a dynamic might help explain why ancient Athens, at the height of its institutional glory, regressed into the great Peloponessian wars with Sparta (from 431 to 421 B.C. and from 414 to 404 B.C.) that ultimately culminated in the fall of Athens.

THE AGE OF REASON

The seventeenth century is referred to as "The Age of Reason" because of the sweeping social, political, religious, scientific, and technological changes that took place, all in the name of rational principles. Towering over the many remarkable individuals of this period was Descartes. In the area of religion, Descartes helped bring about the advent of Deism, discussed later in the chapter. In science he introduced the method of doubt. He played a major role in developing the life sciences, particularly optics, the biological study of the conservation of heat, and the anatomical study of reflexes. In mathematics he created analytic geometry. His views about the nature of the mind and its relationship to the body ultimately gave rise to the science of psychology. He also established epistemology rather than logic as the foundation of philosophy.

That one person could have had this broad an impact on his age is truly extraordinary, but it is also a source of some puzzlement. It is rare for a thinker to have this profound an influence in so many different areas; one would perhaps have to go back to Aristotle to find a comparable example. It was hardly because of his personality, for Descartes was something of a hermit who was quick to take offence and slow to forget a grudge. Nor is it because of the power of his arguments, for the truth is that they are often full of inconsistencies and omissions. Nor is it because his ideas were so novel; on the contrary, apart from his work in mathematics, many of the key ideas that he explored were being discussed by his contemporaries. So we are left with the fundamental question: What exactly was it about Descartes's ideas that accounts for his remarkable impact?

There have been countless attempts to answer this question, in disciplines ranging from philosophy and the history of ideas to theology and the sociology of knowledge.[19] To these we can add some insights provided by our developmental approach. What Descartes represented to his contemporaries, and the eighteenth-century school of Enlightenment philosophers that followed him, was, as many have noted, the conflict between reason and authority. But Descartes went beyond simply championing the power of reason; after all, the great Scholastic and Renaissance thinkers had every bit as much claim to such a title. Rather, Descartes established reason as a revolutionary force, as is reflected in Thomas Paine's revolutionary text of 1793 in which he introduced the title *The Age of Reason*. But then, Descartes was not the political revolutionary Paine was; on the contrary, his actions would suggest quite the opposite.[20] But Descartes did amount to a revolutionary of the intellect, championing a significant advance in the higher reflective capacities that we have been surveying in this chapter.

Descartes did not simply articulate the new ideas that were emerging in the seventeenth century in the diverse scientific fields in which he was active. Above all, Descartes represented the power of applying these new ideas to thinking about the human mind. That is, unlike earlier thinkers, Descartes did not simply look for patterns in mathematics, physics, biology, or anatomy; Descartes looked for patterns in the way we look for such patterns. In other words, he took the leap into metacognition. What Descartes really represented for his contemporaries was, in effect, the sort of intellectual excitement that an adolescent feels as he develops his own internal standards and new reflective abilities in an expanded sphere (the tenth stage). Descartes broadened the higher reflective capacities that the thinkers of his age were mastering in a way that, for the first time, encompassed thinking about thinking.

In this respect, Descartes's *Discourse on Method* is the paradigm of a modern revolutionary text. Not just the content, but the very style in which it is written marks a radical break from the past. Descartes tells of how "returning to the army from the coronation of the Emperor, the onset of winter detained me in quarters where, finding no conversation to divert me and fortunately having no cares or passions to trouble me, I stayed all day shut up alone in a stove-heated room, where I was completely free to converse with myself about my own thoughts."[21] We have grown so accustomed to the voice in which this is written that it requires a conscious effort to appreciate the full significance of this "fragment of autobiography": that is, it is a fragment of *autobiography!*[22]

In what follows, Descartes makes it clear that he is deliberately challenging the central doctrine of Christian cosmology, the orthodox doctrine of the Great Chain of Being. He insists that there is an unbridgeable gap between animals and humans. The human body, according to Descartes, is a machine (which was itself a heretical view); but, unlike animals, humans also possesses the capacities to reason, speak a language, control their actions, and be conscious of their cognitions and sensations. The reason we are said to know that animals don't have minds is because they lack the ability to describe their thoughts. For us to say that an animal is capable of speaking, the animal would have to respond appropriately in varying situations, acquire words spontaneously, and combine words creatively. Only humans, according to Descartes, are born with such a capacity.

That such an argument is so prescient of the skeptical attacks on animal cognition and ape language research that were to flourish in the late twentieth century is an indication of the subtlety of Descartes's thinking about language and cognition, and how these ideas might apply to lower organisms.[23] His polarization in his thinking about animal capacities is a reflection of the regressive elements in the content of his thinking. But in structural terms, even raising such an issue is an indication of the broadening that was taking place in his thinking.

Descartes's universe, unlike that of the Ancients or the Scholastics, is bifurcated. And at its center stands neither the earth nor the sun, but the human being, who makes sense of reality by creating concepts. When Aristotle tells us that "man is by nature a political animal," or

Seneca that "man is a reasoning animal," the emphasis is on "animal": one analyzes the human being as an "animal species."[24] But all this is changed in the *Discourse.* Here we begin with René Descartes: the thoughts of a solitary individual who has come to distrust the teachings of the finest minds of his time; who has renounced the homage to Aristotelian thought that dominated medieval and Renaissance thought; who tells us that he decided to continue his studies by reading from the "great book of the world" rather than from the classics; and whose "real education" has taught him to accept as certain only those ideas that he can see for himself "clearly and distinctly." Spoken like a true adolescent, brimming with confidence in his newfound sense of self.

Indeed, like an adolescent overcome by a sense of the importance of his discovery, Descartes set out to convince his peers that they, too, needed to reform their intellectual habits. His argument is that thinkers are guided by two fundamental cognitive principles: They base their judgments either on the evidence provided by the senses or on the products of the intellect. When Descartes talks about being guided by a principle, he does not mean that one has to have explicitly formulated that principle; rather, whether or not one is guided by a principle is shown by how one reasons. Thinkers need not be aware of the principles that they follow, and probably are not aware of them. Nevertheless, Descartes insists that we are responsible for the principles that guide our thinking insofar as it is possible to find out what principles we unconsciously follow. Hence, Descartes's goal in the *Discourse* is to persuade his contemporaries that they need to divorce their thinking from the immediacy of sense perception.

Descartes is referring here to the importance of factors that are not in conscious awareness. In this regard, he anticipates the discoveries of Schopenhauer and Freud. But on a deeper level, he is appealing to the principle that the permanency or stability of phenomena may be inconsistent with the immediacy of moment-to-moment sensory experience. This sort of sophisticated reasoning, which looks beyond the immediacy of the senses for truths that prevail across the dimensions of time and space, is characteristic of the higher reflective levels of thinking.

For example, Descartes takes the debate over Copernicus's theory to an entirely new level. He is no longer talking about the merits of the geocentric versus heliocentric theory; rather, he transforms the issue into an investigation into the very nature of abstract reasoning. Descartes is saying that to challenge authority we need to understand the unconscious forces that render it so difficult to be guided by the faculty of reason. Such an argument takes us well into the higher stages of f/e development. Descartes is suggesting that because medieval and Renaissance thinkers never considered the operations of the mind—because they were unaware of what modern cognitive psychologists refer to as the "cognitive biases" that guide human thinking—they were drawn to defend the immediacy of their sense perceptions.

The contrast that we have been making between structure and content is particularly illuminating when it comes to assessing Descartes's views about the mind. The basic principle of Descartes's view of mental functioning is that we are perceptually acquainted, through introspection, with the mental causes of our behavior. That is, introspection is an inner sense—a mental analogue of perception—and mental processes and states are said to be transparent (clear and distinct). Our knowledge of other minds, however, is confined to inferences based on the subject's behavior. Indeed, even the belief that there are other subjects is an inference. Thus Descartes argues in the *Second Meditation:*

> If I look out of the window and see men crossing the square, as I just happen to have done, I normally say that I see the men themselves, just as I say that I see the wax. Yet do I see any more than hats and coats which could conceal automatons? I judge that they are men. And so something which I thought I was seeing with my eyes is in fact grasped solely by the faculty of judgement which is in my mind.[25]

Descartes argues that, given that mental states are hidden, that you can only know the mental causes of behavior through behavior, all he actually sees are "mere movements." But he naturally assumes, on the basis of the similarity between these movements and his own, that these are human beings with minds, not automatons. Presumably, this is the result of inferences that he made as an infant, when he was just starting to figure out the way the world works and recording the differences between human beings and objects. Over time, these judgments became so entrained in his thinking that he was no longer aware that he was making a judgment when he observed humans acting. But through the discipline of doing science—by distinguishing between what one actually *sees* and what one *infers*—one is able to disclose these automatic inferential acts.

The upshot of Descartes's argument is that, given that our understanding of mental functioning is grounded in introspection, it follows that I can never know for certain that another human being experiences the same mental states as I do. Even when I see someone writhing on the ground, I can never be certain that the individual feels the same as I feel when I'm in pain. Indeed, I can never even be totally certain that other human beings exist. Hence, it follows that there can never be a science of the mind, comparable to the sciences of matter and motion; for there can never be an objective method for studying mental phenomena.

It took over three centuries before Ludwig Wittgenstein, the great Austrian philosopher, was finally able to overturn this extremely powerful skeptical argument.[26] But far from being an impediment to philosophical and psychological thinking about the mind, Descartes's argument was quite the opposite. For his views about the epistemological asymmetry between first- and third-person psychological knowledge served as a catalyst for philosophical interest in the mind, and ultimately gave birth to the science of psychology. In structural terms, Descartes took a truly enormous step here. In essence, he was attempting to extend scientific principles that had been worked out in the areas of the physical and life sciences to our thinking about thinking. Descartes wasn't guided by some perverse desire to frustrate his contemporaries with what has come to be known as the great Problem of Other Minds purely for the sake of inculcating a skeptical philosophy. Rather, Descartes was attempting to be completely logical in his analysis of the mind. He was trying to think objectively about the sources of our knowledge of the mind and thence establish the limits that such sources impose for our understanding of mental functioning. Descartes was applying more logical models to a range of phenomena that, from the vantage point of his period, had been dealt with in more magical or superstitious modes.

In some respects, the content of Descartes's views of mental functioning was also quite advanced. The idea that when we look at human beings we are constantly making inferences about the causes of their actions without being aware of it anticipates modern ideas about the cognitive unconscious. That is, Descartes envisages a realm of mental functioning of which we are unaware, but of which we can become aware through scientific reflection. Yet, in other respects, the content of Descartes's views of mental functioning lagged considerably behind the structural advance he was making, for despite his revolutionary attack on the Great Chain of Being, Descartes's thinking about human functioning still remained grounded in this medieval dogma.

The fundamental principle of Descartes's view of human behavior is that reason and the emotions are bifurcated and belong to completely different modes of human functioning. Emotions constitute the basest level of human behavior: the instincts, the drives, and the passions that we share with animals. Indeed, emotions, for Descartes, are quite literally a species of involuntary reflex. Reason is completely independent from these mechanical processes, and, one hopes, comes to govern them. Thus, Descartes's view of mental functioning is actually quite limited because he doesn't allow for the role of environmental factors on the *development* of emotions; he doesn't recognize the variability of emotional experience; he doesn't allow for the emotions to have a role in the operations of our cognitive capacities; and he doesn't consider the significance of social factors on the growth of the mind.

This contrast between the developmental levels of different aspects of Descartes's thought provides us with a new level of insight, not only into the reasons Descartes's writings had such a liberating impact on the Age of Reason but, equally, into the reason his views came to have such a constricting influence on subsequent generations. It took a considerable time before the content of philosophical and psychological thinking started to come in line with the structural advances that Descartes instituted. But then, this is the pattern we have seen throughout this chapter: Time and again we see structural advances opening the door for a subsequent broadening in the application of our higher reflective thinking. Indeed, no better example of this phenomenon can be found than in the next major historical epoch, the Enlightenment.

THE ENLIGHTENMENT

The sections into which this chapter is divided are based on the standard view about what constitute the major epochs in Western history. But modern historians of ideas are divided about the legitimacy of looking at history as epochs. The problem is that by assigning a label we tend to focus on a few representative figures and ignore the many who do not conform to our narrow view of what defines the period in question. Yet many remain deeply drawn to the idea of distinctive historical periods, if only because there is a striking difference between periods that are as close in time and culture as, say, the Age of Reason and the

Enlightenment. The problem is how to capture this: how to define the period in question in a way that encompasses as many of the complex social and intellectual cross-currents as possible. But here one runs into the problem that afflicts any attempt at essentialist theories: If the definition is too rigid, we run the risk of excluding far too many elements; and if the definition is too general, it loses any real function. The Enlightenment has come to serve as something of a paradigm example of this problem in the history of ideas.

The Enlightenment was an age of dramatic social change, marked by a continuing decline in the authority of the Church and the continuing social and political ascent of the bourgeoisie. Among the influential figures usually singled out in this era are Adam Smith, who argued for the principle of laissez faire on the grounds that unleashing individuals to pursue their own interests would ensure economic expansion; Charles-Louis de Secondat Montesquieu, whose writings on the division of executive, legislative, and judicial power helped shape the framing of the U.S. Constitution; and, of course, Isaac Newton, whose demonstration that the natural universe is a realm of law and regularity inspired the great scientific revolution.

During these years, a new sense of confidence arose in the powers of the human mind to mold the physical world for the material and moral betterment of humankind. The early exponents of this optimistic outlook, the philosophes popularized the scientific ideas of the seventeenth century and worked to expose contemporary social and political abuses. They argued that reform was necessary and possible and that religion itself should be based on reason and lead to moral behavior. This enlightened form of religious belief, known as Deism, continued to argue for the existence of God, which, according to the philosophes, could be empirically deduced from the contemplation of nature. Indeed, they argued that, because nature provides evidence of a rational God, He must favor rational morality. They consequently argued for the belief in life after death, when rewards and punishments would be meted out according to the virtue a person had shown in this life.

One of the greatest monuments of the philosophes' influence on the Enlightenment was the publication of the *Encyclopedia,* or a *Systematic Dictionary of Science, Arts, and the Trades,* under the leadership of Diderot and d'Alembert. The first volume appeared in 1751 and eventually ran to twenty-eight volumes; it was comprised of articles on all manner of current affairs ranging from politics, ethics, law, religion, and society to manufacturing, canal building, ship construction, and agriculture. The *Encyclopedia* was based on the principle that the average person had every right to the same information as the ruling elite. The government was fully aware of the revolutionary implications of the *Encyclopedia* and tried twice, unsuccessfully, to suppress its publication. The final volume appeared in 1771 and is often seen as one of the critical factors leading up to the French Revolution. As for the *Encyclopedia,* the structure (its being designed for the general public) and the content (the topics surveyed and the manner in which they were expounded) represented a significant developmental advance.

Enlightenment thinkers (e.g., Voltaire and Kant) were themselves aware that a small number of themes running throughout the writings of their era served to distinguish the Enlightenment as a distinctive epoch in Western development. The most important of these themes were the concerns with

1. the autonomy of reason;
2. the dangers of authority;
3. the perfectibility of humanity;
4. the ability of the human mind to discover the mechanical laws governing nature;
5. the ability of the human mind to understand the forces governing man and society;
6. the solidarity of enlightened intellectuals;
7. the dangers of nationalism.

We find writers returning to these issues over and over again, exploring them from every possible angle in polemical debates that, for such an enlightened age, could be remarkably polarized. For that matter, some of the greatest thinkers of the period, among them Rousseau, presented so many different and often conflicting ideas that their views came to inspire diametrically opposed movements. Indeed, the principle reason historians of ideas have balked at trying to define the Enlightenment is because of the dizzying number of different theories and ideas pursued in every conceivable realm of intellectual inquiry.

It is precisely here where distinguishing between the *structure* and *content* of what people were thinking becomes illuminating. To be sure, others have touched on this important theme. For example, Lester Crocker suggests that "the common element" that unified the Enlightenment "lay more in *how* men thought than in *what* they thought."[27] In developmental psychology, Kohlberg based his stage model of moral development on how a child responds to a story about a moral dilemma, where what matters is the structure and not the content of the child's response.[28] What our framework adds to these insights is the first systematic way of looking at the structure of thinking as an integrated emotional/cognitive process. That is, whereas earlier structuralist theorists viewed cognition as a completely self-contained phenomenon, our approach looks at the structure of thinking in terms of its intertwined emotional and cognitive elements.

Such a framework enables us to make sense of what is perhaps the most striking feature of the Enlightenment: the turbulent clash of polarities. The tone of the period is captured in Swift's *Battle of the Books,* where Ancients are pitted against Moderns. Swift's satire is an attack on the endless conflicts that riddled the eighteenth century: In philosophy we see the great battle between Rationalists and Empiricists (the philosophes versus Locke, Berkeley, and Hume); in biology the fight between Vitalists and Materialists (Barthez versus la Mettrie); in politics the clash between republicans and monarchists, utopians and reformists, egalitarians and elitists; in ethics the fight between pantheists and cynics (Shaftsbury versus Mandeville). This is the period that culminates in the U.S. Constitution and the Bill of Rights on the one hand and in the anarchy of the French Revolution on the other.

Yet, as we saw in Chapter 2, it is precisely this capacity to grapple with abstract dichotomies that is one of the hallmarks of the higher stages of f/e development. That is, once individuals have passed beyond the ninth stage, they become increasingly aware of abstract dichotomies and of the connection between the antinomies in their thinking in very different areas. The higher their development, the more individuals begin searching for a resolution of these antinomies that is consistent across multiple domains. It is not until reaching the highest stages, however, that individuals acquire the integrated thinking required for this sort of synthesis.

Our use of the term "antinomy" here is intended to draw attention to the thought of perhaps the greatest of the Enlightenment thinkers, Immanuel Kant. Reflecting the spirit of his age, Kant based his monumental *Critique of Pure Reason* on the significance of four fundamental antinomies—equally rational but contradictory views—dealing with space and time, the principle of atomism, the existence of freedom, and the existence of God. For example, the first antinomy presents the Thesis that "the world has a beginning in time, and is also limited as regards space" and the Anti-Thesis that "the world has no beginning, and no limits in space; it is infinite as regards both time and space." The content of these antinomies is a reflection of just how broad the application of the higher reflective capacities was becoming. But structurally, what is so striking about Kant's argument is his attempt to achieve a critical synthesis by ascending to still a higher level of reflective thinking in which one recognizes the cognitive significance of these antinomies as such.

Kant's famous argument is that when reason attempts to transcend experience, it falls into antinomies. He set out to resolve these antinomies by distinguishing between the limits of what we can know by experience and the a priori structures in the mind that determine the limits of what we can think. Kant's progression to this higher level of reflective thinking is an example of how the highest capacities of reflective thinking emerge only once one has passed beyond the ninth stage of development. For one can only begin looking for a way of reconciling the inconsistencies and dichotomies in one's thinking once one's thinking is grounded in a strong sense of self and internalized standards. This strong sense of self both drives and enables one to search for a way of reconciling the dichotomies in one's thinking, and, in Kant's argument, serves as the emotional/cognitive foundation for his transcendental resolution of the antinomies.

The sort of aspect shift that enables one to transcend a seemingly irreconcilable conflict seen in Kant's self-proclaimed Copernican revolution is the hallmark of ascending to a higher level of reflective thinking. Indeed, one reason Kant had such a profound influence on German philosophy throughout the following century is because a thinker who has reached the highest levels of reflective capacities and is able to look at rational dichotomies orthogonally, as it were, is a source of profound inspiration to those who have advanced far enough to grasp the level of abstraction involved in such a synthesis and proceed from there in their own rational system building.

Not surprisingly, it was Kant who came up with the name "Enlightenment" for this pivotal epoch in human history in his 1784 essay, "Was ist Aufklarung?" ("What Is Enlightenment?"), in which Kant explicitly spelled out that "Sapere aude! 'Have courage to use your own reason'" is the motto of the Enlightenment. This act of self-identification returns us to where we began this section, for it is interesting to reflect on the nature of the terms by which the major epochs in human history are commonly known. Most of the terms that we use are simply descriptive names. Originally the term "renaissance" simply referred to the rediscovery of Roman styles of art (this was the sense in which the Italian writer Giovanni Bocaccio used the term), but through the writings of Voltaire and Jules Michelet (in the eighteenth and nineteenth centuries respectively) the Renaissance came to symbolize not just the extraordinary advances in architecture and art but also the beginnings of the modern age as represented by the great scientific, philosophical, and social revolutions taking place. But of all the titles that we use, "The Age of Reason" and "The Enlightenment" are the only philosophical terms used to designate an historical epoch. Significantly, both were coined by Enlightenment philosophers.

Enlightenment thinkers felt that their era represented an intellectual apex in human history; yet they were deeply concerned, precisely because of this sense of historical importance, about succumbing to hubris. Alexander Pope, for example, in his highly influential *Essay on Man* (1732–1734), warns that human beings are not the center of the universe, that man must learn to accept his intermediary position on the Great Chain of Being. Like so many Enlightenment thinkers, Pope argues that, even though life may seem chaotic and purposeless, it functions in a perfectly rational fashion. However vaunted our intelligence might be, we can perceive only a small portion of this order. Hence, reason ultimately teaches us the limits of reason and the importance of faith.

From our developmental perspective, it is highly interesting to reflect on this interesting tension between medieval and Cartesian ideas. For all the emphasis on free will, Enlightenment thinking remains strongly deterministic. To be sure, the Great Chain of Being was now viewed more as a poetic device than a serious cosmological argument; but the very fact that it could serve this poetic role is a reflection of the strong fatalistic elements that still suffuse eighteenth-century writings. Their thinking is also marked by a new form of determinism, one that is grounded in the science of mechanics rather than in mythology or theology. Indeed, they continued to view history in the same sort of cyclical terms as the Greeks, but with a highly significant modification: Now we encounter the metaphor of the spiral (in the writings of the Italian philosopher Giambattista Vico). The spiral metaphor represents the idea that history does indeed repeat itself, successive generations going through the same cycles of ascent, consolidation, and decline, but with an overall movement upwards and outwards. Thus the Enlightenment view of history introduces an important new dimension in the Western view of *history*, which has become bound up with the idea of progress: a theme that was to dominate nineteenth-century thought. What we see here is how, through an expanding sense of self, Enlightenment thinkers were developing an expanding sense of history: a future filled with unknown possibility.

THE DEVELOPMENTAL MODEL OF HISTORY

In the developmental model of history that we have outlined in this chapter, the sciences of human functioning take a special place in the historical process. It is not an historical accident that psychology did not emerge until the middle of the nineteenth century but rather a reflection of the need to reach a high level of f/e development before we can begin thinking about thinking systematically. Also, science, technology, mathematics, philosophy, and literature had to reach high f/e levels themselves before they could serve as the foundational platform for tackling the ultimate challenge, which is thinking about ourselves.

There are twelve basic principles operating in this f/e view of human history:

1. The dynamics of human history are best understood in structuralist terms.
2. Unlike Piaget's structuralist theory, however, the f/e model places affect at the center of an integrated structuralist theory that accounts for social, emotional, cognitive, and communicative development.
3. The model stresses that individuals, groups, and societies may develop higher reflective capacities in some domains (e.g., science, technology) while remaining at a lower stage of development in others (e.g., human functioning).
4. The model accounts for the overall progressive aspect of human history without postulating a unidirectional process of development. On the contrary, one of the key themes in this approach is that it is always possible for individuals, groups, and societies to regress to a lower developmental level.

5. The advances and the regressions that mark human history are influenced by the stages of affective transformation, including affect gesturing, that characterize early formative and subsequent relationships (as described in Chapters 1 and 2).
6. These advances and regressions are influenced by
   - culturally transmitted practices in co-regulated emotional signaling leading to symbol formation and higher-level reflective capacities, and
   - derivative group structures, institutions, and practices that promote or undermine an individual's continuing development and exercise of these emotional and reflective processes.
7. By not focusing sufficiently on the importance of culturally mediated family practices and structures that promote the processes described in 6 and 7 above, we run the risk of falling victim to the dictum of reliving history as proposed by the Spanish American philosopher George Santayana. That is, by not understanding the engine of human history we are in danger of sabotaging that engine with regressive practices, beliefs, and institutions.
8. Higher-level reflective thinking, including derivative institutions and practices, leads to relatively more stable and technologically more advanced societies, which progressively frees up more time and interest in further improving these formative practices involving families and communities. In turn, these adaptive processes support new technologies as well as even more advanced political and economic opportunities. All these favorable factors maintain a positive progression towards higher developmental levels and the processes that support them.
9. The greater the number of individuals who master these higher reflective abilities, the more widespread the practice of science, literature, and philosophy becomes, and the more stable the society becomes; for the greater the proportion of the society that possesses these abilities, the more stable are the society's political and social organizations or the processes by which the society negotiates conflicts.
10. The more people who master these higher developmental levels, the better able the society is to control its technology. The tendency in human history has been for a society's advances in science and technology to move ahead of its grasp of human emotions and human psychology. A problem occurs when a society's technology is way out in front of its ability to deal with it socially and emotionally (think of the dangers

posed by a fanatical group or country that has acquired nuclear and biological weapons, or the problems relating to cloning).

11. A similar point applies to a society's institutions. Attempts, either by a small group of advanced thinkers or even by a foreign power, to install advanced political and social institutions in a society that has not yet developed the base to support such institutions can also lead to instability or a regressive political movement.
12. The process of f/e development has not only been critical throughout human evolution, up to and including recent human history, but remains critical. That is, we are continuing to develop, to move beyond the high levels we have reached in technology, science, philosophy, and the arts into the realm of human functioning; and we are seeing a broadening in the portion of society that is reaching these higher levels. This is a gradual process, however, and there are all sorts of regressive forces that we must contend with (both internal and external).

It is important to emphasize that what we have presented in this chapter is not a *theory* of human history but rather a *lens* for looking at human history. The importance of this new lens lies in the sorts of questions that it leads us to ask in trying to assess the competing forces and processes involved in the progressive-regressive pattern of human history, which, overall, trends not just towards more sophisticated technologies with a greater capacity to control nature but also towards greater social complexity and differentiation. One of the key features of this new lens is that it prompts us to look at the sorts of problems that earlier cultures were grappling with, for the nature of the solutions provides us with one of the most important tools we have for assessing their stage of f/e development.

—15—

Future Evolution: Toward a Psychology of Global Interdependency

DESPITE DARWIN'S WARNINGS AGAINST looking at evolution in teleological terms, there has been a consistent tendency to view the evolution of human beings as the final step in a long and orderly progression leading from the Common Ancestor up to *Homo sapiens.* On this line of thinking, what marks anatomically modern humans as the terminus of this progression is not our anatomy but our "cognitive architecture": that is, the ways in which we process information and communicate; the basic emotions that we experience and the cluster of physical and behavioral processes linked to these emotions; and the kinds of behaviors in which we are predisposed to engage, some of which may today seem irrational but whose function can be understood when viewed in the context of the Pleistocene conditions in which they were supposedly selected.

To gain an understanding of the nature of human nature, we have also looked at the evolution of humans from nonhuman primates and hominids. But in the theory that we have presented in this book, human evolution is seen as an ongoing, not a completed process, and a child's language, emotional, cognitive, and higher reflective traits are seen not as predetermined but as developing as a function of her "experiences." These experiences involve an ongoing dynamic, largely inseparable, relationship between nature and nurture. Our goal has been not simply to understand the biological and cultural forces that have brought humans to where they are today but, equally important, to understand the nature of these processes so that we can better meet the challenges that confront us.

As societies have grown larger and more diverse, encompassing more and more different cultural subgroups, they have had to develop ever more reflective capacities in meeting the challenges of their increasingly complex social existence. Now as never before, the challenge that humanity faces in this respect is daunting: that of global interdependency. To meet this challenge, humans must ascend to still a higher level of reflection and empathy than has been attained in even the most advanced cultures if we are to avoid the fate of earlier societies that were unable to respond cohesively to the challenges they faced.

As we look back through history, it's possible to observe that in response to new challenges, human groups have evidenced a series of patterns. One of these has been to narrow their focus and become more polarized in their thinking, or to shift into varying degrees of unrealistic or magical thinking, or to become increasingly fragmented and chaotic in their social organization and behavior. On the other hand, another tendency one can observe, a decidedly more adaptive one, is to develop a broader understanding of the challenges and then take an integrated approach based on reflective processes and social and political institutions. As we will discuss later in this chapter, many nations are now being faced with a diminished sense of power relative to other groups, in part related to the growing interdependency of shared dangers as well as economics and communication. Are we currently observing shifts towards polarized and unrealistic thinking? Or are we observing new, more reflective integrated approaches?

To be sure, at present, there is a rapidly growing awareness of the enormous dangers and difficulties posed by global interdependency. But there is also a growing sense of helplessness at the magnitude of these challenges.

We are certainly not the first group to face enormous challenges. Challenge is a cycle that has played out repeatedly throughout human history. Some groups rose to the challenges confronting them by evolving new levels of social organization, and some groups that were psychologically unable to rise to the occasion went into serious decline. What sets us apart today is not the fact that we face such daunting challenges, therefore, but simply that our challenges are far more serious than those faced by earlier societies; indeed, the very continuance of life on this planet is now at stake.

In this chapter we will briefly examine the nature of this social reality, especially the shared dangers involved in nuclear and biological weapons, and then outline the new psychology that is needed to deal in an adaptive and reflective manner with rapidly advancing global interdependency. The developmental model of evolution that we have presented provides us with a new lens to help us understand not only the capacities that need to be inculcated but also the forces that are presently impeding this process. It will help us understand how and why our growing interdependency can lead to greater social fragmentation, more extreme types of polarized beliefs, and greater hostility; or new adaptive levels of personal and social organization.

GLOBAL INTERDEPENDENCY

Terrorist events around the world over the past decade have forced many groups to face more fully a reality that had been emerging for some time: shared dangers—for the first time any one group can, through nuclear, biological, or ecological events, destroy life for all other groups on the planet—as well as shared communications and economies have brought individuals from all parts of the world together into a closer interdependency than at any time before in human history.

Various Western governmental advisory groups warned of terrorist assaults before they became a prominent reality, but with little notice. They cautioned the public that, as weapons of mass destruction become more widely available, these assaults would become more dangerous. These warnings were consistently ignored.

The interdependency of shared dangers, the knowledge that anyone can destroy everyone else, is a difficult concept to grab hold of psychologically. During the Cold War, with only the United States and the Soviet Union in a position to destroy the globe, there was mutual shared deterrence. Where will we be able to find some relative safety when hundreds of groups have massive destructive power?

A small group could set off a biological weapon in its own nation that would destroy life on the entire planet. The seemingly private action of a nation state within its own geographical boundary may, therefore, have far-reaching consequences that are anything but private.

Furthermore, technologies of mass destruction are likely to become more and more accessible, even with our best efforts at limiting them. Over the next fifty years, it is unlikely that the defensive systems we're currently working on—whether immunizations for biological weapons or antiballistic missile systems for nuclear missiles—will keep up with rapidly proliferating offensive technologies. If large numbers of terrorist organizations and outlaw nations gain access to these weapons, traditional military interventions will be less viable due to a new balance of power in which even small groups have world-threatening destructive power.

The current challenge is further complicated by another worrisome fact. Due to the globalization of information and technology, groups no longer have as much autonomy in the evolution of their social, economic, and political structures. In other words, the relative boundaries of groups have gradually been changing with the increased flow of people and ideas across social groups. Groups with very different types of beliefs, different concepts of humanity and reality, and different processes of governance, thinking, and decisionmaking now have access to technologies of mass destruction, and this access will increase in the future. These different beliefs and processes may result in very different ways of regulating or not regulating weapons of mass destruction. This is a frightening reality to confront.

What is the solution to this dilemma? Can destructive technologies be kept out of the hands of groups lacking national and international institutions that can assure safety and global responsibility? It is no wonder that there is a growing sense of numbness in the face of these dangers.

As humans look for a strategy to promote safety and security in the context of the new reality of the interdependency, more can be done in the short term—such as immunization for small pox and better safety procedures on airlines. Improved international collaboration can reduce weapons of mass destruction and strengthen our defensive tactics. But in the long term, a very different process is needed to afford a measure of safety and security.

The first step in this process is to realize that the interdependency of shared dangers means that nations are even more interconnected than ever before. Shared dangers exert an influence on interdependency that adds significantly to economic, social, and cultural factors. Therefore, collective action is more essential than ever before and will be even more so in the future.

How do we create and maintain the collaborations that will be increasingly necessary in the future? How can individuals and groups broaden their shared sense of humanity and reality to encompass a greater range of differences than ever before? How can a greater consensus emerge among peoples on what defines another human being? How can such a consensus accommodate the enormous diversity of the world? How can various nations share a common reality and solve problems together when reflective problem-solving capacities vary, when there is a great deal of polarized thinking ("us" versus "them"), and when thinking often considers only the present rather than the long-term future?

RISING TO THE CHALLENGE OF GLOBAL INTERDEPENDENCY

In Chapter 13, we saw how, at each level of group development, there are adaptive and maladaptive responses to the new challenges that a group faces. In addition to being reflected in the institutions and governmental or social processes embraced by the group, these adaptive versus maladaptive responses have a significant bearing on the psychological functioning of the individuals in that group.

For example, at the first level of group development, a group might respond adaptively to a threat to its security by instituting organizations and practices that maintain order without impinging on the rights and freedoms of the individual; or a group might respond maladaptively, maintaining security through practices that infringe significantly on personal freedoms. In a worse scenario, a group may actually fail to cohere around a common set of practices and decline into a state where the safety of the individual is in constant jeopardy.

At the second level of group development, a group that contains various subgroups—such as we see in modern multicultural societies—can respond adaptively to this internal complexity by cohering around reflective institutions and processes that promote a strong sense of shared identity. Or it can respond maladaptively, unable to attain cohesion at the macro level. It may cohere around superficial traits only, such as skin color or ethnicity, or fall under the sway of a demagogue. In groups that respond adaptively, individuals are able to project themselves into the shoes of another while still maintaining their own sense of personal identity. But individuals in a group that responds maladaptively tend to function as isolated entities. Such individuals typically regress into polarized thinking ("us" versus "them" mentality) when in conflict with other individuals or subgroups.

At the third level of group development, groups faced with the challenge of assimilating a wide range of cultural diversity can strive to maintain a shared sense of reality at the presymbolic level by creating adaptive practices that enable them to develop, through frequent interactions with a wide range of others, an implicit understanding of a broad range of differences and themes. As indicated earlier, this implicit understanding is through a continuous back-and-forth chain of preverbal affect signals. One can readily picture how these types of signals occurring between many members of a group serve as a rapid system of cross-communication regarding the security and norms of that particular group. It's amazing to observe how quickly humans, and even nonhuman primates, respond to the multiple facial expressions, looks, glances, and changes of body posture of several other group members at the same time. Most of this occurs without explicit conscious awareness. However, for such a system to develop and sustain itself and function on behalf of maintaining a group's security and identity, the group members have to have access to one another that allows for this rapid visceral type of communication. Without daily interactions, such implicit understanding is often impossible. Therefore, a group can easily create maladaptive patterns by failing to develop the sorts of programs necessary to promote a great deal of interactions between the groups that need to come together.

In such maladaptive responses, subgroups are likely to disengage from each other and become more and more isolated and self-absorbed. Individuals in a group that responds adaptively experience the immediacy with each other that is needed to develop a shared sense of reality and humanity. Such individuals are better able to engage in cohesive, integrated thought patterns and problem-solving strategies. Individuals in a group that responds maladaptively to the challenge of assimilating a wide range of cultural variations tend to be constricted in the range of differences and themes that they can implicitly communicate. These individuals tend to become more rigid and fragmented in their communication and thinking, and to pursue impulsive, action-oriented attempts at solutions rather than the kinds of reflective, strategic ones seen in groups that respond adaptively.

Groups that respond adaptively to the challenge of achieving a shared sense of reality at the symbolic level are able to recognize and accept the symbols of others as meaningful while not sharing in the belief systems in which those symbols are embedded. Such a group establishes institutions and practices that promote not just tolerance but, even more important, shared understanding amongst its various subgroups. Thus, for example, individuals can understand what is involved in a foreign concept such as voodoo even though they reject some of the basic beliefs enshrined in this religion. But a group that responds maladaptively at this level of group development fails to establish the sorts of institutions and practices that promote mutual understanding, and because of their lack of familiarity with and knowledge of other cultures, individuals are unable to understand the true meaning of others' symbols.

To respond adaptively at the higher levels of group organization, groups have to develop reflective practices and institutions that support logical reality-based problem solving, have stability through change, and promote the development of individuals capable of looking into the future, of embracing a wide range of differences, and of engaging in truly dynamic thinking that involves sizing up a situation in its full complexity and considering and weighing all the relevant factors and options. This is no small task. It demands a variety of emotional and cognitive skills. To begin with, it involves weighing what's important to oneself as well as to other parties. This is a difficult challenge because it demands a high level of empathy (putting oneself in someone else's shoes) as well as considerable knowledge of the other party's background, current situation, and future options. Decisionmaking then involves weighing the different factors, possible options, and related future probabilities. This requires judgment based on considerable emotional experience with and knowledge of the issues at hand. In addition, a high level of reflective thinking employs the processes just described in the face of intense emotions, such as rage, fear, and suspiciousness. Embracing these subtle reflective processes is even more challenging for groups than for individuals because, as suggested earlier, the commonality that organizes large groups can often be associated with more polarized and less reflective practices and institutions.

Groups that respond maladaptively tend to relinquish their capacity for reflective thinking under the pressure of strong feelings and are unable to maintain their institutions and practices at the same level of reflectiveness during periods of acute stress. These groups may be constricted in their range of empathy and frequently regress into polarized, narrow, here-and-now thinking during times of acute stress. For example, such groups may be unable to project sufficiently into the future (i.e., considering only a five-year future consequence rather than a fifty or hundred-year future consequence); unable to characterize the current situation in terms of relevant historical or other current factors (i.e., create a relevant context and internal standards); unable to look at subtlety or gray-area differences in the issues at hand; and unable to look at indirect or multiple influences on the phenomenon at the same time.

The level of individuals, practices, and institutions that characterize a group influences not only the integrity and type of governance structure it can embrace but also its system of justice and its potential for economic growth. Reflective individuals tend to better assess the marginal value of the next unit of investment better than unreflective or polarized individuals who, therefore, tend to precipitate greater economic instability. As we look at how groups organize to meet their developmental needs, it's important to describe a feature of the relationship between the different developmental dimensions of groups. In general, the way in which a group meets its basic developmental needs serves as a foundation for and, to a degree, fuels symbolic group processes.

For example, a group may cohere around its need for security. During the course of evolution, we can readily imagine that a certain group size would enhance the security and survival of the group and its members. But the ability of the group members to feel close to each other, trust one another, and exchange affective signals with each other to negotiate the basic issues of safety, danger, power, relationships, and the like would also, in part, determine the size of the group. For example, beyond a certain size, the intimacy of relationships within the group might be sufficiently compromised as to undermine the relationships and communication patterns that sustain the group. Therefore, the need for security and survival, in part, may motivate the formation of the group. Yet, its ability to maintain relationships and affective signaling to negotiate basic group themes, in part, may determine the structure and size of the group.

The group's capacity to create symbols and ultimately institutions that embrace these symbols may further determine the structure and size of the group. Shared symbols and institutions, such as those that represent concepts like "equality" or "justice," may make it possible for larger and larger groups to form and function. With such shared symbols, individuals are less dependent on direct relationships and communication to negotiate basic group themes such as safety and danger or group norms. Therefore, shared symbols and institutions can, in part, serve the purposes that intimate relationships and direct, implicit affective communication serve.

But shared symbols and the institutions that represent them obtain their saliency, in part, from the emotional investment of group members in them. In times of heightened need for security, for example a threat of another group, a group might invest very intently in its shared symbols and institutions (e.g., a flag). In times where the need for security is not high, the emotional investment in shared symbols may be lower. During these times, the trust between group members, the access they have to each other for implicit communication, and the strength of the traditions behind the shared symbols and institutions, as well as the nature of the symbols and institutions themselves, may be relatively more important for sustaining the group. The larger and more complex the group, and the less the immediate need for security, the more the group may depend on emotional relationships and communication patterns to maintain itself and function.

We can summarize this material with the following table:

TABLE 15.1 Adaptive and Maladaptive Levels of Group Development

	Adaptive	*Maladaptive*
Group security	The group can negotiate security while retaining individual freedoms and reflective institutions.	Group is not able to negotiate security successfully or can only negotiate security at the expense of individual freedoms and reflective institutions.
Cohesion and shared allegiance	Group is able to relate and cohere around reflective institutions and processes.	Group can only cohere around concrete dimensions (such as skin color, polarized beliefs, a charismatic leader); or group is unable to have any cohesion.
Intentions, expectations, and shared assumptions: formation of a collective character	Group possesses communicative abilities that embrace and integrate a broad range of differences and themes.	Group requires daily contact with a wide range of others with a number of programs to develop these implicit communication systems at the level of immediacy that is demanded (see group chapter).

(continued on next page)

TABLE 15.1 *(continued from previous page)*

	Adaptive	*Maladaptive*
Symbolic expression	Symbolic objects seen as representations of important abstract values; differences are reflected and integrated.	Symbolic objects are seen as carrying power in and of themselves; symbol systems are fragmented, inconsistent, idiosyncratic, and discordant with the reality they purportedly represent; they allow for a few strictly defined categories of thought and behavior.
Economic stability and growth and reflective individuals and institutions	Societies characterized by reflective individuals and institutions tend towards more gradual downturns and greater capacity for corrections.	Societies characterized by individuals and groups that are vulnerable to polarized, all-or-nothing thinking are at risk for more sharp downturns, particularly as investment opportunities decrease and those opportunities that are left demand more judgment.
Achieving higher levels of reflective practices and institutions and maintaining stability through change	Stable reflective societies have mechanisms that permit them to change while retaining core values; such societies can handle (literature, art, music, movies, theater, television shows, and news coverage) a wide range of emotional themes. They also continue to grow to higher levels of reflective practices that include taking into account a broader range of themes, longer term projections into the future, and wiser decisions that can support adaptive evolutionary pathways.	In unstable societies that lack high-level reflective institutions, new policies only arise from social upheavals that can result in a new constitution, flag, and sometimes even name; such societies can only handle a limited number of emotional themes, generally in a constricted manner. Such groups also tend to operate in the here-and-now of becoming polarized, fragmented, and/or impulsive.

DEVELOPMENTAL PERSPECTIVES ON CONFLICTS BETWEEN NATIONS

To illustrate the developmental lens we have just outlined, consider its application to one of the most pressing challenges that currently confronts us: that of resolving conflicts among nations in an interdependent world. For it shows us how the level of group functioning influences the way the group deals with conflicts with other groups, as well as new, more adaptive ways worthy of consideration.

As history so well documents, nations often deal with conflict by using strategies such as disengagement, lack of ongoing affective interactions, polarized thinking, and illusions of more definitive power than they actually have, given the interdependency among groups. Diplomats view such actions simply as moves on a chess board. In an interdependent world, however, an effect of a strategy such as disengagement may not be to intimidate or motivate another group to change, but to block communication at a number of levels and increase shared dangers.

As we look at the developmental needs of groups and apply this framework to conflicts within the larger scope of international relationships, we see that an individual group may try to meet its own needs for safety and security and coherence with strategies that diminish the possibility of adaptive patterns in the larger international group. Later we will explore how national groups can meet their developmental needs and resolve conflicts with other groups while supporting, rather than undermining, the functioning of the larger international group. This is especially vital in an interdependent world.

Polarized thinking, disengagement, and other similar strategies are often part of a mindset based on a time in history when there was relatively less interdependency and, perhaps, a greater degree of relative power residing in the hands of one or a few nations. At present, because of the interdependency of shared dangers, such a mindset is no longer realistic. A denial of one's relative vulnerability, however, coupled with an unrealistic perception of one's relative power, may maintain earlier, but now unrealistic, mindsets. It can interfere with accurate current assessments.

For example, an accurate perception of what an adversary needs, can accept, and will do if frustrated, as well as accurate information about oneself, is vital to maintaining a reflective stance and negotiating an agreement. Yielding to the temptation to regard the views and motives of others as all good or all bad prevents one from obtaining realistic information or devising effective strategies. For well into five decades, both sides in the Cold War were drawn into devastating military spending and two actual wars. The relations of the superpowers with countries around the world were distorted by categorizing them as allies or adversaries. This prevented efforts to understand each side's motivations and act on that understanding.

Many nations repeatedly project overly simplified notions onto others and repeatedly discover that foreign plans and aspirations are more complicated than were thought. Oversimplification has the deleterious effect of weakening internal institutions. If leaders give the public pat answers and polarized choices that do not represent the full complexities of challenges, they undermine not only people's confidence but their ability to think reflectively about world affairs.

Moreover, when political leaders use the mass media to feed the public misleading, polarized information, the public loses the capacity to make careful distinctions in the national discourse. Misinformation may serve the short-term political advantage of some individuals, but in the long run it cripples a nation's ability to govern itself.

Modern polling techniques can exacerbate this situation. After public figures have framed issues in a polarized way, pollsters conduct surveys to determine which positions are popular. Leaders then justify their policies and actions by referring to public opinion polls that were initially shaped by their own misinformation. A cycle of escalating misinformation and polarization replaces informed debate.

How can we construct a realistic picture of other groups and use it to deal with conflicts more effectively? Understanding the importance of experience at the level of implicit processes, including emotional exchanges and gestures as outlined earlier, is critical for grasping how to sustain realistic perceptions. Expressions and gestures that convey intentions and values are essential in understanding how to create realistic information and perceptions.

Experience with the broader range of ways we exchange notions of each other can be undermined in five ways:

- First, people often prematurely substitute words for experience before they have developed enough common experience to comprehend each other's real intentions or trust each other's word.
- Second, leaders may rely overly on mere verbal assurances and pay too little attention to actual behavior.
- Third, when people try to force or to manufacture nonexistent accord, their interactions with others may be so structured as to prevent nonverbal cues from betraying their true feelings.
- Fourth, as tensions rise, individuals and nations tend to minimize contact and thus the possibility of many forms of implicit and nonverbal communication.
- Finally, individuals and nations often, and usually futilely, try to set limits for others' behavior without the benefit of a broader relationship within which to mediate disputes and accurately gauge the results of various actions.

When communication is undermined in these ways, a shift occurs toward the tendencies or patterns outlined earlier, that is, for polarized thinking, where power is exaggerated and not seen in its fully relativistic context; where action is valued in its own right, regardless of its impact, because of the unrealistic belief that action and power will lead others to follow the "leader" or "leader group"; and/or where escapist, unrealistic beliefs and practices prevail.

Looking at international affairs in terms of the developmental capacities of groups offers guidelines for evaluating different approaches. Constructive communication means

1. maintaining nonthreatening engagement through international organizations (i.e., not using disengagement or isolation as a signal or sanction—it doesn't work in an interdependent world) even at times of disagreement, stress, and conflict;
2. increasing interactions between leaders and citizens through such avenues as diplomacy and exchange programs;
3. using sanctions and interventions to set limits that are realistically required and implementing these as early as possible and as firmly as needed (but these sanctions and interventions should not be used to isolate an adversary from the world community);
4. tolerating the distortions others make of facts or situations and analyzing these distortions for insights into the aims of others;
5. negotiating differences by using accurate information and realistic assessments of the other party, which includes avoiding personalizing or polarizing the other party's distortions (this includes fuller information about other groups and cultures);
6. keeping a plan on the table towards which moderates can work without becoming impatient and regressive (so that the process can remain reflective); and
7. offering respect and autonomy to other nations.

As the world becomes more and more interdependent, these reflective strategies for dealing with conflict will become more and more essential. The distortions characteristic of existing methods of dealing with conflict will only create greater danger for the entire globe. More than ever we need to understand the nature of our new interdependency and the changes it necessitates in our behavior, perceptions, feelings, and thinking.

THE NEED TO REVIEW OUR FUNDAMENTAL ASSUMPTIONS ABOUT INDIVIDUALS AND GROUPS

Before we can consider what sorts of policies need to be adopted in order to create reflective citizens capable of dealing with the challenges of interdependency, we first need to reassess our fundamental assumptions.

Changing Assumptions About Power

One of the first assumptions we need to come to terms with is how the dynamics of power have changed dramatically. In an interdependent world in which the smallest of splinter groups can wreak terrible havoc on the entire globe, power has become far more relative than it ever was before. Our enormous advantages in weapons systems and technology make it understandably difficult to accept this new reality. Why doesn't enormous military advantage still afford one a definitive advantage in power? But, as indicated earlier, when small groups can set off biological weapons in their own backyards that can engulf the globe, the interdependency of shared dangers makes even overwhelming military advantage much more relative. We may want to deny this fact because it is such a frightening and numbing prospect. Denial, however, can fuel the wish to continue to use intimidation as a major strategy. The problem is that intimidation breeds fear and contempt among those one is trying to intimidate, if not outright rage and hostility. Such feelings have new meaning in an interdependent world that includes shared dangers.

It is not easy to shift from tendencies to use intimidation and maintain the illusion of absolute power. One of the hallmarks of power, however well intended its use, is its inherent tendency to maintain itself. In short, it is hard to give it up. To be sure, power is frequently relinquished in family, community, and even national settings. Families relinquish power for the benefit of a new, more reflective organization as they learn to share decisionmaking with their late adolescent and young adult children. Communities are able to develop flourishing centers that are run as cooperative enterprises. And there are many examples of nations that formed as a result of autonomous groups comprehending and advocating the greater good, relinquishing part of their individual powers to a national entity.

But on the larger world stage, it is much harder to relinquish power, and not surprisingly so. It is easier to share power for a higher purpose with those you love, trust, or at least feel you know, or with whom you share common beliefs and principles. Sharing power is a much more difficult process when dealing with groups that are outside one's orbit of shared humanity and reality. In an interdependent world, however, all groups must share in decisionmaking. To make this possible, power must be used to support reflective institutions that enable collective decisionmaking. Investing in international bodies and relinquishing power after using it to create these reflective representative bodies will take time and will require greater respect for each others' national institutions than now exists.

The principle underpinning this argument is not some utopian vision of international harmony, nor is it an attempt to inculcate classic liberal-democratic ideals. This argument is grounded in an understanding of the changing dynamics of power and the nature of the practices we must embrace *if we are to survive as a global community.* For this same reason, we need to recognize the indirect ways we attempt to maintain power. For example, we need to recognize the shortsightedness of resorting to autocratic tactics when implementing more advanced political ideals. One might argue that the means can justify the ends. But our developmental model suggests that, no matter how lofty the principles, autocratic processes will lead to groups that cohere around polarized beliefs, or feared and aggrandized leaders, or other concrete images. In building more reflective societies, we have to look at the content we're espousing and the processes—that is, the means—we're using to promote these important ideas. For *the means don't justify the ends: The means define the ends.*

What sustains polarized and oversimplified perspectives? We're often unaware of the preconceived frames of reference we use to see the world. A frame of reference literally becomes a lens through which only certain patterns or shades are observed. The competitive lens is an example. Election debates illustrate this example. The media and the candidates are preoccupied with winning or losing the debate rather than explicating the issues. When the media augments this natural tendency with a headline indicating that this or that candidate won the debate rather than that the candidate did or did not fully explicate a complex issue, they are fostering the tendency to view the world through the competitive lens. This reinforces an artificial reality. The public, already predisposed to this lens, follows. It doesn't look for or demand enough substance. As a consequence, politicians, the public, and the media remain preoccupied with winning or losing rather than with creating a long-term framework to resolve conflict.

In resolving conflicts in an interdependent world, we're dealing with issues far more complex than simply the appearance of winning or losing. Those nations and groups that have real power in military and economic spheres are, and will be, perceived with a mixture of envy, suspiciousness, and anger, as well as respect and admiration. We need seriously to consider how mutual respect can be enhanced and suspiciousness and anger lessened.

Attempts at intimidation or destruction without offering hope and opportunities for collaboration will almost always move groups towards the side of the continuum that espouses fear and rage rather than respect. Respect is born from firmness coupled with an understanding of the responsibility of power to work towards fair and reasonable long-term solutions. This is not a utopian vision that ignores the reality of a harsh world populated by groups harboring many different beliefs and goals. Although it sounds utopian, ironically, it's a necessary reality in an interdependent world.

The Global Group: The New Unit of Survival

To resolve conflicts differently we need a new way of thinking about the social group and the survival of the group. A new way of thinking does not involve giving up our individual or group identity; rather it means acquiring the ability to operate at multiple levels of group identification.

As we saw in the preceding chapter, it is the *group,* in its functions of security and its economies of scale, that permits the long-term rearing and education of the young and the creation of new technologies.

The size of the group and the amount of diversity it needs to accommodate is in part based on the challenges surrounding it and what is necessary for survival and adaptation. In an interdependent world, our concept of the group must enlarge beyond previous concepts of a functional group. The unit of survival in an interdependent world is the entire globe. What does the global group require to function adaptively? It would need to embrace similar implicit rules that a family group, a community, or a tribal group embraces in guaranteeing the survival of the group and the individual.

The larger and more diverse the group, however, the harder it is to organize. To organize and maintain such a functional group, one that integrates many diverse elements, requires especially stable, secure, and reflective individuals and institutional processes.

To be successful, the larger group identity needs to be an expression of individual self-interests. An example of this is the family group. Most individuals see their families as part of their own survival. As such, they equate their own interests with those of their families. In working to improve themselves, they improve their families. Similarly, the large group and the global group can be seen as an expression of the individual survival based on the realities of survival.

In other words, one's perception of the group dimensions necessary for survival will determine to what degree the group becomes an expression of individual self-interest. One can imagine different perceptions in different contexts. In a safe world with bountiful food and other basics, individuals and/or small groups may not find the need to congregate in larger groups. On the other hand, with hostile others nearby and/or food and shelter hard to come by, groups of a certain size may be essential for survival. In today's world, because of the degree of interdependency, we have argued that the unit of survival is the global group. Therefore, individuals will, one hopes, over time identify with other human beings across the globe the way they identify with their own individualism and immediate families, communities, and national groups.

One can picture a hierarchical relationship where individual self-interest relates to the family, the community, the nation, and the global group. Each boundary, which is real and has its own integrity, also has a relationship with the next boundary, which provides another level of survival protection. For the large group to work, this hierarchical relationship must be strong at every level or boundary. Yet, it is expected that very different feelings will exist for one's family in comparison to one's nation and, in turn, for one's nation in comparison to the global group. While different, if there is true recognition of interdependency and a new unit of survival (the globe), we would expect a new ethic to emerge regarding others. In other words, individual self-interest is not to be given up. It's to be transformed into a hierarchical set of relationships where individual and collective needs are part of one integrated pattern. This will require highly reflective individuals who can attend to their unique individual and family patterns and the larger group patterns at the same time.

THE MORALITY OF INTERDEPENDENCY: SUPPORTING A NEW ETHIC OF SURVIVAL OR INCREASING THE LIKELIHOOD OF RAGE AND DESTRUCTION

The global group will require a new morality consistent with changing global relationships. In an interdependent world, we all share each other's wish and need to survive. Respect for survival, however, cannot be limited to outlawing terrorism. As discussed, it must involve respect for the life and well-being of all members of the world, including those who are poor, hungry, or ill. If a new morality of survival is not applied to all peoples equally, those excluded and those who identify with the excluded may decide not to embrace the collective ideal of survival. They may support efforts to undermine it. In other words, in an interdependent world, all human survival must be the goal of all groups and become part of a global effort.

In this context, the failure of an interdependent partner to acknowledge the importance of one's own existence may fuel rage and destructive action. This sense of disavowal had a very different meaning in a world where groups were relatively isolated from one another. When there weren't shared dangers, communications, or economies, peoples of one group didn't expect much from the peoples of another. In an interdependent world, where cultures are intertwined and where it's nearly impossible to escape from the influence of "others," even if one wishes to, the feelings are quite different. Now, one's needs and very existence are being disavowed by another with whom one has a strong relationship—whose movies and books are part of one's daily life, whose businesses are reaping profits in one's economy, and whose relative prosperity is being witnessed every day.

Why should this closer, more intimate relationship among the world's peoples breed more rage and destructiveness? To understand this phenomenon, consider the metaphor of a family: How does the member of a family whose basic needs are not attended to begin to feel? There are usually two feelings: enormous rage and enormous helplessness, sometimes coupled with depression. These feelings can fuel the most destructive acts. The disavowal of another human being in an intimate context is perhaps one of the strongest insults and humiliations human beings can suffer.

In addition to the disavowal of the other's basic needs, the disregard of the other's culture, religion, or beliefs can also fuel outrage and destructive acts. The paradox is that the more interdependent the world, the greater the need to respect each peoples' unique needs and characteristics while at the same time supporting a shared sense of humanity and reality through global reflective processes and institutions.

The morality of interdependency can be characterized by five basic principles:

1. The world is interdependent by the nature of its shared dangers as well as its shared economies and communications.
2. Interdependency alters the expectations and the emotional reactions of individuals and groups to one another.
3. In an interdependent world, one of the largest insults and humiliations an individual, community, or culture can suffer is the lack of acknowledgment of its very existence (in terms of basic survival needs).
4. Such disavowal breeds rage and hopelessness and increases the likelihood of destructive acts and the support of destructive acts by even moderate groups.
5. A new ethic and morality for an interdependent world must be based on the premise that the viability of the world—that is, its health, security, and very survival—is measured by how well it tends to the needs of its weakest members. This ethic may establish a more difficult standard than is apparent. Helping others cannot be based on handouts or even the most enlightened forms of altruism. Rather, it must stem from the basic realization of the interdependency of all individuals and groups to share in the survival of the world and each of its inhabitants. In addition, this new ethic must go beyond dealing with poverty and enhancing the survival of individuals and groups. It must also provide opportunities for individuals and groups to grow economically and socially. This includes stable families and communities, educational opportunities, and economic opportunities.

Working together on behalf of survival will, in the short run, certainly not immediately reduce all the reasons for conflict or terrorist activity. Different belief systems, a lack of economic growth and opportunities for meaningful personal and family improvement and careers, and poverty all need to be considered. Even if these factors are dealt with, challenges will obviously still exist. If, over time, however, a new morality of interdependency can support the moderate elements in most groups, extremists will have a harder and harder time finding support and the moderates will be relatively better able to maintain safety and security. In short, the greater the interdependency, the greater the needed collaboration. There must also, however, be recognition that the equilibrium between thoughtful collective action on behalf of safety and security and global danger is a precarious one.

BUILDING THE CITIZENS OF THE FUTURE

Currently, much of the world is still organized around simpler types of group structures in which power is concentrated in the hands of one individual or a small ruling elite. Such power structures seem to be a natural feature, not just of human, but indeed, of primate social groups. We saw in Part II how nonhuman primate societies are invariably dominated by a physically dominating figure, an alpha male whose power is further buttressed by a small number of slightly less powerful allies. His control over the group only subsides when he can no longer fend off the physical challenge of a powerful rival. But the social structure of the group remains the same during such transitions; only the alpha himself changes, and perhaps his allies. Interestingly, a similar pattern can be observed in children's "playground politics," when the most competent kid at the activity at hand assumes the role of the leader who oversees a rigid hierarchical structure; for as long as he remains the strongest, he will be the one who decides who gets to play where or on which team.[1]

As we saw in Chapter 14, this kind of simple power hierarchy was the mainstay of early human communities and civilizations, and still remains characteristic of many tribal cultures that are ruled over by a powerful chief or cadre. Interestingly, such a form of social organization is also widely seen in the business world, particularly in the early days of large companies that are built up and run by a powerful dominating figure. As can be clearly seen in the business example, in certain conditions such a structure is highly effective insofar as it provides for rapid and centralized decisionmaking. Some of the most powerful corporations in America today were originally built up by charismatic figures that inspired a workforce with their vision and work ethic. But when these highly successful national companies started to make the transition to multinationals, the old style of autocratic governance proved to be much less adaptive; these companies then had to undergo the difficult adjustments involved in developing the kind of diffused decision-making structure that is needed to deal with the exigencies of running an international conglomerate.

We are faced with a very similar issue today on the political front: the problem of whether the simpler kinds of tribal structures that are very effective for certain kinds of social or economic conditions, such as when groups are highly homogenous, are sufficiently adaptive to deal with the challenges of global interdependency. If, as we have argued above, we need to develop more reflective kinds of structures to deal with the challenges of interdependency, then we need to consider how we can enable a significant proportion of the populace to develop the sorts of reflective capacities that are necessary to sustain the kinds of institutions and decisionmaking processes that are necessary to deal with the exigencies of global interdependency.

For more basic kinds of social structures, the individual only need be capable of relating to the other members of the group by cohering around some concrete dimension: the charismatic leader, a dominant belief. But to support and participate in much more reflective types of group structure, including groups that cohere around abstract principles like justice or equality, the individual must be able to operate at a reflective level that permits thinking off an internal sense of self and an internal standard. As we described in Chapter 2, this involves the ability to operate in two frames of reference at the same time, one involving an awareness of one's own inner sense of self and the evolving standards or ideals associated with that. The other frame of reference involves one's immediate moment-to-moment experiences, which are compared to and judged or interpreted according to one's internal sense of self and standards. For example, in order to invest in a concept such as justice and an institutional court system that supports the concept of justice, one must be able to have developed an internal sense of self and standards that can resonate with these abstract concepts and institutions.

In other words, to invest in institutions—the consent of the governed, according to Jefferson—requires a sophisticated level of reflective thinking that's considerably more complex than simply believing in a strong leader or a rigid set of principles.

What is often not clear to students of group behavior is the degree to which this level of reflective thinking or the departure from reflective thinking is an indication not just of the intellectual processes used by the individual or group but of their emotional processes. Reflective thinking of the type we have been describing requires well-developed forms of empathy (being able to project oneself into another person's shoes and understand a wide range of feelings and views without distorting them), coupled with the ability for being realistic and having a firm assessment of reality. The ability to hold on to reality in the face of intense feelings and complex circumstances and to be able to balance realistic appraisals with empathetic understanding is a complex emotional and intellectual task. It is especially difficult for groups and individuals to carry out this task when under extreme stress and/or influenced by strong emotions. When done well, however, these thinking processes seamlessly harness and integrate our emotional and cognitive capacities. But this integrated development needs to reach a level where individuals can think off an internal standard, and a sense of self, even when there are strong emotions, if the group is to have any chance of supporting concepts such as justice and equality or other elements of a representative government. Although one may argue from a purely theoretical point of view about the value of reflective institutions and processes versus less reflective ones in terms of a group's coherence and ability to carry out fundamental goals, it is clear that in an interdependent world the need for high-level reflective institutions to support collective problem solving is highly adaptive.

From such a perspective, we have just as strong a vested interest in the emotional well-being of children in distant parts of the globe as we have in the emotional well-being of our own children and the children of our neighborhoods. Consider, for example, the reports that Palestinian children as young as eight are expressing a keen desire for martyrdom. This development has been seized upon as proof that pleas for a negotiated settlement of the Middle East conflict are delusional because such young children are being indoctrinated with a terrorist mindset. But, of course, an eight-year-old child is incapable of grasping the concept of martyrdom. In fact, the studies by Palestinian psychologists on which these reports are based[2] point to a very different phenomenon that needs to be addressed. Some of the children in question demonstrate alarmingly high levels of distress. In other words, the evidence suggests that some of these children may already be potentially suicidal and that all they need is a pretext to trigger their self-destructive inclinations.

Similar reports are coming in from all over the world: stories of young children who, after being abducted and trained in guerrilla armies, take part in heinous genocidal acts without a trace of remorse; or children raised in a climate of violence who then participate enthusiastically in acts of sectarian violence. In an interdependent world, we need to take such reports as seriously as if they were happening in our own backyard. Indeed, we are seeing alarmingly high rates of difficulties in our own backyard as well. For example, from 1950 through 1990, the rate of suicide for persons fifteen to twenty-four years of age increased from 4.5 to 13.5 per 100,000 in the United States.[3] With such a contagion of adolescent distress spreading around the world, it is clear that urgent action must be taken if the next generation is to develop into the sort of reflective society that will be capable of supporting the kinds of institutions and decisionmaking processes that are necessary to deal with the exigencies of global interdependency.

THE PATHWAY TO EMPATHY, REFLECTIVENESS, AND INTERDEPENDENCY

As can be seen, many new policies need to be explored. These policies, however, must not only reduce distress in children and families, but facilitate needed coping skills, especially for empathy and reflective thinking. Our recent understanding of the steps leading to high levels of reflection and empathy suggests the types of new policies that are needed for families and communities. Let's take a look at the steps leading to higher and higher levels of empathy and reflective thinking and explore the types of family patterns and educational practices that will support this progression.

To begin with, children must learn *to engage emotionally with "another."* This is the basis for a shared sense of humanity. It is promoted by consistent, nurturing care from loving caregivers who will be part of the child's life for years to come. It is undermined by full-time care from a series of caregivers and by caregivers having to care for too many babies (as happens in many busy daycare centers). It is also undermined by stressed caregivers and/or families, inadequate or inappropriate caregiving, and/or exposure to dysregulating environments or toxic substances that affect the central nervous system.

Next, the child must learn *to interact and to signal with emotions and broaden their emotional range.* This is the basis for being aware of the intentions and feelings of "another." It is promoted by nurturing interactions and lots of emotional exchanges: long, emotional dialogues involving such basic emotional themes of life as love, anger, and curiosity. It is undermined by insufficient interactions, the misreading of a baby's emotional signals, or interactions that are too intense, constricted, or short.

Following this, the child must learn *to care for, protect, share with, and behave altruistically toward another.* This is the basis for cooperation. It is promoted by opportunities for long, playful, preverbal interactions and emotional dialogues with caregivers and peers, as well as supportive, firm limits. The limits should use lots of communicative gestures that teach a child to cope with frustration and learn patience and respect for others. It is undermined by insufficient interactive opportunities and overly punitive or inconsistent limits.

Next the child must learn *to broaden the range of "others" with whom one can relate, share, and respect.* This is the basis for broadening one's future empathetic range and identifications. It is promoted by play and interactive opportunities with children from different cultural and ethnic backgrounds. These interactions need to occur, however, as part of one-on-one or very small group play opportunities in a context of ongoing nurturing care with one or a few consistent caregivers (who provide a great deal of security). It is undermined by isolation, very busy child care settings that keep children in large groups (more than three or four children with a caregiver), or dysregulating environments.

After this, the child must *share emotions and ideas with others.* This is the basis for feeling a part of another person's feelings and ideas, or inner world. It is promoted by shared pretend play with caregivers and peers and lots of verbal or pictorial "discussions" with others, and by bringing in the ideas and customs, including dress, music, games, and activities, of children from other cultures. It is undermined when children play mostly alone, engage only in action-oriented rather than imaginative activities, and encounter overly rigid child care attitudes and practices.

Sharing emotions and ideas enables the child to take the next step: that of *seeing the world from another's perspective.* This is the basis for identifying with someone else (putting oneself in someone else's shoes).

It is promoted by challenging children to connect ideas together logically through opinion-oriented conversations and logical pretend dramas with others. It is also promoted by setting firm but gentle limits on aggression while promoting its verbal or imaginative expression and understanding. It is promoted by teaching the value of respect and concern for others with actions (helping others, sharing in chores), as well as words. It is supported by relationships with peers from different cultures and opportunities to expand on the ideas and customs of others in play, stories, songs, and activities. This capacity is undermined by factual and memory-based dialogue, self-centered types of experiences, and narrow, rigid child care practices.

A child is now in a position to *understand the needs and wishes of others.* This is the basis for understanding that others may have different needs and feelings than we do and the reasons for these differences. It is promoted by logical discussions that look at the multiple reasons for events or feelings; compare people, objects, or feelings (different friends or foods); and explore the degrees to which one feels or believes a certain way ("I like Sarah much better than Allison because she is much nicer when I'm upset"). It is also promoted by the experience of being empathized with when angry, competitive, needy, or sad. It is also promoted by real friendships with children from other cultures. (These can be promoted through increased diversity in schools and neighborhoods, as well as through relationships that result from communication and exchange programs with "brother" and "sister" communities and schools in other parts of the world.) It is also promoted by learning about other cultures and experiences with "others" that promote friendship, respect, and an understanding of fundamental similarities as well as differences. It is undermined by interactions that promote action over reasoning and polarized thinking over reflection, as well as punitive school or family interactions.

A child now begins to develop *reflective empathy: experiencing how another person feels and comparing it to one's own feelings.* This is the basis for emotionally understanding the feelings of others and reasoning about the feelings of others in relationship to one's own personal experiences. It is promoted by experiences that lead to a sense of self, comfort with a broad range of feelings, and the capacity to construct internal standards. It is also promoted by experiences and friendships with a range of "others" that challenge stereotyped views, stretch one's emotional range, and promote reflective thinking. It is further promoted by factual knowledge of other cultures through broad-based school programs and personal experience of other cultures through school and community diversity and special international programs. This capacity is undermined by experiences that lead to a fragmented sense of self, weak internal standards, impulsive or polarized thinking, and comfort with only a narrow range of emotions.

Finally, the individual can progress *from reflective empathy to interdependency.* This is the basis for increasing reflective empathy to include broader and broader identifications with others from around the globe. It is promoted by experiences that increase knowledge of the cultural, social, historical, geographical, and economic dimensions of other peoples, coupled with increasing interaction with other peoples in a variety of contexts. These contexts may include work and service programs in other cultures, integrated schools, communities, community centers, and cross-cultural work and exchange programs. It is also promoted by the mastery of the developmental challenges during adolescence and adulthood that leads to higher and higher levels of reflective thinking. It is undermined by limited knowledge and experiences of other peoples, weak reflective thinking capacities, and poor mastery of the stages of adolescence and adulthood.

The kinds of programs that we are outlining here have, of course, been the subject of lively political debate in recent years. What has been missing in these deliberations, however, is a larger purpose to determine the direction in which we should be moving. Our developmental model of evolution provides us not only with just such a goal but also with a psychological framework for considering the kinds of family and educational practices necessary to create a reflective citizenry and the reflective institutions that can rise to the challenge of global interdependency.

THE ROAD TO DEMOCRACY—PAVED WITH AGGRESSION AND RESENTMENT?

The levels of social groups described earlier can help us understand the challenges faced in trying to facilitate democratic forms of government. Nation building can be successful only if the underlying social and psychological needs for a dictatorial regime are understood and the multiple steps required for democracy are anticipated. It is often taken for granted that dictatorships are somehow imposed on a populace, generally against their will, and then kept in power through the agencies of a secret police force. But groups with warring factions often lend their tacit if not outright support to a dictatorial regime in their efforts to contain emergent hostilities. Initially, such groups often do not have the collaborative relationship, trust, and reflective processes that are needed to invest in shared democratic practices and institutions.

Consider two situations. In the first, a nation is made up of a cohesive social and cultural group where individuals have some degree of trust in one another and a history of democratic-type institutions. When such a group is liberated from a dictatorial regime, it can come together to support democratic reforms readily, as was observed in post-Communist Poland.

In the second, a group has a long history of many separate factions. Each faction distrusts the others, and they all harbor polarized views about each other. What happens when this type of group is freed from a dictatorial regime? To answer this question, we first must look at why a group would embrace or tolerate a dictatorial regime in the first place.

As discussed earlier, social groups have basic needs, including safety, cohesion, implicit survival-based communication, shared symbols, and a relatively shared sense of reality and humanity. Different social and political groups, however, have different ways of coming together to meet these basic needs. Social groups that are chaotic and fragmented and invested in polarized beliefs meet their basic needs in ways that are different from those of social groups that are relatively integrated and share a sense of reality and humanity.

For example, the fragmented and/or polarized social organizations often can come together only by adhering to a powerful feared leader or set of polarized beliefs or images. In a sense, the group feels that security is attainable only through a feared leader because no other structure could contain the imminent hostility and danger from the different factions. Paradoxically, fear of hostility is used to contain hostility and provide security for the group.

The dictatorial regime, ironically, may also provide a familiar social structure. Dictatorial regimes tend to operate at a level of group functioning dominated by "all-or-nothing" thinking, concrete rules, extreme punishments, and little tolerance for flexible reflective thinking that would challenge an authoritative structure.

Warring factions also tend to operate with "all-or-nothing" thinking ("The other factions are all bad") and rule by severe authority figures.

Therefore, the polarized thinking that characterizes a dictatorial regime feels familiar. In contrast, reflective thinking embodied in representative forms of government and concepts such as justice, equality, debate, and compromise feel alien.

What is likely to occur when groups that have been fragmented or polarized into different warring factions are suddenly freed from the seeming tyranny of a dictator? Will they be appreciative? Will they feel a sense of security from a potential governing process that seems alien to their basic needs? Or will they experience a collective anxiety and uncertainty because the proposed new governing structure, although supporting individual freedoms, does not provide an immediate sense of protection against factional hostilities or a familiar social organization? Not surprisingly in such a circumstance, we can expect certain reactions: frank hostility by those with very different objectives; anxiety, fear, and resentment by those for whom a new structure does not provide the security and cohesion of the former one; and a lack of appreciation by those who do not intuitively embrace the liberating power as representing goals that are familiar and attainable.

The road towards democracy, therefore, often involves a considerable period of time and a number of steps. The different subgroups have to meet the challenge of beginning to work together and resolve underlying hostilities. They have to trust each other enough to experience some relative degree of a shared sense of reality and humanity and invest in some type of representative governing structure. The groups have to develop the psychological and social capacities to embrace reflective integrated compromises rather than polarized beliefs. Achieving the psychological level where differences are evaluated and debated is a long and arduous task.

What policies will promote these processes? In addition to established policies such as preparing for the long haul; establishing safety, security, and basic necessities; investing in schools and health care; and working with moderates toward self-governance, there are additional objectives. One is to anticipate the hostile reactions described earlier. Another is to realize that to cope with the anxiety of losing the relative security of a dictatorial regime, the liberated group needs a greater sense of safety than might be anticipated. Without this first step of "safety," there will be an enormous temptation to return to autocratic practices. In addition, the investment in human capital and family and educational programs to enhance the capacities of future generations for reflective thinking, collaboration, and empathy will have to be far greater than is usually planned.

As the United States was forming, Thomas Jefferson said that democratic processes can be only as strong as the capacity of the governed to invest in these processes. To invest in abstract democratic principles, however, requires high levels of personal and group functioning.

BUILDING FOR THE FUTURE: FAMILY AND INSTITUTIONAL POLICIES

To help new generations become reflective and empathic at levels that support interdependency, we need to explore new national and international policies. Creating the conditions for healthy human development and adaptive social organization requires enlightened policies. What are the societal policies and practices needed to promote the development of individuals and groups capable of coping with growing global interdependency?

In formulating these policies, it is important to realize that currently we are shepherding a delicate evolutionary process of group formation. Therefore, practical decisions must always take into account the long-term needs of the global group.

To support adaptive interdependency in both the short and the long term, we have to invest in a new type of capital, namely, human capital. Interestingly, many of the skills required for a successful labor force are the same ones required for supporting the infrastructures of complex adaptive social organizations and a shared sense of humanity and reality. But our developmental model of group development prompts us to consider whether we should be considering a whole host of programs geared towards promoting the psychology of interdependency.

To begin with, should we institute *a public education campaign* on the nature of global interdependency and the changes it means for individuals and groups?

We certainly need to pursue policies that will promote *the development of safe and caring communities.* Adaptive social organizations and political structures depend on our capacities for sustained relationships, comprehending our own and other's emotions, engaging in reflective thinking (even in the face of intense feelings), and supporting institutions that can interpret reality and mediate and resolve conflicts. As is well known (but often overlooked), these critical capacities grow from nurturing human relationships between children and their caregivers, and can occur only in stable, secure families who live in safe, well-organized communities. This means that greater emphasis and, where necessary, appropriate types of support must be given to human development, families, and communities. The problem is, it may take many generations of significantly enhanced child care, family functioning, and social policies to reach these goals. How, then, do we design and implement policies that employ the necessary long-term perspective?

On still a more basic level, do we need to institute policies to support *safety and protection?* The problem here is that individuals or nations who are hungry, ill, in constant danger, or at immediate risk can hardly be expected to invest emotionally in others or the collective processes that sustain large groups. Therefore, one of the most basic principles in a psychology of interdependency should be that of making sure that every child and every family on the globe have the basics of physical protection and care. This includes adequate food and shelter, healthy water supplies, appropriate medical care, and safe environments. We have to be as concerned for a child in Africa who is afflicted with AIDS, or who lives in poverty, or who is suffering from neglect, abuse, or anxiety, as we would be for a child in our own extended family or neighborhood; for we cannot be at the highest level of personal, intrapsychic development if we turn a blind eye to impoverished child-rearing conditions in other parts of the world.

It also seems vital that we address the nature of *educational policies* from the perspective of global interdependency. To be sure, educational policies, like culture and the media, have long been regarded as the private preserve of autonomous nation states. But isn't it imperative that we strengthen international organizations and exchange programs that cultivate thinking skills and that look at issues from a broad geopolitical perspective? To foster a more shared sense of a common humanity, that is, a global identification where the primary unit is the global group rather than the individual, family, community, or national group (although these will continue to exist as foundations), isn't it critical that we help children see events from multiple contexts and frames of reference? Indeed, the foregoing argument suggests that it is imperative we develop educational policies that support reflective thinking skills. In an interdependent world, "intellectual isolation" and rote rather than reflective learning would pose a risk to future collaboration and understanding. As part of a thinking approach to literacy, educational programs will need to help children further understand who they are as human beings and the psychological, social, cultural, and governmental processes that explain their behavior.

One of the key issues raised by our developmental model is how children deal with cultural and/or religious experiences. At their personal core, children will require a very strong grounding in their immediate cultural and/or religious values. This core will need to be strong enough to sustain and support cultural and/or religious values that foster a sense of shared humanity and reality as well as an ethic of survival that is shared with other peoples of the world. Because a shared sense of humanity is only learned through personal experience with a range of "others," should we not be pursuing a number of programs that will foster this broader ethic?

It goes without saying, perhaps, that we need to recognize the overwhelming importance of providing economic opportunities to all groups. In an interdependent world, having the motivation and opportunity to succeed fuels all the initiatives we have been describing. What is not always appreciated, however, is how related these goals are. With growing technologies, there is a general consensus that the ability to take advantage of economic opportunities will depend more and more on being a "reflective" individual who can work with others as well as solve new challenges. Nations without large numbers of logical, reflective individuals who can learn and master new challenges generally do not progress economically in the same way as nations with large numbers of individuals who have these capacities. Therefore, we need to consider initiatives in which economic opportunities are coupled with an infrastructure of nurturing families, healthy environments, safe communities, and cross-cultural, thinking-based, educational initiatives.

But the heart of a family and children's program must be those child-caregiver interactions, as described in the previous section, that lead to the types of personal characteristics that can support the higher level of reflection needed to meet the challenges of interdependency.

CONCLUSION

In an interdependent world, the unit of survival is no longer the individual or even the small group, but the global group. Therefore, self-interest and the large group interests are the same. But the dynamics of groups and individuals are such that groups and individuals operate at a number of levels ranging from fragmented patterns and polarized thinking ("us" versus "them") to cohesive reflective thinking where problem solving and a truly shared sense of humanity and reality is possible at the individual, group, and institutional level. Understanding how groups meet their developmental needs for safety and security, cohesion, communication, and the creation of stable, reflective institutional processes to solve problems and adapt to the future can enable us to move toward these more adaptive organizations. In an interdependent world that includes weapons of mass destruction, only shared reflective institutions can enable collective action to support safety and security. This will require an unprecedented investment in individuals, families, and communities around the globe; that is, it will require human capital. It will also require changes in the assumptions that guide economic, political, and military goals. For example, to embrace the global group and unite the world, individuals will need to broaden their identifications to include not only their own families, cultures, or religions but also "others" across the globe who subscribe to different beliefs and practices.

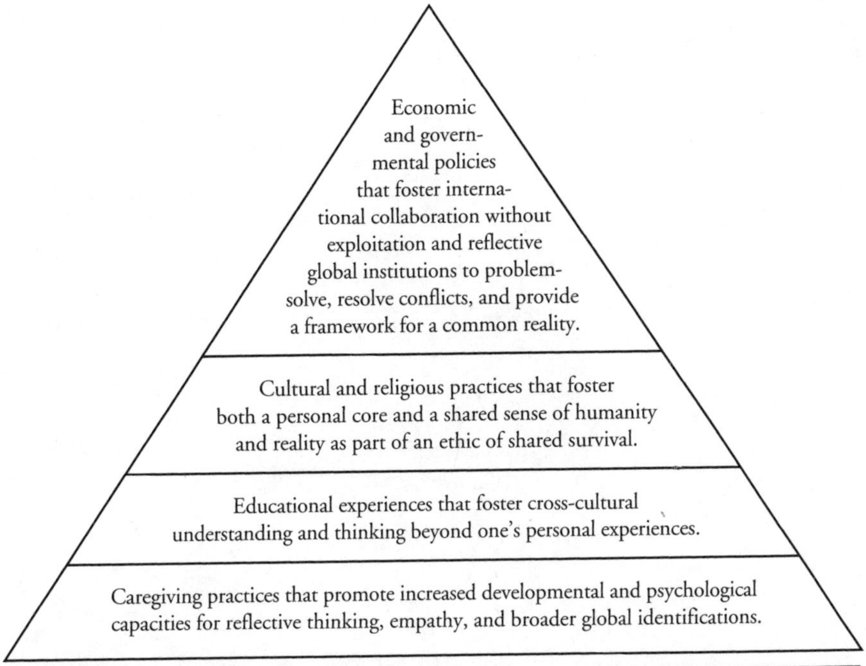

Economic and governmental policies that foster international collaboration without exploitation and reflective global institutions to problem-solve, resolve conflicts, and provide a framework for a common reality.

Cultural and religious practices that foster both a personal core and a shared sense of humanity and reality as part of an ethic of shared survival.

Educational experiences that foster cross-cultural understanding and thinking beyond one's personal experiences.

Caregiving practices that promote increased developmental and psychological capacities for reflective thinking, empathy, and broader global identifications.

FIGURE 15.1 Foundations for Adaptive Global Interdependency

A new morality of interdependency will need to be defined by how well the global community deals with its weakest or neediest members. Therefore, the marginal value of the world's weakest members will define the new morality. However, helping others with the basics of survival cannot be based on the notion of "handouts," or even of enlightened altruism. The morality of interdependency must be based on a fundamental commitment and responsibility to work for shared survival.

Interdependency can breed either destructive or constructive patterns. Destructive patterns are likely to the degree that the psychological processes that come into play in an interdependent world are ignored, and rage, impulsivity, and polarized thinking are fomented. Constructive patterns are likely to the degree that a shared sense of humanity and reality is constructed through the formation of global identifications, reflective institutions, and a true morality of interdependency.

The support and maintenance of reflective individuals, processes, and institutions that contribute to a shared sense of reality and humanity depend on a series of critical developmental processes that characterize individuals and groups, as outlined earlier. These developmental processes require significant investments in present and future "human capital," and educational, political, and economic initiatives.

Notes

INTRODUCTION

1. Boyd and Richerson 1985.
2. Based on James Mark Baldwin's "social hereditary" argument; see Baldwin 1896; Weber and Depew 2003.
3. See, for example, Weingart et al. 1997; Maynard, Smith, and Szathmáry 1999; Ridley 2003.
4. What she refers to as "acquired/innate" dualism (see Oyama 2003).
5. Gottlieb 1997.
6. Gottlieb, Wahlsten, and Lickliter 1998.
7. Gottlieb 1963.
8. Kandel 1989.
9. See Spitz 1965; Bowlby 1969; Greenspan et al. 1987; Harlow, Gluck, and Suomi 1972.
10. It's Nathan.

CHAPTER ONE

1. Chomsky 1966, 1980.
2. Pinker 2002.
3. LeDoux 1996, 2001.
4. Allen et al. 1974.
5. Greenspan 1979, and Chapter 10 of this work.
6. Volosinov 1973.
7. Vygotsky 1934.
8. Spitz 1945; Bowlby 1951; Hunt 1941; Harlow 1953.
9. Vandell and Wolfe 2000; Peisner-Feinberg et al. 1999; NICHD Early Child Care Research Network 1998, 1999.
10. NICHD Early Child Care Research Network 2000.
11. Greenspan et al. 1987.
12. Sameroff et al. 1986.
13. Sameroff et al. 1993.
14. Jaffe et al. 2001.

15. Greenspan et al. 2003.

16. Shonkoff and Phillips 2000.

17. Ramey and Ramey 1998; Provence and Naylor 1983; Lally et al. 1988; Schweinhart, Barnes, and Weikart 1993.

18. LeDoux 1996, 2001; Damasio 1994.

19. Goleman 1995; Gardner 1983; Greenspan, Jacokes, and Cassily 2003; Shonkoff et al. 2000; Damasio 1994.

20. Greenspan and Wieder 1998, 1999; Interdisciplinary Council on Developmental and Learning Disorders Clinical Practice Guidelines Workgroup 2000.

21. Greenspan 1979, 1981, 1989, 1997a; Interdisciplinary Council on Developmental and Learning Disorders Clinical Practice Guidelines Workgroup 2000; Greenspan and Lourie 1981; Greenspan et al. 1987; Greenspan and Wieder 1998, 1999; Greenspan, DeGangi, and Wieder 2001; Greenspan et al. 2003.

22. Greenspan 1997b.

23. Vygotsky 1934, 1978; Piaget 1955.

24. Shanker 2002.

25. Bateson 1975; Tevarthen 1979.

26. Fogel 1993.

CHAPTER TWO

1. Ross 1974.

2. McLaughlin 1974.

3. Dennis 1960.

4. So, for example, in the paintings of the great French baroque painter, Nicholas Poussin, the figures are visibly involved with and responding to each other in the context of the event in which they are mutually engaged.

5. Tomkins 1963b, 1963a; Izard 1979; Ekma et al. 1980; LeDoux 1996; Schacter 1997.

6. Damasio 1994.

7. Descartes 1637.

8. Izard 1993.

9. Ekman et al. 1972; Ekman 1992; Izard 1977.

10. Boas 1982; Lutz 1998.

11. Izard 1997.

12. See Erikson 1940; Freud 1965; Spitz 1945; Bowlby, 1951.

13. See, for example, Ainsworth, Bell, and Stayton 1974; Bruner 1983; Brazelton, Koslowski, and Main 1974; Hofer 1988; Sander 1962; Sroufe, Waters, and Matas 1974; Stern 1974.

14. Klinnert et al. 1983.

15. This reaction has been observed in a cross-cultural study of American, Canadian, and Chinese babies (Tronick, Ricks, and Cohn 1982).

16. See Witherington, Campos, and Hertenstein 2001.

17. Messinger, Fogel, and Dickson 1997, 1999; Dickson, Fogel, and Messinger 1998.

18. Fogel et al. 2000.

19. Greenspan 1979, 1989, 1992, 1997b, 2001, 1997a; Interdisciplinary Council on Developmental and Learning Disorders Clinical Practice Guidelines Workgroup 2000;

Greenspan and Wieder 1998; Greenspan and Lewis 2002; Greenspan and Glovinsky 2002.

20. Greenspan 1997b.

21. Erikson 1963; Freud 1965; Spitz 1945; Bowlby 1951; Hunt 1941; Murphy 1974; Escalona 1968; Provence and Lipton 1962; Fraiberg 1979; Sander 1962; Brazelton and Cramer 1990; Ainsworth et al. 1974; Thomas, Chess, and Birch 1968; Yarrow 1975; Sroufe 1979; Emde, Gaensbauer, and Harmon 1976; Stern 1974.

22. The field studies referred to in the text have been conducted as part of studies that are contributing to the revisions of the Bayley Developmental Scales for Infants and Young Children. A functional emotional developmental questionnaire that characterizes the early capacities has been studied as part of the Bayley Scales package and will be available, both as an independent assessment of emotional functioning and as part of the new Bayley kit. The results of field studies, including studies that demonstrate the validity of the age prediction and that show that the functional emotional developmental capacities can discriminate between healthy children and children with developmental and emotional challenges, are available on request from the author and will soon be published.

23. Klaus and Kennell 1976.

24. Klaus and Kennell 2000.

25. Greenspan 1979, 1997b, 1989.

26. Ayres 1964.

27. Greenspan and Press 1985.

28. Greenspan 2001.

29. Greenspan and Wieder 1997, 1998, 1999; Greenspan 1992.

30. Erikson 1959, 1963, 1964, 1968; Michels 1993; Greenspan and Polk 1980.

31. Jaques 1990.

32. Erikson 1963.

PART TWO INTRODUCTION

1. Pinto-Correia 1997.
2. Gould and Lewontin 1979.
3. Dawkins 1986.
4. Barkow, Cosmides, and Tooby 1992.
5. Ibid.
6. Shanker 1998.
7. Dennett 1978.

CHAPTER THREE

1. We are indebted to Barbara King for this point.
2. As Sue Carter showed in her research on voles; see Carter 1992.
3. Small 1998; Dissanayake 2000.
4. Plooij 1984; Goodall 1986.
5. Goodall 1986; Kano 1992.
6. Savage-Rumbaugh et al. 1996.
7. Denison 1995.
8. Falk 2003.

9. Culminating in a book that we wrote, with Talbot Taylor, titled *Apes, Language and the Human Mind* (Savage-Rumbaugh, Shanker, and Taylor 1998).
10. King and Shanker 2003.
11. I have also been deeply influenced by the work of my colleague, Joanna Blake; see Blake 2000.
12. Wittgenstein 1953.
13. See "Mother and Child," in Savage-Rumbaugh et al. 1998.
14. King 2002.
15. King 1999, 2002; King and Shanker 2003.
16. Goodall 1976.
17. Lorenz 1952; Hinde 1974.
18. Bowlby 1951.
19. Ainsworth conducted extensive home studies of such factors as how promptly the mother responded to her infant's signals, how smooth feeding was, how much synchrony the dyad exhibited in play interactions, and how animated were the caregiver's facial expressions of affect (see Bretherton 2002).
20. See, for example, Polan and Hofer 1999.
21. Savage-Rumbaugh and Lewin 1992; King (in press); De Waal 2001.
22. Falk 2000.
23. Berman, Rasmussen, and Suomi 1997.
24. Blum 2002.
25. Ibid.
26. Gibson 1986, 2001; Falk 2000.
27. Russon, Bard, and Parker 1996.
28. Hofer 1988.
29. Savage-Rumbaugh and Lewin 1994.
30. Toth 1987.
31. Wrangham 2000.
32. King and Shanker 2003.
33. Savage-Rumbaugh et al. 1996.
34. Ibid.
35. Waal 1996, 1998; Waal and Lanting 1997; Waal and Tyack 2003; Kano 1992.
36. McGrew 1992; Wrangham 1996; Noble and Davidson 1996; Falk 2003.
37. Dunbar, Knight, and Power 1999; Dunbar and Barrett 2001.
38. Falk 2000.

CHAPTER FOUR

1. Shanker 1998.
2. Pinker 2002.
3. Russell 1919.
4. Hayes and Hayes 1952.
5. Fouts 1997.
6. Rumbaugh 1977.
7. Stanford 1999.
8. Sabater 1992.
9. McConnell and Moses 1995.
10. Boesch and Boesch-Achermann 2000.

11. Walker and Leaky 1993.

12. In the display at the American Museum of Natural History in New York, this pair of individuals is depicted as a prehistoric married couple out for an evening *passagiata,* the male's arm casually draped around the shoulder of his mate.

13. Toth and Schik 1993.

14. Mercader, Panger, and Boesch 2002.

15. Boesch et al. 2000.

16. Roche et al. 1999.

17. See, for example, Klein 1989, Walker et al. 1993, and Tattersall 1995.

18. Leakey 1979, 1994.

19. Greenspan 1997b.

20. Tronick 1982.

21. Bialystok 1999.

22. Diamond 2002.

23. Falk 2003.

CHAPTER FIVE

1. Waal and Lanting 1997.
2. Boesch and Boesch-Achermann 2000.
3. Ibid., 235–236.
4. Ibid., 15.
5. Ibid., 249.
6. Goodall 1986.
7. King 2003.
8. Templeton 2002.
9. Gottlieb 1997; Small 1998.
10. Trinkhaus 1993.
11. Balter 1996; Berkowitz 1996.
12. Klein 1989, 512.
13. Barham 2000.
14. Bickerton 1995.
15. Russon, Bard, and Parker 1996; Hauser 1996.
16. Chomsky 1998.
17. Pinker 1994.
18. Chomsky 1966.
19. Ibid.; Savage-Rumbaugh, Shanker, and Taylor 1998.
20. Chomsky 1975.
21. Shanker 2002.
22. Noble and Davidson 1996.
23. Lieberman 1984; Laitman 1984.
24. Lieberman 1991.
25. Lieberman 1984; Laitman 1984.
26. Arensburg et al. 1989.
27. Ibid.
28. Savage-Rumbaugh, Fields, and Taglialatela 2001. Interestingly, it appears that those with a strong musical or linguistic background are better able to understand what Kanzi is saying than those with no such background.

29. Armstrong, Stokoe, and Wilcox 1995; Corballis 2002.
30. Segerdahl, Fields, and Savage-Rumbaugh, unpublished manuscript.
31. Reé 1999; Shanker 2000.
32. Taylor 1997.
33. King 2002; Falk 2003.

CHAPTER SIX

1. Bahn 1998; Hadingham 1979; Leroi-Gourhan 1982; Golomb 2002.
2. Bahn 1998.
3. Bednarik 1998.
4. Bednarik 2001.
5. Marschak 1995.
6. Bahn 1998.
7. Mithen 1996.
8. Dickson 1990; Henshilwood 1999.
9. Henshilwood 1999.
10. Henshilwood et al. 2002.
11. Bahn 1998.
12. Golomb 1992.
13. Ibid.
14. Berk 2003; Martlew 1996.
15. Bahn 1998.
16. Noble and Davidson 1996.
17. There is considerable controversy over the dating of the so-called Paleolithic lighting system. A fuel lamp was found at La Chaire that dates between 20,000 and 15,000 years ago (Bahn 1998).
18. Burenhult 1995.
19. Note that a similar phenomenon has been found in Balzi Rossi, Italy, where two children, seven and thirteen years old, were buried with elaborate adornments.
20. For example, fish traps were used to catch salmon in western Europe (Wenke 1999).
21. Thorndike 1911.
22. Köhler 1951.
23. For anyone who has worked on attachment, the parallels here to Tronick's "still-face" experiments are remarkably striking; see Tronick 1980.

CHAPTER SEVEN

1. Darwin 1859, 345.
2. McShea 1996.
3. Ayala and Dobzhansky 1974.
4. McShea 1996.

CHAPTER EIGHT

1. Bickerton 1995.
2. Chomsky 1998.
3. Ibid.

4. Falk 2003.
5. Harlow 1971; Suomi 1999; Blum 2002.
6. Savage-Rumbaugh and Lewin 1994; King 2002.
7. Spitz 1965; Bowlby 1969; Karen 1998.
8. Harris 1980.
9. Wittgenstein 1953.
10. Pinker 1994.
11. Armstrong, Stokoe, and Wilcox 1995; Bickerton 1995; Deacon 1997; Donald 1991; King 1999; Lieberman 1984; Noble and Davidson 1996; Pinker 1994.
12. Chomsky 1998; Bickerton 1995; Piatelli-Palmerini 1979.
13. Armstrong 1998; King 1999.
14. Chomsky argued that the ability to understand and produce infinitely many sentences that one has never heard before can only be explained if the "deep" structure of language is a *generative* system, in the mathematical sense that, like the axioms in an axiomatic system, a limited number of basic "principles" generate infinitely many "values" (i.e., sentences). See Chomsky 1966.
15. Lenneberg 1967.
16. Barkow, Cosmides, and Tooby 1992.
17. Shanker 1998.
18. de Groot 1965.
19. Young 1978.
20. Such a mechanist paradigm was clearly needed to explain how we acquire our higher capacities. The problem was, however, that AI scientists were notoriously silent on developmental questions. But Chomsky's vision of a "language gene" offered a model for how a child acquires his higher capacities *"automatically, unconsciously, and without effort"*; that is, like a computer processing language. And the reasoning behind this thesis was, first, that just as a computer program cannot figure out language patterns from scratch but needs certain fundamental processing "heuristics," so, too, a child must have what Chomsky referred to as the "built-in structure of an information-processing (hypothesis-forming) system" that enables him "to arrive at the grammar of a language from the available data in the available time" (Chomsky 1959); second, that children actually know things about language that they could not possibly have learned; third, that all known languages have the same abstract structure, and that all children go through the same maturational sequence when acquiring language; and finally, that children acquire language without explicit instruction from their caregivers, and, indeed, despite the flagrant language mistakes that they are exposed to from birth. Hence, the capacity to speak must be innate in the sense that the child is born with a genetic program—what Chomsky called Universal Grammar—that controls the process of language acquisition.
21. Hockett 1960.
22. E.g., semanticity; arbitrariness (noniconic representations); displacement (communicate about events displaced in space or time); duality of patterning (different rules for combining sound units and meaning units); discreteness (speech is composed of a small set of acoustically distinct units).
23. King and Shanker 2003.
24. Chomsky 1967, 11–12.
25. Bruner 1983.
26. Bateson 1975.
27. Trevarthan 1979.

28. Shanker and Taylor 2001.
29. Savage-Rumbaugh 1986.
30. Greenspan 1992, 1997.
31. King and Shanker 2003.
32. Fenson et al. 1994.
33. Johnson 1997.
34. Armstrong et al. 1995.
35. Corballis 2002.
36. Condillac 2001.
37. Taylor 1992.
38. Quoted in Harris and Taylor 1989, 120.
39. Reé 1999; Shanker 2000.
40. Juliard 1970, 42f.
41. Aitcheson 2003; Lieberman 1998.
42. Petitto 1997.
43. Gallaway and Richards 1994.
44. Groce 1985.
45. Savage-Rumbaugh, Shanker, and Taylor 1998; Shanker et al. 2001.
46. Greenspan and Porges 1984.

CHAPTER NINE

1. Chomsky 1972.
2. Tomasello 2003.
3. Shanker and Taylor 2001.
4. Brown 1973.
5. Tomasello 2003.
6. Shanker 2002.
7. Wellman 1990.
8. Savage-Rumbaugh, Shanker, and Taylor 1998.
9. Pinker 1994b.
10. Chomsky 1980, 134.
11. Pinker 1994b, 18.
12. Pinker 1994b.
13. Shanker 2002.
14. Baker and Hacker 1980; Taylor 1997; Wittgenstein 1953.
15. Owens 2001.
16. Ibid.
17. Kuhl and Meltzoff 1997, 7.
18. Petitto 1997, 51.
19. Fenson et al. 1994.
20. So, for example, Temple Grandin reports that "during the last couple of years, I have become more aware of a kind of electricity that goes on between people. I have observed that when several people are together and having a good time, their speech and laughter follow a rhythm. They will all laugh together and then talk quietly until the next laughing cycle. I have always had a hard time fitting in with this rhythm, and I usually interrupt conversations without realizing my mistake. The problem is that I can't follow the rhythm" (Grandin 1960).

21. Fiezet al. 1996.
22. Wallace et al. 2001.
23. Fraisse 1963; Whitrow 1988.
24. DeCasper and Carstens 1980; DeCasper and Fifer 1980.
25. Ramus et al. 2000.
26. Greenspan 2001, 1997.
27. Jaffe et al. 2001.
28. Shaffer et al. 2001.
29. Drake, Jones, and Baruch 2000; Greenspan et al. 2003.
30. Shanker 1998, 2002.
31. Shanker 1998.
32. Chomsky 1957, 54.
33. Chomsky 1957.
34. Ibid., 58.
35. Shannon 1949.
36. Pinker 1994b.
37. Gesell 1933.
38. Chomsky 1966, 65.
39. Ibid.
40. Bruner 1983.
41. Saffran, Aslin, and Newport 1996.
42. Bates and Elman 1996.

CHAPTER TEN

1. Greenspan and Wieder, 1998.
2. Gardner 1983; Sternberg and Berg 1992.
3. Studies and overviews of the importance of early emotional interactions for development include the following: Shonkoff and Phillips 2000; Sameroff et al. 1986; Sameroff et al. 1993; NICHD Early Child Care Research Network 1998, 1999, 2000a, 2000b, 2003; Greenspan et al. (unpublished work); Vandell and Wolfe 2000; Peisner-Feinberg et al. 1999.

In addition, it is important to note that, historically, emotions or affects have been viewed in various ways: as outlets for extreme passion, as physiologic reactions, as subjective states of feeling, as interpersonal social cues (Campos, Campos, and Barrett 1989; Greenspan 1979, 1997; Young 1943). Beginning nearly six decades ago, the importance of emotions for aspects of learning was documented by psychoanalytic observers such as René Spitz and John Bowlby, who described the effects of emotional deprivation, and Heinz Hartmann and David Rappaport, who explored clinical and theoretical relationships (Bowlby 1951; Hartmann 1939; Rappaport 1960; Spitz 1945). Sibylle Escalona (1968) and Lois Murphy (1974) further explored affective development and described individual differences in infants and their relationship to psychopathology. Building on this work, the pediatrician T. Berry Brazelton systematized the observation of the infant's social and emotional repertoire (Brazelton and Cramer 1990).

Neuroscience research has further documented the importance of the environment and "experiences" for the development of the central nervous system and learning. During the formative years there is a sensitive interaction between genetic proclivities and environmental experience. Experience appears to adapt the infant's biology to his or her environment (Greenough and Black 1992; Hein and Diamond 1983; Hofer 1988, 1995; Holloway 1966; Rakic, Bourgeois, and Goldman-Rakic 1994; Schanberg and Field 1987; Singer 1986; Thinus-Blanc 1981; Turner and Greenough 1983, 1985; Weiler, Hawrylak, and Greenough 1995; Wiesel and Hubel 1963). In this process, however, not all experiences are the same. As described in the prior section, children seem to require certain types of experiences involving a series of specific types of emotional interactions geared to their particular developmental needs. For example, as we will shortly explore, in the early months of life, babies require soothing, regulating synchronous interactions. They can be observed to move their arms and legs in rhythm to their mothers' voices (Condon 1975; Condon and Sander 1974). These types of interactions enable them to begin integrating what they hear and see (Spelke 1976). By four to five months, one can readily observe synchronous movement in rhythm with the mother's affective communication via her voice, facial expressions, or body movements. As development proceeds, reciprocal gestural, vocal, and verbal communication generally occurs in an interactive rhythm (Greenspan, Jacokes, and Cassily 2003; NICHD Early Child Care Research Network 1998, 1999, 2000a, 2000b, 2003; Peisner-Feinberg et al. 1999; Sameroff et al. 1993; Sameroff et al. 1986; Vandell and Wolfe 2000).

4. Greenspan 1979, 1997.
5. Polanyi 1967.
6. Astington 1993.
7. Goleman 1995.
8. LeDoux 1996.

CHAPTER ELEVEN

1. Damasio 1994, 1999; Glenberg 1997; Hurley 1998; Jeannerod 1997.
2. LeDoux 2001.
3. Ibid.
4. Greenspan 1997a.
5. Winnicott 1957.
6. Greenspan 1997b.
7. Damasio 1998.
8. Ibid.
9. Greenspan 1997b, 1989.
10. Kandel 1989.
11. Greenough, Black, and Wallace 1987.
12. Ibid.
13. LeDoux 2001, 1996.
14. Greenspan 1997b, 1979.
15. Bachevalier 1998.
16. Davidson 1998.
17. Carlson 1998.
18. Gunnar and Vazquez 2001.
19. Bachevalier 1998.
20. Bloom and Capatides 1987.
21. Capatides and Bloom 1993.
22. Siller and Sigman 2002.
23. Greenspan and Wieder 1997, 1999.

24. For a more complete discussion on Piaget, see Greenspan 1979.
25. Greenspan 1979 and Chapter 10 on intelligence in this work.
26. Greenspan 1979.
27. Greenspan 1979, 1981, 1989, 1992, 1997a; Greenspan and Wieder 1998, 1999; Interdisciplinary Council on Developmental and Learning Disorders Clinical Practice Guidelines Workgroup, 2000, 1981, 1987, 1992, 1997a; Greenspan et al. 1998, 1999.
28. Greenspan 1979.
29. Piaget and Inhelder 1966.
30. See Greenspan 1979 for more discussion.
31. Greenspan 1997b.
32. Gazzaniga et al. 1998.
33. Greenspan 1979, 1989.
34. Piaget 1962; Emde, Gaensbauer, and Harmon 1976; Campos, Campos, and Barrett 1989; Stern 1974; Greenspan 1979, 1989, 1997b.

CHAPTER TWELVE

1. Greenspan et al. 1987; Greenspan 1992; Greenspan and Wieder 1998, 1999.
2. Greenspan and Wieder 1997.
3. Greenspan et al. 1997, 1999.
4. Bertrand et al. 2001.
5. Greenspan and Bauman (unpublished work); Greenspan (work in progress).
6. Baron-Cohen 1994.
7. Minshew and Goldstei 1998.
8. Mundy, Sigman, and Kasari 1990.
9. Baranek 1999; Dawson and Galpert 1990; Osterling and Dawson 1994; Tanguay 1999; Tanguay, Robertson, and Derrick 1998.
10. Wetherby and Prizant 1993.
11. Greenspan 2001; Sperry 1985; Baron-Cohen 1989; Baron-Cohen, Leslie, and Frith 1985; Bowler 1992; Dahlgren and Trillingsgaard 1996; Dawson et al. 1998; Frith 1989; Klin, Volkmar, and Sparrow 1992; Ozonoff 1997; Pennington and Ozonoff 1996.
12. Greenspan 1979, 1989, 1997b.
13. See Chapter 2 and Greenspan 1997b.
14. Greenspan et al. 1997.
15. Bell 1970; Emde et al. 1991; Greenspan 1979, 1997b; Kagan 1981; Piaget 1981; Werner and Kaplan 1963; Winnicott 1931.
16. Condon 1975; Condon and Sander 1974.
17. Greenspan 1997b, and Chapter 2.
18. See Chapter 2.
19. Greenspan and Wieder 1997.
20. Zimmerman and Gordon 2000.
21. Ibid.
22. Minshew et al. 1998.
23. Benson and Zaidel 1985; Courchesne et al. 1994; Dawson, Warrenburg, and Fuller 1982; Greenspan and Wieder 1997; Greenspan 1997b; Sperry 1985.
24. Bauman 2000.
25. Benson et al. 1985; Courchesne et al. 1994; Dawson et al. 1982; Greenspan 1997b; Sperry 1985; Wetherby et al. 1981.

26. Baron-Cohen, Frith, and Leslie 1988; Baron-Cohen, Tager-Flusberg, and Cohen 1993; Frith 1993.
27. Klin et al. 1995; Schultz et al. 2000.
28. Source: Morton Gernsbacher, Ph.D., University of Wisconsin-Madison (magernsb@facstaff.wisc.edu).
29. Greenspan 1992; Greenspan, DeGangi, and Wieder 2001.
30. Greenspan 1992; Greenspan et al. 1999; Interdisciplinary Council on Developmental and Learning Disorders Clinical Practice Guidelines Workgroup 2000.
31. Greenspan 1989, 1992.
32. Greenspan et al. 1999.
33. See Committee on Educational Interventions for Children with Autism 2001. For relationship-based approaches, see the Web site of the Interdisciplinary Council for Developmental and Learning Disorders at http://www.icdl.com.
34. Hunt 1941; Spitz 1945.
35. Greenspan et al. 1987, 1992.
36. Greenspan 1992; Greenspan et al. 1998.
37. Greenspan et al. 1987.

CHAPTER THIRTEEN

1. See Tomasello 1999; Wellman 1990.
2. Nelson 1973.
3. Bowerman and Levinson 2001; Toren 1998.
4. Brown and Levinson 2000. We are indebted to Talbot Taylor (personal communication) for this way of illustrating Brown and Levinson's argument.
5. Hall 1994.
6. Ibid., 89–90.
7. Ibid., 17.
8. Ibid., 25.
9. Ibid., 86–87.
10. Ibid., 26.
11. Shanker 2000.
12. Hall 1994, 26.
13. We are deeply indebted to Dr. Willow Power in this discussion of Navajo enculturation.
14. Turquet 1975.
15. Cheney and Seyfarth 1990.
16. Plomin 1997.
17. Gottlieb 1997.
18. Pinker 2002.
19. White 1799, 134.
20. Quine 1960.
21. Lukes, Gellner, and Gellner 1998.
22. Shanker 1986.
23. Leach 1970.
24. One of Levi-Strauss's famous examples of such a universal structural pattern is his explanation of family systems. The essential drive behind marriage is said to be to establish interdependence between biological units. These rules express a basic human refusal

to acknowledge the family as an exclusive or isolated reality. All systems, no matter how complicated they might be (e.g., by distinctions of terminology, prohibitions, prescriptions, or preferences, are no more than a process for dividing families into rival or allied camps who can and must take part in marriage. Regardless of the society studied, the family construct is grounded in several invariable properties: The family originates in marriage. It includes a husband, wife, and children born from their union, forming a nucleus around which other relatives gather. The members of the family are united among themselves by legal bonds. Rights and obligations are of an economic, religious, or some other nature. Incestuous marriages are prohibited.

25. Chomsky 1967.
26. Brown 1991.
27. Morris 1967.
28. McNeill 1995.
29. Tronick and Adamson 1980.
30. Bowlby 1951, 1958, 1969; Greenspan et al. 1987, 1997; Harlow, Gluck, and Suomi 1972; Harlow 1958; Hunt 1941; Spitz 1945.
31. Greenspan 1992; Greenspan and Wieder 1998.
32. Levi-Strauss 1967.
33. Freud 1985.
34. Malinowski 1929.
35. Boesch and Boesch-Achermann 2000.
36. Tomasello 2003.
37. Winch 1958.
38. Deacon 1997.
39. Ragir 2002.
40. Parsons 1999.
41. Lipuna, Postone, and Calhoun 1993.

CHAPTER FOURTEEN

1. We are grateful to Robert Thomas for this mathematical metaphor.
2. Pinker 2002.
3. Barkow, Cosmides, and Tooby 1992.
4. As Gould pointed out on numerous occasions.
5. Pinker 2002.
6. Tomasello 2003.
7. Baron-Cohen 1994.
8. Rasmussen 1908.
9. Briggs 1970.
10. Rasmussen 1946.
11. Tomasello 2003.
12. An interesting speculation based on this model relates to the relative roles of men and women in the transformation leading to more reflective ways of thinking. There has been a striking tendency in the history of science to treat the growth of science as primarily a male phenomenon. Indeed, there is even a widespread assumption that history reveals that, in general, women were not provided with the opportunity to develop fully their creative and logical faculties. But as we consider the types of emotional interactions that lead to critical thinking and problem-solving in scientific discoveries, which we

described in Chapter 2, it is clear that mothers or other female caregivers were often providing these formative learning interactions. To accomplish this task, the caregivers must themselves have mastered these critical stages and have embraced these varying capacities and attitudes. We might speculate that through their vital role as caregivers, they were the hidden engines behind the great cultural and scientific revolutions occurring at key junctures in history. In fact, now that men and women are beginning to share the more public roles as leading scientists, business people, and so forth, we may be facing a crisis, in some families at least, in which this role involving formative caregiving is being left to others who lack adequate training or the intimacy of a loving relationship with the child that is necessary to develop these critical thinking and problem-solving capacities.

13. Johnson and Wright 1975.
14. Gluckman 1964.
15. Taylor 1997, 67–77.
16. Assmann 2002, 67.
17. The only historical writing before Herodotus in any other culture is the narrative portions of the Hebrew Books of Samuel and Kings. The first Egyptian writer of history is Manetho (third century B.C.), on whose king lists, preserved in Diodorus, all our Pharaonic chronology depends. In the third century, there were an increasing number of historical writers in Egypt, including some on Jewish history. Berossus, who is contemporary with Manetho, wrote a history of Babylon, which is lost, but seems to have moved seamlessly from mythological times to historical chronicle. We are deeply indebted to Richard Stoneman for this information.
18. Snell 1960.
19. See, for example, Cottingham 1999; Gaukroger 1995; Watson 2002.
20. Watson 2002.
21. Descartes (1637), 1996.
22. Cameron 1987.
23. Savage-Rumbaugh, Shanker, and Taylor 1998.
24. See the opening chapter of Aristotle 1984.
25. Descartes 1641.
26. Wittgenstein 1953.
27. Crocker 1991.
28. Kohlberg 1984.

CHAPTER FIFTEEN

1. Greenspan 1997b.
2. See *Child Martyrs and Lovers of Their Homeland*, a film broadcast on Palestinian Television on June 27, 2002.
3. See "Suicide Contagion and the Reporting of Suicide: Recommendations from a National Workshop," CDC Reports, April 22, 1994, 43 (RR–6).

References

Ainsworth, M., S. M. Bell, and D. Stayton. 1974. Infant-mother attachment and social development: Socialization as a product of reciprocal responsiveness to signals. In M. Richards, ed., *The Integration of the child into a social world* (pp. 99–135). Cambridge: Cambridge University Press.

Aitcheson, J. 2003. *A glossary of language and mind.* Oxford: Oxford University Press.

Allen, M. G., S. Cohen, W. Pollin, and S. I. Greenspan. 1974. Affective illness in the NAS-NRC registry of 15,909 veteran twin pairs. *American Journal of Psychiatry* 131:1234–1239.

Arensburg, B., A. M. Tillier, B. Vandermeersch, H. Duday, L. A. Scheparts, and Y. Rak. 1989. A middle Palaeolithic human hyoid bone. *Nature* 338:758–760.

Aristotle. 1984. Metaphysics. In J. Barnes, ed., *The complete works of Aristotle: The revised Oxford translation* (1984). Princeton: Princeton University Press.

Armstrong, D. F. 1998. *Original Signs.* Washington, D.C.: Gallaudet University Press.

Armstrong, D. F., W. C. Stokoe, and S. Wilcox. 1995. *Gesture and the nature of language.* New York: Cambridge University Press.

Assmann, J. 2002. *The mind of Egypt: History and meaning in the time of the Pharaohs.* Translated and edited by A. Jenkins. New York: Metropolitan Books.

Astington, J. W. 1993. *The child's discovery of the mind.* Cambridge, Mass.: Harvard University Press.

Ayala, F. J., and T. Dobzhansky. 1974. *Studies in the philosophy of biology: Reduction and related problems.* Berkeley: University of California Press.

Ayres, J. 1964. Tactile functions: Their relation to hyperactive and perceptual motor behavior. *The American Journal of Occupational Therapy* 18:6–11.

Bachevalier, J. H. 1998. How developing memory systems affect emotion and behavior. Presentation at the NIH, NIMH Conference: Discovering Ourselves: The Science of Emotion, Library of Congress, Washington, D.C.

Bahn, P. 1998. *Journey through the Ice Age.* London: Weidenfeld and Nicholson.

Baldwin, M. 1896. A new factor in evolution. *American Naturalist* 30: 441–451.

Baker, G. P., and P. M. S. Hacker. 1980. *Wittgenstein, understanding and meaning: An analytical commentary on the Philosophical Investigations.* Oxford: Blackwell.

Balter, M. 1996. Cave structure boosts Neanderthal image. *Science* 271:449.

Baranek, G. T. 1999. Autism during infancy: A retrospective video analysis of sensory-motor and social behaviors at 9–12 months of age. *Journal of Autism and Developmental Disorders* 29:213–224.

Barham, A. J. 2000. Prehistoric body paint. *Archaeology,* vol. 53.

Barkow, J., L. Cosmides, and J. Tooby. 1992. *The adapted mind: Evolutionary psychology and the generation of culture.* New York: Oxford University Press.

Baron-Cohen, S. 1988. Autistic children's understanding of seeing, knowing, and believing. *British Journal of Developmental Psychology* 4:315–324.

______. 1989. The theory of mind hypothesis of autism: A reply to Boucher. *British Journal of Disorders of Communication* 24:199–200.

______. 1994. *Mindblindness: An essay on autism and theories of mind.* Cambridge, Mass.: MIT Press.

Baron-Cohen, S., A. M. Leslie, and U. Frith. 1985. Does the autistic child have a "theory of mind"? *Cognition* 21:37–46.

Baron-Cohen, S., H. Tager-Flusberg, and D. Cohen. 1993. *Understanding other minds: Perspectives from autism.* London: Oxford University Press.

Bates, E., and R. Elman. 1996. Learning rediscovered. *Science* 274:1849–1850.

Bateson, M. C. 1975. Mother-infant exchanges: The epigensis of conversation interaction. *Annals of the New York Academy of Science* 263:101–113.

Bauman, M. 2000. Autism: Clinical features and neurobiological observations. In Stanley I. Greenspan, chairman and ed., *Clinical practice guidelines: Redefining the standards of care for infants, children, and families with special needs* (pp. 689–704). Interdisciplinary Council on Developmental and Learning Disorders Clinical Practice Guidelines Workgroup. Bethesda, Md.: Interdisciplinary Council on Developmental and Learning Disorders Clinical Practice Guidelines Workgroup.

Bednarik, R. G. 1998. The first stirring of creation. *Unesco Courier* 51:4.

______. 2001. Beads and pendants of the Pleistocene. *Anthropos* 96:545–555.

Bell, S. M. 1970. The development of the concept of the object as related to infant-mother attachment. *Child Development* 41:219–311.

Benson, F., and E. Zaidel. 1985. *The dual brain.* New York: Guilford.

Berk, L. 2003. *Child Development.* Canadian edition, adapted and edited by Elizabeth A. Levin. Toronto: Allyn and Bacon.

Berkowitz, M. 1996. Neanderthal news. *Archaeology,* vol. 22.

Berman, C. C. M., K. L. R. Rasmussen, and S. J. Suomi. 1997. Group size, maternal behavior and social networks: A natural experiment with free-ranging monkeys. *Animal Behaviour* 53: 421.

Bertrand, J., A. Mars, C. Boyle, F. Bove, M. Yeargin-Allsopp, and P. Decoufle. 2001. Prevalence of autism in a United States population: The Brick Township, New Jersey, investigation. *Pediatrics* 108: 1155–1161.

Bialystok, E. 1999. Cognitive complexity and attentional control in the bilingual mind. *Child Development* 70: 636–644.

Bickerton, D. 1995. *Language and human behavior.* Seattle: University of Washington Press.

Bloom, L., and J. B. Capatides. 1987. Sources of meaning in the acquisition of complex syntax: The sample case of causality. *Journal of Experimental Child Psychology* 43:112–128.

Blum, Deborah. 2002. *Love at Goon Park: Harry Harlow and the science of affection.* Cambridge, Mass.: Perseus Publishing.

Boas, F. 1982. *A Frans Boas Reader: The shaping of American anthropology, 1883–1911.* Chicago: University of Chicago Press.

Boesch, C., and H. Boesch-Achermann. 2000. *The chimpanzees of the Taï forest: Behavioral, ecology and evolution.* Oxford: Oxford University Press.

Bowerman, M., and S. Levinson. 2001. *Language acquisition and conceptual development.* Cambridge: Cambridge University Press.

Bowler, D. M. 1992. Theory of mind in Asperger's syndrome. *Journal of Child Psychology and Psychiatry* 33:893.

Bowlby, J. 1951. *Maternal care and mental health.* World Health Organization (WHO) Monograph Series, no. 51. Geneva: World Health Organization.

______. 1958. The nature of the child's tie to its mother. *International Journal of Psychoanalysis* 39:350–373.

______. 1969. *Attachment and loss.* Vol. 1. London: Hogarth Press.

Boyd, R., and P. Richerson. 1985. *Culture and the evolutionary process.* Chicago: University of Chicago Press.

Brazelton, T. B., B. Koslowski, and M. Main. 1974. The origins of reciprocity; the early mother-infant interaction. In M. Lewis and L. Rosenblum, eds., *The Effect of the infant on its caregiver.* New York: John Wiley & Sons, Inc.

Brazelton, T. B., and B. Cramer. 1990. *The earliest relationship: Parents, infants, and the drama of early attachment.* Reading, Mass.: Addison-Wesley Publishing Co.

Briggs, J. 1970. *Never in anger: Portrait of an Eskimo family.* Cambridge, Mass.: Harvard University Press.

Brown, D. E. 1991. *Human universals.* New York: McGraw Hill.

Brown, P., and S. Levinson. 2000. Frames of spatial reference and the acquisition in Tenejapan Tzeltal. In G. Nucci, G. Saxe, and E. Turiel, eds., *Culture, thought and development* (pp. 167–197). Mahwah, N.J.: Lawrence Erlbaum.

Brown, R. 1973. *A first language: The early stages.* Cambridge, Mass.: Harvard University Press.

Bruner, J. S. 1983. *Child's talk: Learning to use language.* New York: W. W. Norton & Co.

Burenhult, G. 1995. *The first human: Origins and history to 10,000.* Vol. 1. New York: Harper Collins.

Cameron, J. M. 1987. The theory and practice of autobiography. In B. Davies, ed., *Language, Meaning and God.* London: Geoffrey Chapman.

Campos, J., R. Campos, and K. Barrett. 1989. Emergent themes in the study of emotional development and emotion regulation. *Developmental Psychology* 25:394–402.

Capatides, J. B., and L. Bloom. 1993. Underlying process in the socialization of emotion. In L. Lipsitt and C. Rovee-Collier, eds., *Advances in infancy research.* Norwood, N.J.: Ablex.

Carlson, M. 1998. Developing self and emotion in extreme social deprivation. Presentation at the NIH, NIMH Conference: Discovering Ourselves: The Science of Emotion. Library of Congress, Washington, D.C.

Carter, C. S. 1992. Oxytocin and sexual behavior. *Neuroscience Biobehaviour Review* 16:131–144.

Cheney, D., and R. Seyfarth. 1990. *How monkeys see the world: Inside the mind of another species.* Chicago: University of Chicago Press.

Chomsky, N. 1957. *Syntactic structures.* The Hague: Mounton.

______. 1959. Review of *Verbal behavior* by B. F. Skinner. *Language* 35:26–58.
______. 1966. *Cartesian linguistics.* New York: Harper & Row.
______. 1967. Recent contributions to the theory of innate ideas. *Synthese: An International Journal for Epistemology* 17:2–11.
______. 1972. *Language and mind.* New York: Harcourt Brace Jovanovich.
______. 1975. *The logical structure of linguistic theory.* New York: Plenum Press.
______. 1980. *Rules and representations.* New York: Columbia University Press.
______. 1998. *On language.* New York: The New Free Press.
Condillac, E. 2001. *Essay on the origin of human knowledge.* Translated and edited by Hans Aarsleff. New York: Cambridge University Press.
Condon, W. S. 1975. Multiple response to sound in dysfunctional children. *Journal of Autism and Developmental Disorders* 5:43.
Condon, W. S., and L. Sander. 1974. Neonate movement as synchronized with adult speech: Interactional participation and language acquisition. *Science* 183:99–101.
Corballis, M. C. 2002. *From hand to mouth: The origins of language.* Princeton: Princeton University Press.
Cottingham, J. 1999. *Descartes.* New York: Routledge.
Courchesne, E., N. Akshoomoff, B. Egaas, A. J. Lincoln, O. Saitoh, L. Schreibman, J. Townsend, and R. Yeung-Courchesne. 1994. *Role of cerebellar and parietal dysfunction in the social and cognitive deficits in patients with infantile autism.* Paper presented at the Autism Society of America Annual Conference, Las Vegas, Nevada.
Crocker, M. 1991. Multiple meta-interpreters in a logical model of sentence processing. In C. G. Brown and G. Koch, eds., *Natural language: Understanding and logic* (pp. 127–145). Amsterdam: Elsevier.
Dahlgren, S. O., and A. Trillingsgaard. 1996. Theory of mind in non-retarded children with autism and Asperger's syndrome: A research note. *Journal of Child Psychology and Psychiatry* 37: 763.
Damasio, A. R. 1994. *DeCartes' error: Emotion, reason, and the human brain.* New York: Putnam.
______. 1998. The science of emotion. Keynote address at the NIH, NIMH Conference: Discovering Ourselves: The Science of Emotion. Library of Congress, Washington, D.C.
______. 1999. *The feeling of what happens.* New York: Harcourt Brace.
Darwin, C. 1859. *On the origin of species.* London: John Murray.
Davidson, R. J. 1998. Understanding positive and negative emotion. Presentation at the NIH, NIMH Conference: Discovering Ourselves: The Science of Emotion. Library of Congress, Washington, D.C.
Dawkins, R. 1986. *The blind watchmaker.* New York: W. W. Norton & Co.
Dawson, G., S. Warrenburg, and P. Fuller. 1982. Cerebral lateralization in individuals diagnosed as autistic in early childhood. *Brain Language* 15:353–368.
Dawson, G., and I. Galpert. 1990. Mother's use of imitative play for facilitating social responsiveness and toy play in young autistic children. *Development and Psychopathology* 2:151–162.
Dawson, G., A. Meltzoff, J. Osterling, and J. Rinaldi. 1998. Neuropsychological correlates of early symptoms of autism. *Child Development* 69:1276–1285.
de Groot, J. 1965. *The development of the mind: Psychoanalytic papers on clinical and theoretical problems.* New York: International Universities Press.

Deacon, T. W. 1997. *The symbolic species: The co-evolution of language and the brain.* New York: W. W. Norton & Co.
DeCasper, A. J., and A. Carstens. 1980. Contingencies of stimulation: Effects on learning and emotion in neonates. *Infant Behaviour and Development* 4:19–35.
DeCasper, A. J., and W. P. Fifer. 1980. Of human bonding: Newborns prefer their mothers' voices. *Science* 208:1174–1176.
Denison, S. 1995. From modern apes to human origins. *British Archaeology,* vol. 8.
Dennett, D. 1978. *Brainstorms.* Montgomery, Vt.: Bradford Books.
Descartes, R. 1637. *Discours de la methode: pour bien conduire sa raison & character veinte dans la sciences.* Leyde: Ian Maire.
______. 1641; 1996. *Meditations on discourse on the method*; and *Meditations on first philosophy.* Edited by D. Weissman. New Haven: Yale University Press.
Diamond, A. 2002. Normal development of prefrontal cortex from birth to young adulthood: Cognitive functions, anatomy, and biochemistry. In D. T. Stuss and R. T. Knight, eds., *Principles of frontal lobe function* (pp. 466–503). London: Oxford University Press.
Dickson, D. B. 1990. *The dawn of belief: Religion in the upper Paleolithic of southwestern Europe.* Tucson: University of Arizona Press.
Dickson, K. L., A. Fogel, and D. Messinger. 1998. The development of emotion from a social process view. In F. Mascolo and S. Griffen, eds., *What develops in emotional development?* (pp. 253–271). New York: Plenum Press.
Dissanayake, E. 2000. Antecedents of the temporal arts in early mother-infant interaction. In B. Wallin, B. Merker, and S. Brown, eds., *The origins of music.* Cambridge, Mass.: MIT Press.
Donald, M. 1991. *Origins of the modern mind.* Boston: Harvard University Press.
Drake, C., M. Jones, C. and Baruch. 2000. The development of rhythmic attending in auditory sequences: Attunement, referent period, focal attending. *Cognition* 77:251–288.
Dunbar, R., C. Knight, and C. Power. 1999. *The Evolution of culture: An interdisciplinary view.* Edinburgh: Edinburgh University Press.
Dunbar, R., and L. Barrett. 2001. *Cousins: Our primate relatives.* London: DK Publishing, Inc.
Ekman, P. 1992. Facial expression of emotion: New findings, new questions. *Psychological Science* 3:34–38.
Ekman, P., W. Friesen, and P. Ellsworth. 1972. *Emotions in the human face.* Elmsford, N.Y.: Pergamon.
Ekman, P., W. Friesen, M. O'sullivan, and K. Scherer. 1980. Relative importance of face, body, and speech in judgments of personality and affect. *Journal of Personality and Social Psychology* 38:270–277.
Emde, R. N., T. J. Gaensbauer, and R. J. Harmon. 1976. *Emotional expression in infancy: A biobehavioral study.* Psychological Issues Monograph Series, no. 37. New York: International Universities Press.
Emde, R. N., Z. Biringen, R. B. Clyman, and D. Openheim. 1991. The moral self of infancy: Affective core and procedural knowledge. *Developmental Review* 11:251–270.
Erikson, E. H. 1959. *Identity and the life cycle: Selected papers.* With a historical introduction by David Rapaport. New York: International Universities Press.
______. 1963. *Childhood and society.* Rev. ed. New York: W. W. Norton & Co.
______. 1964. *Insight and responsibility.* New York: W. W. Norton & Co.

______. 1968. *Identity: Youth and crisis.* New York: W. W. Norton & Co.
Escalona, S. 1968. *The roots of individuality.* Chicago: Aldine.
Falk, D. 2000. *Primate diversity.* New York: W. W. Norton & Co.
______. In press. Prelinguistic evolution in early hominins: Whence motherese? *Brain and Behavioral Sciences.*
Fenson, L., P. Dale, S. Reznick, E. Bates, D. Thal, and S. Pethick. 1994. *Variability in early communicative development.* Chicago, Ill.: Society for Research in Child Development.
Fiez, J., M. Raichle, D. Balota, P. Tallal, and S. Petersen. 1996. PET activation of posterior temporal regions during passive auditory word presentation and verb generation. *Cerebral Cortex* 6:10.
Fogel, A. 1993. *Developing through relationships.* Chicago: University of Chicago Press.
Fogel, A., G. C. Nelson-Goens, H. C. Hsu, and A. F. Shapiro. 2000. Do different infant smiles reflect different positive emotions? *Social Development* 9:497–520.
Fouts, R. 1997. *Next of kin: What chimpanzees have taught me about who we are.* New York: William Morrow.
Fraiberg, S. 1979. Treatment modalities in an infant mental health program. Paper presented at the National center for Clinical Infant Programs, December 5–7, Washington, D.C.
Fraisse, P. 1963. *The psychology of time.* New York: Harper & Row, Inc.
Freud, A. 1965. *Normality and pathology in childhood: Assessments of development.* New York: International Universities Press.
Freud, S. 1911. *Formulations on the two principles of mental functioning.* Standard Edition. London: Hogarth Press.
______. 1985. *Civilization, society and religion: Group psychology; civilization and its discontents.* Harmondsworth: Penguin.
Frith, U. 1989. *Autism: Explaining the enigma.* London: Blackwell.
______. 1993. Autism. *Scientific American* 268 (June):108–114.
Gallaway, C., and B. Richards. 1994. *Input and interaction in language acquisition.* Cambridge: Cambridge University Press.
Gardner, H. 1983. *Frames of mind: The theory of multiple intelligences.* New York: Basic Books.
Gaukroger, S. 1995. *Descartes: An intellectual biography.* Oxford: Clarendon Press.
Gazzaniga, M., R. Ivry, and G. R. Mangun. 1998. *Fundamentals of cognitive neuroscience.* New York: W. W. Norton & Co.
Gesell, A. 1933. Maturation and patterning of behaviour. In C. Murchison, ed., *A handbook of child psychology.* Worcester, Mass.: Clark University Press.
Gibson, K. 1986. Cognition, brain size and the extraction of embedded food resources. In J. Else and P. Lee, eds., *Primate ontogeny, cognition and social behaviour.* Cambridge: Cambridge University Press.
Gibson, K. 2001. Bigger is better: Primate brain size in relationship to cognition. In D. Falk and K. Gibson, eds., *Evolutionary anatomy of the primate cerebral cortex.* Cambridge and New York: Cambridge University Press.
Glenberg, A. M. 1997. What memory is for. *Behavioural and Brain Sciences* 20:1–55.
Gluckman, M. 1964. *Closed systems and open minds.* Chicago: Aldine Publishing Co.
Goleman, D. 1995. *Emotional intelligence.* New York: Bantam.
Golomb, C. 1992. *The child's creation of a pictorial world.* Berkeley: University of California Press.

______. 2002. *Child art in context: A cultural and comparative perspective.* Washington, D.C.: American Psychological Association.

Goodall, J. 1976. Infant development. Video Recording. Washington, D.C., National Geographic Society.

______. 1986. *The chimpanzees of Gombe: Patterns of behavior.* Cambridge, Mass.: Belknap Press of Harvard University Press.

Gottlieb, G. 1963. A naturalistic study of imprinting in wood ducklings (Aix sponsa). *Journal of Comparative Physiology and Psychology* 56:86–91.

______. 1997. *Synthesizing nature-nurture: Prenatal roots of instinctive behavior.* Mahwah, N.J.: Lawrence Erlbaum.

Gottlieb, G., D. Wahlsten, and R. Lickliter. 1998. The significance of biology for human development: A developmental psychobiological systems view. In R. Lerner, ed., *Handbook of child psychology.* Vol. 1 of *Theory* (pp. 233–273). New York: John Wiley & Sons, Inc.

Gould, S. J., and R. C. Lewontin. 1979. The spandrels of San Marco and the Panglossian paradigm: A critique of the adaptationist programme. *Proceedings of the Royal Society of London* 205:281–288.

Greenough, W. T., and J. E. Black. 1992. Induction of brain structure by experience: Substrates for cognitive development. In M. R. Gunnar and C. Nelson, eds., *Developmental Behavioral Neuroscience* (pp. 155–299). Mahwah, N.J.: Lawrence Erlbaum.

Greenough, W. T., J. E. Black, and C. S. Wallace. 1987. Experience and brain development. *Child Development* 38:539–559.

Greenspan, S. I. 1979. *Intelligence and adaptation: An integration of psychoanalytic and Piagetian developmental psychology.* Psychological Issues Monograph Series, nos. 47–48. New York: International Universities Press.

______. 1981. Psychopathology and adaptation in infancy and early childhood: Principles of clinical diagnosis and preventive intervention. Clinical Infant Reports, no. 1. New York: International Universities Press.

______. 1989. *The development of the ego: Implications for personality theory, psychopathology, and the psychotherapeutic process.* New York: International Universities Press.

______. 1992. *Infancy and early childhood: The practice of clinical assessment and intervention with emotional and developmental challenges.* Madison, Conn.: International Universities Press.

______. 1993. *Playground politics: Understanding the emotional life of your school-age child.* Reading, Mass.: Addison Wesley.

______. 1997a. *Developmentally based psychotherapy.* Madison, Conn.: International Universities Press.

______. 1997b. *The growth of the mind and the endangered origins of intelligence.* Reading, Mass.: Addison Wesley Longman.

______. 1999. *Building healthy minds: The six experiences that create intelligence and emotional growth in babies and young children.* Cambridge, Mass.: Perseus Publishing.

______. 2001. The affect diathesis hypothesis: The role of emotions in the core deficit in autism and the development of intelligence and social skills. *Journal of Developmental and Learning Disorders* 5:1–45.

______. 2005. *Autistic spectrum disorders.* Work in progress.

Greenspan, S. I., L. E. Jacokes, and J. F. Cassily. Unpublished paper. Timing and rhythmicity and cognitive-academic performance.

Greenspan, S. I., and W. J. Polk. 1980. A developmental approach to the assessment of adult personality functioning and psychopathology. In S. I. Greenspan and G. H. Pollock, eds., *Toward understanding personality development* (pp. 255–297). N.Y.: International Universities Press.

Greenspan, S. I., and R. S. Lourie. 1981. Developmental structuralist approach to the classification of adaptive and pathologic personality organizations: Infancy and early childhood. *American Journal of Psychiatry* 138:725–735.

Greenspan, S. I., and S. W. Porges. 1984. Psychopathology in infancy and early childhood: Clinical perspectives on the organization of sensory and affective-thematic experience. *Child Development* 55:49–70.

Greenspan, S. I., and B. K. Press. 1985. The toddler group: A setting for adaptive social/emotional development of disadvantaged 1- and 2-year-olds in a peer group. *Zero to Three* (Bulletin of the National Center for Clinical Infant Programs) 5:6–8.

Greenspan, S. I., S. Wieder, A. Lieberman, R. Nover, R. Lourie, and M. Robinson. 1987. Infants in multirisk families: Case studies in preventive intervention. *Clinical Infant Reports.* New York: International Universities Press.

Greenspan, S. I., and S. Wieder. 1997. Developmental patterns and outcomes in infants and children with disorders in relating and communicating: A chart review of 200 cases of children with autistic spectrum diagnoses. *Journal of Developmental and Learning Disorders* 1:87–141.

______. 1998. *The child with special needs: Encouraging intellectual and emotional growth.* Reading, Mass.: Perseus Publishing.

______. 1999. A functional developmental approach to autism spectrum disorders. *Journal of the Association for Persons with Severe Handicaps* 24:147–161.

Greenspan, S. I., G. A. DeGangi, and S. Wieder. 2001. *The functional emotional assessment scale (FEAS) for infancy and early childhood: Clinical & research applications.* Bethesda, Md.: Interdisciplinary Council on Developmental and Learning Disorders.

Greenspan, S. I., and I. Glovinsky. 2002. *Children with bipolar patterns of dysregulation: New perspectives on developmental pathways and a comprehensive approach to prevention and treatment.* Bethesda, Md.: The Interdisciplinary Council on Developmental and Learning Disorders.

Greenspan, S. I., and D. Lewis. 2002. *The affect-based language curriculum: An intensive program for families, therapists and teachers.* Bethesda, Md.: The Interdisciplinary Council on Developmental and Learning Disorders.

Greenspan, S. I., and M. Bauman. (Submitted for publication). *Autistic spectrum disorders: a multipath, cumulative risk model.*

Groce, N. E. 1985. *Everyone here spoke sign language: Hereditary deafness on Martha's Vineyard.* Cambridge, Mass.: Harvard University Press.

Gunnar, M. R., and D. M. Vazquez. 2001. Low cortisol and a flattening of expected daytime rhythm: Potential indices of risk in human development. *Development and Psychopathology* 13:516–538.

Hadingham, E. 1979. *Secrets of the Ice Age: The world of the cave artists.* New York: Walker.

Hall, E. T. 1994. *West of the thirties: Discoveries among the Navajo and Hopi.* New York: Doubleday.

Harlow, H. F. 1953. Motivation as a factor in the acquisition of new responses. In *Current theory and research in motivation: A symposium* (pp. 24–29). Lincoln: University of Nebraska Press.

______. 1958. The nature of love. *American Psychologist* 13:573–685.

______. 1971. *Learning to love.* San Francisco: Albion Publishing Company.

Harlow, H. F., J. P. Gluck, and S. J. Suomi. 1972. Generalization of behavioral data between nonhuman and human animals. *American Psychologist* (August):709–716.

Harris, R. 1980. *The language-makers.* London: Duckworth.

Harris, R., and T. J. Taylor. 1989. *Landmarks in linguistic thought.* Vol. 1 of *The Western tradition from Socrates to Saussure.* London: Routledge.

Hartmann, H. 1939. *Ego psychology and the problem of adaptation.* New York: International Universities Press.

Hauser, M. 1996. *The evolution of communication.* Cambridge, Mass.: MIT Press.

Hayes, K. J., and C. Hayes. 1952. Imitation in a home-raised chimpanzee. *Journal of Comparative and Physiological Psychology* 45:450–459.

Hein A., and R. M. Diamond. 1983. Contribution of eye movement to the representation of space. In A. Hein and M. Jeannerod, eds., *Spatially oriented behavior* (pp. 119–134). New York: Springer.

Henshilwood, C. S. 1999. Blombos cave: Early evidence for modern human behaviour. *Museum News,* vol. 13.

Henshilwood, C. S., F. d'Errico, R. Yates, Z. Jacobs, C. Tribolo, G. A. T. Duller, N. Mercier, J. C. Sealy, H. Valladas, I. Watts, and A. G. Wintle. 2002. Emergence of modern human behaviour: Middle stone age engravings from South Africa. *Science* 295:1278–1280.

Hockett C. 1960. *The origin of speech.* San Francisco: W. H. Freeman.

Hofer, M. A. 1988. On the nature and function of prenatal behavior. In W. Smotherman and S. Robinson, eds., *Behavior of the fetus.* Caldwell, N.J.: Telford.

Hofer, M. A. 1995. Hidden regulators: Implications for a new understanding of attachment, separation, and loss. In S. Goldberg, R. Muir, and J. Err, eds., *Attachment theory: Social, developmental, and clinical perspectives* (pp. 203–230). Hillsdale, N.J.: Analytic Press.

Holloway, R. L. 1966. Dendritic branching: Some preliminary results of training and complexity in rat visual cortex. *Brain Research* 2:393–396.

Hunt, J. M. 1941. Infants in an orphanage. *Journal of Abnormal and Social Psychology* 36:338.

Hurley, S. 1998. *Consciousness in action.* Cambridge, Mass.: Harvard University Press.

Izard, C. E. 1971. *The face of emotion.* New York: Meredith & Appleton-Century-Crofts.

______. 1977. *Human emotions.* New York: Plenum Press.

______. 1979. Emotions as motivation: An evolutionary-developmental perspective. *Evolutionary and developmental perspective* 26:163–200.

______. 1993. Four systems for emotion activation: Cognitive and noncognitive processes. *Psychological Review* 100:68–90.

______. 1997. Emotions and facial expressions: A perspective from differential emotions theory. In J. Russell and J. Fernández-Dols, eds., *The psychology of facial expression.* Cambridge: Cambridge University Press.

Jacques, E. 1990. The midlife crisis. In S. I. Greenspan and G. H. Pollock, eds., *The course of life: Early adulthood.* Vol 5. New York: International Universities Press.

Jaffe, J., B. Beebe, S. Feldstein, C. L. Crown, and M. D. Jasnow. 2001. *Rhythms of dialogue in infancy.* Monographs of the Society for Research in Child Development 66, no. 2, serial no. 265. With commentaries by Philippe Rochat and Daniel N. Stern. Boston: Blackwell.

Jaques, E. 1990. The midlife crisis. In S. I. Greenspan and G. H. Pollock, eds., *The course of life: Early adulthood.* Vol. 5. New York: International Universities Press.

Jeannerod, M. 1997. *The cognitive neuroscience of action.* Oxford: Blackwell.

Johnson, M. H., and H. Wright. 1975. Population, exchange, and early state formation in southwestern Iran. *American Anthropologist* 77:267–289.

Juliard, P. 1970. *Philosophies of language in eighteenth-century France.* The Hague: Mouton.

Kandel, E. R. 1989. Genes, nerve cells, and the remembrance of things past. *Journal of Neuropsychiatry and Clinical Neuroscience* 1:103–125.

Kano, T. 1992. *The last ape: Pygmy chimpanzee behavior and ecology.* Stanford: Stanford University Press.

Karen R. 1998. *Becoming attached: First relationships and how they shape our capacity to love.* New York: Oxford University Press.

King, B. 1999. *The origins of language: What nonhuman primates can tell us.* Santa Fe, N.M.: School of American Research Press.

______. 2002. On patterned interactions and culture in great apes. In R. Fox and B. King, eds., *Anthropology beyond culture.* Oxford: Berg.

______. Unpublished work. *The dynamic dance: Nonvocal social communication in the African great apes.*

King, B., and S. Shanker. 2003. The expulsion of primates from the garden of language. *Evolution of Communication* 1:59–99.

______. 2003. How can we know the dancer from the dance? The dynamic nature of African great ape social communication. *Anthropological Theory* 3:5–26.

Klaus, M., and J. Kennell. 1976. *Maternal-infant bonding: The impact of early separation or loss on family development.* St. Louis: C. V. Mosby.

Klaus, M. H., and P. H. Klaus. 2000. *Your amazing newborn.* Cambridge, Mass.: Perseus Publishing.

Klein, R. 1989. *The human career: Human biological and cultural origins.* Chicago: University of Chicago Press.

Klin, A., F. R. Volkmar, and S. Sparrow. 1992. Autistic social dysfunction: some limitations of the theory of mind hypothesis. *Journal of Child Psychology and Psychiatry* 33:861–876.

Klin, A., F. R. Volkmar, S. Sparrow, D. V. Cicchetti, and B. P. Rourke. 1995. Validity and neuropsychological characterization of Asperger syndrome: Convergence with nonverbal learning disabilities syndrome. *Journal of Child Psychology and Psychiatry* 36:1127–1140.

Klinnert, M. D., J. Campos, F. J. Sorce, R. N. Emde, and M. J. Svejda. 1983. Emotions as behavior regulators: Social referencing in infancy. In R. Plutchik and H. Kellerman, eds., *Emotion: Theory, research, and experience* (pp. 57–86). New York: Academic.

Kohlberg, L. 1984. *The psychology of moral development: The nature and validity of moral stages.* San Francisco: Harper & Row.

Köhler, W. 1951. *The Mentality of apes.* Translated by Ella Winter. New York: The Humanities Press Inc.

Kuhl, P. K., and A. N. Meltzoff. 1997. Evolution, nativism and learning in the development of language and speech. In M. Gopnik, ed., *The inheritance and innateness of grammars* (pp. 7–44). New York: Oxford University Press.

Laitman, J. T. 1984. Evolution of the hominid upper respiratory tract: the fossil evidence. In P. Tobias, ed., *Hominid evolution: Past, present and future* (pp. 281–286). New York: John Wiley & Sons, Inc.

Lally, J. R., P. L. Mangione, A. S. Honig, and D. S. Wittner. 1988. More pride, less delinquency: Findings from the ten-year follow-up study of the Syracuse University Family Development Research Program. *Zero to Three* (Bulletin of the National Center for Clinical Infant Programs) 4:13–18.

Leach, E. 1970. *Lévi-Strauss.* London: Fontana/Collins.

Leakey, R. 1979. *Origins: What new discoveries reveal about the emergence of our species and its possible future.* New York: Dutton.

______. 1994. *The origin of humankind.* New York: Basic Books.

LeDoux, J. E. 1996. *The emotional brain.* New York: Simon & Schuster.

______. 2001. *Synaptic self.* New York: Viking.

Lenneberg E. 1967. *Biological foundations of language.* New York: John Wiley & Sons, Inc.

Leroi-Gourhan, A. 1982. *The dawn of European art: An introduction to Paleolithic cave painting.* Translated by Sara Champion. New York: Academic.

Levi-Strauss, C. 1967. *Structural anthropology.* Garden City, N.Y.: Doubleday Books.

Lieberman, P. 1984. *The biology and evolution of language.* Cambridge, Mass.: Harvard University Press.

______. 1991. *Uniquely human: The evolution of speech, thought, and selfless behavior.* Cambridge, Mass.: Harvard University Press.

______. 1998. *Eve spoke: Human language and human evolution.* New York: W. W. Norton & Co.

Lipuna, E., M. Postone, and C. Calhoun, eds. 1993. *Bourdieu: Critical perspectives.* Chicago: University of Chicago Press.

Lukes, S., E. Gellner, and D. Gellner. 1998. *Language and solitude: Wittgenstein, Malinowski, and the Habsburg dilemma.* Cambridge: Cambridge University Press.

Lutz, C. 1998. *Unnatural emotions: Everyday sentiments on a Micronesian atoll and their challenge to Western theory.* Chicago and London: University of Chicago Press.

Mahler, M. S., F. Pine, and A. Bergman. 1975. *The psychological birth of the human infant: Symbiosis and individuation.* New York: Basic Books.

Malinowski, B. 1929. *The sexual life of savages in north-western Melanesia: An ethnographic account of courtship, marriage and family life among the natives of the Trobriand Islands.* London: G. Routledge.

Marks, J. 2002. *What it means to be 98% chimpanzee: Apes, people, and their genes.* Berkeley: University of California Press.

Marschak, A. 1995. A Middle Paleolithic symbolic composition from the Golan Heights: The earliest known depictive image. *Current Anthropology* 37:356–365.

Martlew, M. 1996. Human figure drawings by schooled and unschooled children in Papua New Guinea. *Child Development* 67:2743–2762.

Maynard Smith, J., and E. Szathmáry. 1999. *The origins of life.* Oxford: Oxford University Press.

McConnell, C., and C. Moses. 1995. *The new chimpanzees.* Washington, D.C.: National Geographic Society. Video recording.

McGrew, W. C. 1992. *Chimpanzee material culture: implications for human evolution.* Cambridge: Cambridge University Press.

McLaughlin, M. 1974 Survivors and surrogates: Children and parents from the ninth to the thirteenth centuries. In L. deMause, ed., *The history of childhood.* Northvale, New Jersey: Jason Aronson, Inc.

McNeill, D. 1995. *Hand and mind: What gestures reveal about thought.* Chicago: University of Chicago Press.

McShea, D. W. 1996. Metazoan complexity and evolution: Is there a trend? *Evolution* 50:477–492.

Mercader, J., M. Panger, and C. Boesch. 2002. Excavation of a chimpanzee stone tool site in the African rainforest. *Science* 296:1452–1455.

Messinger, D., A. Fogel, and K. Dickson. 1997. A dynamic systems approach to infant facial action. In J. Russell, ed., *New directions in the study of facial expression.* New York: Cambridge University Press.

______. 1999. What's in a smile? *Developmental Psychology* 35:701–708.

Michels, R. 1993. Adulthood. In S. I. Greenspan and G. H. Pollock, eds., *The course of life: Early adulthood.* Vol. 5 (pp. 1–14). New York: International Universities Press.

Minshew, N., and G. Goldstein. 1998. Autism as a disorder of complex information processing. *Mental Retardation and Developmental Disabilities* 4:129–136.

Mithen, S. 1996. *The prehistory of the mind: The cognitive origins of art, religion and science.* London, New York: Thames and Hudson.

Morris, D. 1967. *The naked ape.* Toronto: Bantam Books.

Mundy, P., M. Sigman, and C. Kasari. 1990. A longitudinal study of joint attention and language development in autistic children. *Journal of Autism and Developmental Disorders* 20:115–128.

Murphy, L. B. 1974. *The individual child.* Publication no. OCD 74–1032. Washington, D.C.: U.S. Government Printing Office, Department of Health, Education and Welfare.

Nelson, K. 1973. *Structure and strategy in learning to talk.* Monographs of the Society for Research in Child Development 38, nos. 1–2, serial no. 149.

NICHD Early Child Care Research Network. 1998. Early child care and self-control, compliance and problem behavior at twenty-four and thirty-six months. *Child Development* 69:1145–1170.

______. 1999. Effect sizes from the NICHD Study of Early Child Care. Paper presented at the Biennial Meeting of the Society for Research in Child Development, Albuquerque, New Mexico.

______. 2000a. Characteristics and quality of child care for toddlers and preschoolers. *Applied Developmental Science* 4:116–135.

______. 2000b. The relation of child care to cognitive and language development. *Child Development* 71:960–980.

______. 2003. Does amount of time spent in child care predict socioemotional adjustment during the transition to kindergarten? *Child Development* 74:976–1005.

Noble, W., and I. Davidson. 1996. *Human evolution, language, and mind: A psychological and archaeological inquiry.* Cambridge: Cambridge University Press.

Owens, R. 2001. *Language development: An introduction.* Boston: Allyn and Bacon.

Oyama, S. 2003. On having a hammer. In B. Weber and D. Depew, eds., *Evolution and learning: The Baldwin Effect reconsidered.* Cambridge, Mass.: MIT Press.

Ozonoff, S. 1997. Causal mechanisms of autism: Unifying perspectives from an information-processing framework. In D. Cohen and F. Volkmar, eds., *Handbook of autism and pervasive developmental disorders.* New York: John Wiley & Sons, Inc.
Parsons, T. 1999. *The Talcott Parsons reader.* Edited by B. S. Turner. Oxford: Blackwell.
Peisner-Feinjerg, E. S., M. R. Burchinal, R. M. Clifford, M. L. Culkin, C. Howes, S. L. Kagan, N. Yazejian, P. Byler, J. Rustici, and J. Zelazo. 1999. *The children of the cost, quality, and outcomes study go to school: Technical report.* Chapel Hill, N.C.: FPG Child Development Center, University of North Carolina-Chapel Hill.
Pennington, J., and S. Ozonoff. 1996. Executive functions and developmental psychopathology. *Journal of Child Psychology and Psychiatry* 37:51–87.
Petitto, L. 1997. In the beginning: On the genetic and environmental factors that make early language acquisition possible. In M. Gopnik, ed., *The inheritance and innateness of grammars* (pp. 45–69). New York: Oxford University Press.
Piaget, J. 1952. *The origins of intelligence in children.* New York: International Universities Press.
______. 1955. *The language and thought of the child.* New York: Meridian Books.
______. 1962. The stages of intellectual development of the child. In S. Harrison and J. McDermott, eds., *Childhood psychopathology* (pp. 157–166). New York: International Universities Press.
Piaget, J., and B. Inhelder. 1966. *The psychology of the child.* New York: Basic Books.
Piatelli-Palmerini, M. 1979. *Language and learning.* London: Routledge and Kegan Paul.
Pinker, S. 1994a. *The language instinct.* New York: William Morrow.
______. 1994b. *The language instinct: The new science of language and mind.* London: Allen Lane.
______. 2002. *The blank slate: The modern denial of human nature.* New York: Viking.
Pinto-Correia, C. 1997. *The ovary of Eve: Egg and sperm and preformation.* Chicago: University of Chicago Press.
Plomin, R. 1997. Nature, nuture, and cognitive development from 1 to 16 years: A parent offspring adoption study. *Psychological Science* 8:442–447.
Plooij, F. 1984. *The behavioral development of free-living chimpanzee babies and infants.* Norwood, N.J.: Ablex.
Polanyi, M. 1967. *The tacit dimension.* London: Routlege and Kegan Paul.
Provence, S., and R. C. Lipton. 1962. *Infants in institutions.* New York: International Universities Press.
Provence, S., and A. Naylor. 1983. *Working with disadvantaged parents and their children: Scientific and practical issues.* New Haven: Yale University Press.
Quine, W. V. O. 1960. *Methods of logic.* New York: Holt.
Ragir, S. 2002. Constraints on communities with indigenous sign languages: Clues to the dynamics of language genesis. In A. Wray, ed., *The Transition to Language.* Oxford: Oxford University Press.
Rakic, P., J. Bourgeois, and P. Goldman-Rakic. 1994. Synaptic development of the cerebral cortex: Implications for learning, memory, and mental illness. In J. van Pelt, M. A. Corner, H. B. M. Uylings, and F. H. Lopes da Silver, *The self-organizing brain: From growth cones to functional networks* (pp. 227–243). New York: Elsevier Science.
Ramey, C. T., and S. L. Ramey. 1998. Early intervention and early experience. *American Psychologist* 58:109–120.

Ramus, F., M. Hauser, C. Miller, D. Morris, and J. Mehler. 2000. Language discrimination by human newborns and by cotton-top tamarin monkeys. *Science* 288:349–351.

Rappaport, D. 1960. *The structure of psychoanalytic theory: A systematizing attempt.* New York: International Universities Press.

Rasmussen, K. 1908. *The people of the polar north: A record.* London: K. Paul, Trench, Trübner.

______ 1946. *Report of the fifth Thule expedition, 1921–1924: The Danish expedition to Arctic North America in charge of Knud Rasmussen.* Copenhagen: Gyldendal 1946.

Reé, J. 1999. *I see a voice: A philosophical history of language, deafness and the senses.* London: Harper Collins.

Ridley, M. 2003. *Nature via nurture.* New York: Harper Collins.

Roche, H., A. Delagnes, J. P. Brugal, C. Feibel, M. Kibunjia, V. Mourre, and P. J. Texier. 1999. Early hominid stone tool production and technical skill 2.34 myr ago in West Turkana, Kenya. *Nature* 399:57–60.

Ross, J. 1974. The middle-class child in urban Italy, fourteenth to early sixteenth century. In L. deMause, ed., *The history of childhood.* Northvale, New Jesey: Jason Aronson, Inc.

Rumbaugh, D., ed. 1977. *Language learning by a chimpanzee: The Lana project.* New York: Academic Press.

Russell, B. 1919. *Introduction to mathematical philosophy.* London: G. Allen and Unwin.

Russon, A. E., K. A. Bard, and S. T. Parker. 1996. *Reaching into thought: the minds of the great apes.* Cambridge, Mass.: Cambridge University Press.

Sabater, P. 1992. *Los origenes de la Cultura.* Mexico: Fondo de Cultura.

Saffran, A., E. Newport, R. Aslin, R. Tunick, and S. Barrueco. 1997. Incidental language learning: Listening (and learning) out of the corner of your ear. *Psychological Science* 8:101–105.

Sameroff, A., R. Seifer, R., Barocas, M. Zax, and S. I. Greenspan. 1986. IQ scores of 4-year-old children: Social-environmental risk factors. *Pediatrics* 29:343–350.

Sameroff, A., A. Seifer, A. Baldwin, and C. Baldwin. 1993. Stability of intelligence from preschool to adolescence: The influence of social and family risk factors. *Child Development* 64:97.

Sander, L. 1962. Issues in early mother-child interaction. *Journal of the American Academy of Child Adolescent Psychiatry* 1:141–166.

Savage-Rumbaugh, S. 1986. *Ape language: From conditioned response to symbol.* New York: Columbia University Press.

Savage-Rumbaugh, S., and R. Lewin. 1994. *Kanzi: the ape at the brink of the human mind.* New York: John Wiley & Sons, Inc.

Savage-Rumbaugh, S., S. L. Williams, T. Furuichi, and T. Kano. 1996. Language perceived: Pan paniscus branches out. In W. C. McGrew, L. R. Marchant, and T. Nishida, eds., *Great ape societies.* Cambridge: Cambridge University.

Savage-Rumbaugh, S., S. G. Shanker, and T. J. Taylor. 1998. *Apes, language, and the human mind.* New York: Oxford University Press.

Savage-Rumbaugh, S., W. M. Fields, and J. P. Taglialatela. 2001. Language, speech, tools, and writing. *Journal of Consciousness Studies* 8:273–292.

Schacter, D. L. 1997. The neuropsychology of false recognition. *Current Directions in Psychological Science* 6:65–70.

Schanberg, S. M., and T. M. Field. 1987. Sensory deprivation stress and supplemental stimulation in the rat pup and preterm human neonate. *Child Development* 58:1431–1447.

Schultz, R., I. Gauthier, A. Klin, R. Fulbright, A. Anderson, F. Volkmar, P. Skudlarski, C. Lacadie, D. Cohen, and J. Gore. 2000. Abnormal ventral temporal cortical activity during face discrimination among individuals with autism and Asperger syndrome. *Archives of General Psychiatry* 57:331–340.

Schweinhart, L. J., H. V. Barnes, and D. P. Weikart. 1993. Significant benefits: The High/Scope Perry Preschool study through age 27. Ypsilanti, Mich., High/Scope Press. High/Scope Educational Research Foundation, No. 10.

Shaffer, R. J., L. E. Jacokes, J. F. Cassily, S. I. Greenspan, R. F. Tuchman, and P. J. Stemmer, Jr. 2001. Effect of Interactive Metronome training on children with ADHD. *American Journal of Occupational Therapy* 55:155–161.

Shanker, S. 1986. Computer vision or mechanist myopia. In S. G. Shanker, ed., *Philosophy in Britain today* (pp. 213–266). Albany, N.Y.: State University of New York Press; London: Croom Helm.

______. 1998. *Wittgenstein's remarks on the foundations of AI.* London: Routledge.

______. 2000. Review of Jonathan Reé: I see a voice. *Sign Language Studies* 1, no. 1:93–102.

______. 2002. The generativist-interactionist debate over specific language impairment: Psycholinguistics at a crossroads. *American Journal of Psychology* 115:415–450.

Shanker, S., and T. Taylor. 2001. The house that Bruner built. In D. Bakhurst and S. Shanker, eds., *Language, culture, self: The philosophical psychology of Jerome Bruner* (pp. 50–70). London: Sage.

Shannon, C. 1949. *The mathematical theory of communication.* Urbana, Ill.: University of Illinois Press.

Shonkoff, J. P., and D. A. Phillips. 2000. *From neurons to neighborhoods: The science of early childhood development.* National Research Council, Committee on Integrating the Science of Early Childhood Development. Washington, D.C.: National Academy Press.

Siller M., and M. Sigman. 2002. The behaviors of parents of children with autism predict the subsequent development of their children's communication. *Journal of Autism and Developmental Disorders* 32:77–89.

Singer, W. 1986. Neuronal activity as a shaping factor in postnatal development of visual cortex. In W. T. Greenough and J. M. Juraska, eds., *Developmental neuropsychobiology* (pp. 271–293). Orlando, Fla.: Academic Press.

Small, M. 1998. *Our babies, ourselves: How biology and culture shape the way we parent.* New York: Anchor Books.

Smith, M., and E. Szathmáry. 1997. *The major transitions in evolution.* New York: Oxford University Press.

Snell, B. 1960. *The discovery of the mind.* New York: Harper.

Spelke, E. S. 1976. Infants' intermodal perception of events. *Cognitive Psychology* 8:553–560.

Sperry, R. W. 1985. Consciousness, personal identity, and the divided brain. In F. Benson and E. Zaidel, eds., *The dual brain* (pp. 11–27). New York: Guilford.

Spitz, R. A. 1945. Hospitalism: An inquiry into the genesis of psychiatric conditions in early childhood. *The Psychoanalytic Study of the Child* 1:53–74.

______. 1965. *The first year of life: A psychoanalytic study of normal and deviant development of object relations.* New York: International Universities Press.

Sroufe, L. A. 1979. Socioemotional development. In J. Osofsky, ed., *Handbook of Infant Development.* New York: John Wiley & Sons, Inc.

Sroufe, L. A., E. Waters, and L. Matas. 1974. Contextual determinants of infant affective response. In M. Lewis and L. Rosenblum, eds., *The origins of fear* (pp. 49–72). New York: John Wiley & Sons, Inc.

Stanford, C. 2000. *Chimpanzee and red colobus: The ecology of predator and prey.* Cambridge, Mass.: Harvard University Press.

Stanford, G. 1999. *The hunting apes: Meat eating and the origins of human behavior.* Princeton: Princeton University Press.

Stern, D. 1974a. Mother and infant at play: The dyadic interaction involving facial, vocal, and gaze behaviors. In M. Lewis and L. Rosenblum, eds., *The effect of the infant on its caregiver.* New York: John Wiley & Sons, Inc.

______. 1974b. The goal and structure of mother-infant play. *Journal of the American Academy of Child Psychiatry* 13:402–421.

Sternberg, R. J., and C. Berg. 1992. *Intellectual development.* Cambridge: Cambridge University Press.

Suomi, S. J. 1976. Mechanisms underlying social development: A reexamination of mother-infant interactions in monkeys. In A. Pick, ed., *Minnesota Symposium on Child Psychology.* Vol. 10. Minneapolis: University of Minnesota Press.

______. 1999. Attachment in rhesus monkeys. In J. Cassidy and P. R. Shaver, eds., *Handbook of attachment: Theory, research, and clinical applications* (pp. 181–197). New York: Guilford Press.

Tattersall, I. 1995. *The fossil trail: How we know what we think we know about human evolution.* New York: Oxford University Press.

Taylor, T. 1992. *Mutual misunderstanding: Skepticism and the theorizing of language and interpretation.* Durham, N.C.: Duke University Press.

______. 1997. *Theorizing language: Analysis, normativity, rhetoric, history.* New York: Pergamon.

Templeton, A. 2002. Out of Africa again and again. *Nature* 416:45–51.

Thinus-Blanc, C. 1981. Volume discrimination learning in golden hamsters: Effects of the structure of complex rearing cages. *Developmental Psychobiology* 14:397–403.

Thomas, A., S. Chess, and H. G. Birch. 1968. *Temperament and behavior disorders in children.* New York: New York University Press.

Thorndike, E. 1911. *Animal Intelligence.* New Brunswick, N.J.: Transaction Publishers.

Tomasello, M. 1999. *The cultural origins of human cognition.* Cambridge, Mass.: Harvard University Press.

______. 2003. *Constructing a language: A usage-based theory of language acquisition.* Cambridge, Mass.: Harvard University Press.

Tomkins, S. 1963a. *Affect, imagery, consciousness.* Vol. 1. New York: Springer Publishing.

______. 1963b. *Affect, imagery, consciousness.* Vol. 2. New York: Springer Publishing.

Toren, C. 1998. Cannibalism and compassion: transformations in Fijian notions of the person. In V. Keck, ed., *Common worlds and single lives: Constituting knowledge in Pacific societies* (pp. 95–115). London: Berg.

Toth, N. 1987. The first technology. *Scientific American* 4:121.

Toth, N., and K. Schik. 1993. *Making silent stones speak: Human evolution and the dawn of technology.* New York: Simon & Schuster.

Trevarthen, C. 1979. Communication and cooperation in early infancy: A description of primary intersubjectivity. In M. Bullowa, ed., *Before speech: The beginning of interpersonal communication* (pp. 321–347). Cambridge: Cambridge University Press.

Trinkhaus, E. 1993. *The Neanderthals: Of skeletons, scientists, and scandal.* Atlanta: Vintage Books.

Tronick, E. 1982. *Social interchange in infancy: Affect, cognition, and communication.* Baltimore: University Park Press.

Tronick, E., and L. Adamson. 1980. *Babies as people: New findings on our social beginnings.* New York: Collier Books.

Tronick, E., M. Ricks, and J. Cohn. 1982. Maternal and infant affective exchange: patterns of adaptation. In T. Field and A. Fogel, eds., *Emotion and early interaction.* Mahwah, N.J.: Lawrence Erlbaum.

Turner, A. M., and W. T. Greenough. 1983. Synapses per neuron and synaptic dimensions in occipital cortex of rats reared in complex, social, or isolation housing. *Acta Stereologica* 2:239–244.

______. 1985. Differential rearing effects on rat visual cortex synapses. I: Synaptic and neuronal density and synapses per neuron. *Brain Research* 329:195–203.

Turquet, P. 1975. Threats to identity in the large group: A study in the phenomenology of the individual's experiences of changing membership status in a large group. In L. Kreeger, ed., *The large group: Dynamics and therapy* (pp. 87–158). London: Constable.

Vandell, D. L., and B. Wolfe. 2000. *Child care quality: Does it matter and does it need to be improved?* Washington, D.C.: Office of the Assistant Secretary for Planning and Evaluation, U.S. Department of Health and Human Services.

Volosinov, V. N. 1973. *The philosophy of language.* New York: Seminar Press.

Vygotsky, L. S. 1962. *Thought and language.* 1934. Edited and translated by Eugenia Hanfmann and Gertude Vakar. Cambridge, Mass.: MIT Press.

______. 1978. *Mind in society: The development of higher psychological processes.* Cambridge, Mass.: Harvard University Press.

Waal, F. 1996. *Good natured: The origins of right and wrong in humans and other animals.* Cambridge, Mass.: Harvard University Press.

______. 1998. *Chimpanzee politics: Power and sex among apes.* Baltimore: Johns Hopkins University Press.

Waal, F., and F. Lanting. 1997. *Bonobo: The forgotten ape.* Berkeley: University of California Press.

Waal, F., and P. Tyack. 2003. *Animal social complexity: Intelligence, culture, and individualized societies.* Cambridge, Mass.: Harvard University Press.

Walker, A., and R. E. F. Leaky. 1993. *The Nariokotome homo erectus skeleton.* Cambridge, Mass.: Harvard University Press.

Wallace, W. P., T. R. Shaffer, M. D. Amberg, and V. L. Silvers. 2001. Divided attention and prerecognition processing of spoken words and nonwords. *Memory and Cognition* 29:1102–1110.

Watson, R. 2002. *Cogito, ergo sum: The life of René Descartes.* Boston: D. R. Godine.

Weber, B., and D. Depew, eds. 2003. *Evolution and learning: The Baldwin Effect reconsidered.* Cambridge, Mass.: MIT Press.

Weiler, I. J., N. Hawrylak, and W. T. Greenough. 1995. Morphogenesis in memory formation: Synaptic and cellular mechanisms. *Behavioural Brain Research* 66:6.

Weingart, P., S. Richerson, P. Mitchell, and S. Maasen, eds. 1997. *Human by nature: Between biology and the social sciences.* Mahwah, N.J.: Lawrence Erlbaum.

Wellman, H. M. 1990. *The child's theory of mind.* Cambridge, Mass.: M.I.T. Press.

Wenke, R. 1999. *Patterns in prehistory: Humankind's first three million years.* New York and Oxford: Oxford University Press.

Wetherby, A., R. L. Koegel, and M. Mendel. 1981. Central auditory nervous system dysfunction in echolalic autistic individuals. *Journal of Speech and Hearing Research* 24:420–429.

Wetherby, A. M., and B. M. Prizant. 1993. Profiling communication and symbolic abilities in young children. *Journal of Childhood Communication Disorders* 15:23–32.

White, C. 1799. *An account of the regular gradation in man and in different animals and vegetables.* London: C. Dilly.

Whitrow, G. J. 1988. *Time in history.* Oxford: Oxford University Press.

Wiesel, T. N., and D. H. Hubel. 1963. Single-cell responses in striate cortex of kittens deprived of vision in one eye. *Journal of Neurophysiology* 26:1003–1017.

Winch, P. 1958. *The idea of a social science and its relation to philosophy.* London: Routledge.

Winnicott, D. W. 1987. *The child, the family and the outside world.* 1957. Reading, Mass.: Addison-Wesley.

Witherington, D. C., J. J. Campos, and M. J. Hertenstein. 2001. Principles of emotion and its development. In G. Bremmer and A. Fogel, eds., *Blackwell handbook of infant development* (pp. 427–464). Oxford: Blackwell.

Wittgenstein, L. 1953. *Philosophical investigations.* New York: Macmillan.

Wrangham, R. 1996. *Demonic males: Apes and the origins of human violence.* Boston: Houghton Mifflin.

Yarrow, L. 1975. *Infant and environment: Early cognitive and motivational development.* New York: John Wiley & Sons, Inc.

Young, J. A. 1978. *Patterns of the brain.* Oxford: Oxford University Press.

Young, T. T. 1943. *Emotions in man and animal.* New York: John Wiley & Sons, Inc.

Zimmerman, A., and B. Gordon. 2000. Neuromechanisms in autism. In Interdisciplinary Council on Developmental and Learning Disorders Clinical Practice Guidelines, *Redefining the standards of care for infants, children, and families with special needs.* Bethesda, Md.: ICDL.

Index

About the Authors

STANLEY I. GREENSPAN, M.D. is Clinical Professor of Psychiatry and Pediatrics at George Washington University Medical School and Chairman of the Interdisciplinary Council on Developmental and Learning Disorders. Regarded as the world's leading authority on clinical work with infants and young children, he is a former director of the NIMH's Mental Health Studies Center and Clinical Infant Development Program, a founding president of Zero to Three: The National Center for Infants, Toddlers and Families, and a supervising child psychoanalyst at the Washington Psychoanalytic Institute.

Dr. Greenspan is the author or editor of more than 37 influential books, translated into over a dozen languages, including *The Growth of the Mind* (with Beryl Lieff Benderly), *Building Healthy Minds, The Challenging Child, The Child with Special Needs* (co-authored by Serena Wieder, Ph.D.); *Infancy and Early Childhood,* and, together with T. Berry Brazelton M.D., *The Irreducible Needs of Children.*

STUART G. SHANKER, D. Phil. (Oxon), is Distinguished Research Professor at York University in Toronto and Co-Chair of the Council of Human Development. One of the world's leading authorities on the philosophy of Ludwig Wittgenstein, he has been at the forefront of ape-language research and child-language studies.

Dr. Shanker, who was the Marian Buck Scholar at Christ Church, Oxford and a Canada Council Postdoctoral Fellow, has received the highest research award at his university, the Walter L. Gordon Fellowship. His twenty highly praised books, including *Wittgenstein and the Turning-Point in the Philosophy of Mathematics, Wittgenstein's Remarks on the Foundations of AI,* and *Apes, Language and the Human Mind* (with Sue Savage-Rumbaugh and Talbot Taylor) have established him as a leading figure in the philosophy of psychology.